Novel Thermoelectric Materials and Device Design Concepts

T0172141

Sergey Skipidarov • Mikhail Nikitin
Editors

Novel Thermoelectric Materials and Device Design Concepts

 Springer

Editors
Sergey Skipidarov
RusTec LLC
Moscow, Russia

Mikhail Nikitin
RusTec LLC
Moscow, Russia

Thermoelectric Power Generation
ISBN 978-3-030-12059-7   ISBN 978-3-030-12057-3   (eBook)
https://doi.org/10.1007/978-3-030-12057-3

This Springer imprint is published by the registered company Springer Nature Switzerland AG
The registered company address is: Gewerbestrasse 11, 6330 Cham, Switzerland

Preface

Book 1 of book series Thermoelectric Power Generation includes chapters that discussed ways on how to create effective thermoelectric materials for manufacturing high-performance thermoelectric generators (TEGs) operating in low-, mid-, and high-temperature ranges. The problem is very old and very difficult to solve. Unfortunately, nowadays, we see a lack of innovative and affordable n-type and p-type thermoelectric materials and TEGs on the market.

The indomitable consumption of fossil fuels has resulted in gigajoules of low-potential waste heat and huge amount of greenhouse gases. Waste heat energy, which is conditionally a free-of-charge energy, is estimated to be from 50 to 70% of the primary energy produced from the burning of fossil fuels. Low-potential heat can be generated due to absorption of sunlight as well. So, it is easy to generate waste heat, but difficult to convert it into electrical energy.

The efficient recovery of low-potential heat is an important and nontrivial task. Sources of such heat are plentifully everywhere, and, as a rule, it is simply dissipated without any benefit to the people. This is caused by the fact that low-potential waste heat is strongly localized near heat sources; therefore, that heat is difficult to use in a cost-effective manner for an intended purpose. TEGs are small-sized items that can be placed as close as possible to heat (thermal) energy sources, which can have temperatures of hundreds of degrees Celsius. It can operate anywhere (including indoors) and at any time of the day. These factors are decisive for applying TEGs to recover low-potential heat. Therefore, in practical applications, TEG's structure will be exposed to systematic long-term heavy temperature gradients, mechanical stresses (thermomechanical stresses), and high temperatures on one (hot) side. TEGs, and hence the thermoelectric materials forming legs of thermocouples, must withstand the abovementioned shock.

Obviously, only materials based on pressed powders and composites can withstand long time in real heavy thermal and mechanical attacks during TEG exploitation.

In TEGs, due to high temperatures (hundreds of degrees Celsius) on the hot side and heavy thermomechanical stresses in module, many processes become active, leading to a quick or gradual degradation in the performance of the thermoelectric materials and TEG itself. These degradation processes are, namely, interdiffusion, recrystallization, alloying, dissolution, phase transitions, phase separation, phase segregation, sublimation, oxidation, mechanical damage of legs, commutation and interconnections, and other phenomena.

Authors of chapters present their look on modern solutions of considered problem including microstructural manipulation (alloying, nanoprecipitates and strains, composites, nanoinclusions, multiphase and all-scale nanocomposites), optimizing concentration of charge carriers (deep-level doping, dynamic doping), band engineering (band convergence, resonant states, low effective mass, and deformation potential coefficient), crystal structure defect engineering, potential interface barriers, and solubility manipulation.

To become attractive and affordable to customers, TEGs should have a service life of at least 5000 h, with thousand cycles on-off, and, of course, be cheap as well.

This book is an attempt to arrange the interchange of research and development results concerned with hot topics in TEGs research, development, and production, including:

1. Trends in traditional inorganic materials
2. Novel inorganic materials
3. Research results in innovative composite nanomaterials
4. Novel methods and measurement techniques for performance evaluation of thermoelectric materials and TEGs
5. Thermoelectric power generators simulation, modeling, and design

Moscow, Russia   Sergey Skipidarov
                                                            Mikhail Nikitin

Contents

Part I Trends in Traditional Inorganic Materials

1 **Investigating the Performance of Bismuth-Antimony Telluride**   3
   Zinovi Dashevsky and Sergey Skipidarov

2 **SnSe: Breakthrough or Not Breakthrough?**   23
   Christophe Candolfi, Dorra Ibrahim, Jean-Baptiste Vaney,
   Selma Sassi, Philippe Masschelein, Anne Dauscher,
   and Bertrand Lenoir

3 **Tin Sulfide: A New Nontoxic Earth-Abundant Thermoelectric
   Material**   47
   Hong Wu, Xu Lu, Xiaodong Han, and Xiaoyuan Zhou

4 **SnTe-Based Thermoelectrics**   63
   Wen Li, Jing Tang, Xinyue Zhang, and Yanzhong Pei

5 **Lead Chalcogenide Thermoelectric Materials**   83
   Shan Li, Xinyue Zhang, Yucheng Lan, Jun Mao,
   Yanzhong Pei, and Qian Zhang

6 **High Thermoelectric Performance due to Nanoprecipitation,
   Band Convergence, and Interface Potential Barrier
   in PbTe-PbSe-PbS Quaternary Alloys and Composites**   105
   Dianta Ginting and Jong-Soo Rhyee

7 **Multicomponent Chalcogenides with Diamond-Like
   Structure as Thermoelectrics**   137
   Dan Zhang, Guangsheng Fu, and Shufang Wang

8 **1-2-2 Layered Zintl-Phase Thermoelectric Materials**   159
   Jing Shuai, Shan Li, Chen Chen, Xiaofang Li, Jun Mao,
   and Qian Zhang

9   Skutterudites: Progress and Challenges 177
Gerda Rogl and Peter Rogl

10 Half-Heusler Thermoelectrics 203
Ran He, Hangtian Zhu, and Shuo Chen

Part II   Novel Inorganic Materials

11 Polymer-Derived Ceramics: A Novel Inorganic
Thermoelectric Material System 229
Rakesh Krishnamoorthy Iyer, Adhimoolam Bakthavachalam
Kousaalya, and Srikanth Pilla

Part III   Performance Evaluation and Measurement Techniques

12 Grain Boundary Engineering for Thermal Conductivity
Reduction in Bulk Nanostructured Thermoelectric Materials 255
Adam A. Wilson, Patrick J. Taylor, Daniel S. Choi,
and Shashi P. Karna

13 Novel Measurements and Analysis for Thermoelectric Devices 277
Patrick J. Taylor, Adam A. Wilson, Terry Hendricks,
Fivos Drymiotis, Obed Villalpando, and Jean-Pierre Fleurial

Part IV   Device Design, Modeling and Simulation

14 Modeling and Optimization of Thermoelectric Modules
for Radiant Heat Recovery 297
Je-Hyeong Bahk and Kazuaki Yazawa

Index 325

Part I
Trends in Traditional Inorganic Materials

Chapter 1
Investigating the Performance of Bismuth-Antimony Telluride

Zinovi Dashevsky and Sergey Skipidarov

Abstract We provide the rationale for possible significant improving efficiency of low-temperature thermoelectric generators (TEGs) based on bismuth-antimony telluride $(Bi_2Te_3)_x(Sb_2Te_3)_{1-x}$ ternary alloys.

It has been shown by experiments that using in TEGs of p-type legs made of $(Bi_2Te_3)_x(Sb_2Te_3)_{1-x}$ material with orientation alternative to traditional, i.e., when cleavage planes of legs are transverse to heat flux direction, results in increasing in thermoelectric efficiency by an average of 25% in the temperature range from 100 °C to 350 °C.

1.1 Introduction

Development and wide application of thermoelectric generation as user-friendly direct energy conversion technology are limited mainly by two factors:

- Relatively low conversion efficiency of thermoelectric generators (TEGs)
- Limited resources of thermoelectric materials for large-scale production of high-performance TEGs for industrial applications

Researchers and engineers focus their efforts on solving these problems by:

- Increasing in thermoelectric efficiency Z in a wide range of operating temperatures 50–1000 °C
- Research and development activity and arranging production of novel high-performance thermoelectric materials consisting of elements which are in abundance on the Earth

Unfortunately, situation with candidates for new high-performance materials is far from satisfactory: there are some potential effective candidates for using in mid-temperature range (300–550 °C), but there are no currently high-performance candidates for low-temperature range (below 300 °C) which are able to replace

Z. Dashevsky · S. Skipidarov (✉)
RusTec LLC, Moscow, Russia
e-mail: skipidarov@rustec-msk.com

© Springer Nature Switzerland AG 2019
S. Skipidarov, M. Nikitin (eds.), *Novel Thermoelectric Materials and Device Design Concepts*, https://doi.org/10.1007/978-3-030-12057-3_1

traditional materials based on bismuth-antimony telluride $(Bi_2Te_3)_x(Sb_2Te_3)_{1-x}$ ternary alloy. The main problem is high content of deficit tellurium in these materials reaching about 50%. There are no deposits of tellurium ores in the Earth; tellurium is extracted as a by-product in the production of copper. From one side, consumption of tellurium in industry is low, but from another side, due to low content in the Earth, production volume of tellurium is very limited. Till so far, tellurium has been used as doping additive in alloy steel production. However, production of solar photovoltaic (PV) panels based on CdTe began recently, and, if production will grow in the near future, then affordability of tellurium will be a serious challenge for thermoelectric applications, due to both possible sharp rise in tellurium prices and shortage on the market.

It all stimulates research works for increasing in thermoelectric efficiency of well-known bismuth and antimony chalcogenides that, albeit indirectly, can reduce the consumption of tellurium.

We will consider evident, but nontrivial, approach of improving the efficiency of low-temperature thermoelectric p-type $(Bi_2Te_3)_x(Sb_2Te_3)_{1-x}$ ternary alloys. We made a choice on p-type materials deliberately. If new effective tellurium-free thermoelectric materials will be discovered, then, likely, those will be n-type materials, and p-type thermoelectric materials available now will be irreplaceable for a long time. It's because electron mobility μ_e in conduction band of semiconductor materials (Si, Ge, GaAs, etc.) is practically always much higher than heavy hole mobility μ_h in valence band in entire reasonable temperature range. And, as the result, thermoelectric efficiency, which is proportional to mobility of major charge carriers, should generally be higher in materials of n-type legs (major charge carriers are electrons) than in materials of p-type legs (major charge carriers are heavy holes).

In addition, lower heavy hole mobility in comparison with electron mobility should cause a sharp drop of Z value of p-type legs with an increase in temperature and approach to intrinsic conductivity when contributions of holes and electrons to total conductivity value become comparable. It is well known that sign of Seebeck coefficient is positive for holes and negative for electrons. Therefore, thermal generation of electrons in p-type material, even in low concentration, will sharply decrease in total value of Seebeck coefficient, which, eventually, tends to zero as the concentration of minority charge carriers (electrons) increases.

The structures of the valence and the conduction bands of $(Bi_2Te_3)_x(Sb_2Te_3)_{1-x}$ ternary alloys are practically similar (that is an essential advantage) which provides highly competitive and unrivaled performance till so far.

1.2 Selection Criteria of Effective Thermoelectric Material

A.F. Ioffe has stated selection criteria of effective thermoelectric material as "the material with the electrical conductivity like a good metal and the thermal conductivity like good isolator." A.F. Ioffe deserves a full credit for understanding that, namely, "middle" class of materials, semiconductors can provide the best thermo-electric performance as thermoelectric materials.

The conversion efficiency of heat energy to electric power depends on figure of merit of thermoelectric material—factor Z, which is expressed as:

$$Z = \frac{S^2\sigma}{\kappa},$$

(1.1)

where S is Seebeck coefficient, σ is electrical conductivity, and κ is thermal conductivity of thermoelectric material.

At acoustic phonon scattering of charge carriers (charge carrier's mean free path $l = l_0 E^r$, where l_0 is constant, E is charge carrier energy, and scattering parameter $r = 0$), Seebeck coefficient is written as:

$$S = \frac{k_0}{e}\left[2\frac{F_1(\mu^*)}{F_0(\mu^*)} - \mu^*\right],$$

(1.2)

where k_0 is Boltzmann constant, e is electron charge, F_i are Fermi integrals, and μ^* is reduced Fermi level (for conduction band $\mu^* = \frac{E_c - E_F}{k_0 T}$, E_c is the bottom of conduction band and E_F is Fermi level; for valence band $\mu^* = \frac{E_F - E_v}{k_0 T}$, E_v is the top of valence band).

The electrical conductivity σ of a semiconductor with concentration of charge carriers $n(p)$ is given by well-known formula:

$$\sigma = en(p)\mu_n(\mu_p),$$

(1.3)

where $n(p)$ and $\mu_n(\mu_p)$ are electron (hole) concentration and mobility. Electron concentration in the conduction band is determined by expression:

$$n = 4\pi\frac{\left(2m_n^* k_0 T\right)^{3/2}}{h^2}F_{1/2}(\mu^*),$$

(1.4)

where m_n^* is the density of states (DOS) effective mass of electrons, and in the case of many-ellipsoid (N) model, it is defined as:

$$m_n^* = N^{2/3}(m_1 m_2 m_3)^{1/3},$$

(1.5)

where m_1, m_2, m_3 are the components of effective mass along the main axis of ellipsoids.

The hole concentrations in the valence band are determined by expression:

$$p = 4\pi\frac{\left(2m_p^* k_0 T\right)^{3/2}}{h^2}F_{1/2}(\mu^*),$$

(1.6)

where m_p^* is the density of states (DOS) effective mass of heavy holes. The mobility in a semiconductor with a simple parabolic band is written as:

$$\mu = e\tau/m_c^*, \qquad (1.7)$$

where τ is the average pulse relaxation time (the time defining a mean free path) and m_c^* is effective mass of charge carrier.

Heat transport is determined by total thermal conductivity κ of semiconductor which consists of three components—lattice thermal conductivity κ_L by phonons, electronic thermal conductivity κ_e by free electrons or holes, and ambipolar thermal conductivity κ_a by electron-hole pairs in the intrinsic conduction region:

$$\kappa = \kappa_L + \kappa_e + \kappa_a. \qquad (1.8)$$

The heat transport by elementary excitations (phonons) can be described by formula:

$$\kappa_L = \frac{1}{3} C_v \nu^3, \qquad (1.9)$$

where C_v is the specific heat at constant volume and ν is the velocity of propagation of elementary excitation.

Electronic thermal conductivity can be described by Wiedemann-Franz law:

$$\kappa_e = L_0 \sigma T, \qquad (1.10)$$

where L_0 is constant known as Lorenz number. For scattering by the acoustic phonons, Lorenz number is given by:

$$L_0 = 2 \left(\frac{k_0}{e} \right)^2. \qquad (1.11)$$

Seebeck coefficient S decreases with a growth of charge carriers concentration $n(p)$, and in a metal with charge carriers concentration $\sim 10^{22}$ cm^{-3}, S is close to zero. At the same time, electrical conductivity σ increases with growing concentration of charge carriers $n(p)$. Therefore, dependence of product $S^2\sigma$ as a function of Fermi level E_F has the bell-shaped form. The maximum $S^2\sigma$ is achieved when E_F is close to the bottom of conduction band for n-type semiconductor or close to the top of valence band for p-type semiconductor. In this case, the value of Seebeck coefficient $S \approx 180 \pm 10$ μV/K and concentration $n(p) \sim 5 \times 10^{19}$ cm^{-3}. At concentration $n(p) \sim 5 \times 10^{19}$ cm^{-3}, heat transport is determined mainly by lattice thermal conductivity κ_L through phonons. For optimal concentration of charge carriers (reduced Fermi level $\mu^* \approx 0$), the figure of merit Z practically depends on three parameters:

$$Z \sim \left(m_n^*\right)^{3/2}(\mu/\kappa_{\mathrm{L}}). \qquad (1.12)$$

Now we can formulate the selection criteria of the effective thermoelectric material:

1. Semiconductor material (elemental, compound, alloy, composite, multilayer, superlattice, low-dimensional) should preferably consist of heavy atoms; it will cause a low frequency of thermal vibrations of lattice and, properly, significant decrease in lattice thermal conductivity κ_{L}.
2. Semiconductor material should have high dielectric constant $\varepsilon \geq 100$, that reduces significantly the scattering of charge carriers by impurity ions, and small effective mass of major charge carriers (m_n^* of electrons for n-type and m_p^* of heavy holes for p-type). These two criteria provide high mobility of charge carriers μ.

   Note that high dielectric constant ε leads to decrease in ionization energy of the impurity atoms (which is practically approaching zero). As a result, energy level of impurity merges with the conduction or valence band, and concentration of electrons in the conduction band of n-type material or concentration of heavy holes in the valence band of p-type material becomes constant (similar to metal) and equals to dopant concentration in temperature range from 0 K to the onset of intrinsic conductivity.
3. Semiconductor material should have multi-valley structure of the conduction band for n-type and the same structure of the valence band for p-type. In this case, concentration of charge carriers increases many times without changing energy of Fermi level E_{F}, because it is simply proportional to the number of ellipsoids and, properly, the electrical conductivity σ rises seriously as well.
4. Semiconductor material should allow heavy doping with impurities, ensuring the achievement of high concentrations of charge carriers $n(p) \geq 10^{19}$ cm^{-3} that will result in optimal Fermi level energy E_{F}, and, having in mind maximum value of $S^2\sigma$, it will mean $(E_{\mathrm{F}} - E_{\mathrm{c}}) \sim 0$ eV for n-type and $(E_{\mathrm{v}} - E_{\mathrm{F}}) \sim 0$ eV for p-type.
5. Bandgap of semiconductor E_{g} in operating temperature range of thermoelectric material should be $E_{\mathrm{g}} \geq 8k_0T$, to minimize contribution of thermally generated minority charge carriers to total electrical conductivity, leading to decrease in thermoelectric efficiency.

1.3 Basic Properties of Thermoelectric Materials Based on Bismuth Telluride

Bismuth telluride Bi_2Te_3 consists of heavy atoms; that is one of the reasons for its high thermoelectric efficiency. Crystal structure of Bi_2Te_3 (Fig. 1.1, left) is characterized by rhombohedral actual primitive cell (Fig. 1.1, top right). Lattice constant c of the rhombohedral cell (between bottom and top points) equals to approximately 5.08 nm. This is maximal size of cell which is the actual measure for comparison

Fig. 1.1 (Left) Crystal structure of Bi_2Te_3 with rhombohedral unit cell embedded inside. Bismuth and tellurium atoms are shown as yellow and green circles, respectively. (Top right) A separate rhombohedral unit cell structure, generated from XCrySDen. (Bottom right) Side view of a quintuple layer QL $(Te' - Bi - Te - Bi - Te')$ with each layer separated from the next by the real-ratio distance. The superscripts are used to distinguish two types of differently bonded Te atoms [1]

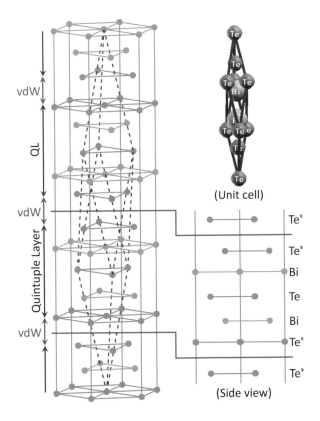

with size of nanoparticles when we will consider nanotechnology approaches. The bulk structure consists of orderly stacked alternating hexagonal monatomic crystal planes. Stacks of five neighboring monoatomic layers $Te' - Bi - Te - Bi - Te'$ form quintuple layers (QLs) (Fig. 1.1, bottom right). Bonding between atomic planes within a QL is strong covalent with a small fraction of ionic bonding. Ionic bond component is explained by different valence of Te (6) and Bi (5). The outer adjacent atomic layers of QLs are attracted by weak van der Waals (vdW) forces. This weak vdW bonding between QLs allows the crystal to be easily cleaved along an inter-QL plane (vdW interface), so crystals of Bi_2Te_3 have clearly defined cleavage planes. For convenience, crystal structure of Bi_2Te_3 and related alloys is often considered in the hexagonal basis. A unit cell in the hexagonal basis contains three QLs; lattice constant c equals to 3.049 nm.

As already mentioned, Z is determined by the ratio of the mobility of charge carriers to lattice thermal conductivity. Let's consider the case when the mobility is determined by scattering of charge carriers due to thermal vibrations of lattice only. It should be understood here that electrons or holes are scattered not by thermal vibrations of the atoms but by electric fields which occur due to distortion of crystal periodic potential because of these vibrations. The stronger the distortion of the periodic potential, the higher is the intensity of local fields caused by thermal vibrations. Therefore, mobility of charge carriers is usually low in ionic crystals, where relief of crystal potential is most pronounced. In contrast, in crystals with predominant covalent bonds, the crystal potential is smooth (in this case, there is no alternation of positively and negatively charged ions), and the mobility of charge carriers is much higher. As for lattice thermal conductivity, bismuth telluride consists of heavy Bi and Te atoms that cause a low frequency of thermal vibrations of lattice and, properly, a low lattice thermal conductivity. Basic parameters of Bi_2Te_3 are given in Table 1.1.

Specific crystalline structure of Bi_2Te_3 semiconductor gives in the result strong anisotropy of basic properties. As a result, values of electric and thermal conductivity in direction parallel to cleavage planes are much higher than transverse values (Table 1.1). Van der Waals (vdW) interfaces between quintuple outer adjacent atomic layers play a significant role in forming anisotropy of electric and thermal conductivity (Fig. 1.2) [1].

Table 1.1 Basic properties of bismuth telluride [2]

	Parameter	Symbol	Value
1	Melting point	T_m	585 °C
2	Bandgap at 300 K	E_g	0.13 eV
3	Temperature dependence of bandgap	dE_g/dT	-9×10^{-5} eV/K
4	Debye temperature	T_D	155.5 K
5	Number of ellipsoids:		
	– Valence band	N_c	6
	– Conduction band	N_v	6
6	Density of states (DOS) effective mass:		
	– Electrons	m_n^*	0.45 m_0
	– Heavy holes	m_p^*	0.69 m_0
7	Mobility of charge carriers:		
	– Electrons	μ_e	1200 cm^2/(V × s)
	– Heavy holes	μ_p	510 cm^2/(V × s)
8	Temperature dependence of mobility:		
	– Electrons	$d\mu_e/dT$	$T^{-1.7}$
	– Heavy holes	$d\mu_p/dT$	$T^{-2.0}$
9	Scattering parameter:		
	– Electrons	r_e	0
	– Holes	r_p	0
10	Dielectric constant	ε_0	400
11	Concentration of charge carriers at 300 K	n_i	10^{18} cm^{-3}
12	Lattice thermal conductivity along cleavage planes	$\kappa^\parallel$	14.5×10^{-3} W/(cm × K)
13	Anisotropy of thermal conductivity	$\kappa^\parallel/\kappa^\perp$	~2–3
14	Anisotropy of electrical conductivity:		
	n-type	$(\sigma^\parallel/\sigma^\perp)_n$	4–6
	p-type	$(\sigma^\parallel/\sigma^\perp)_p$	2.7

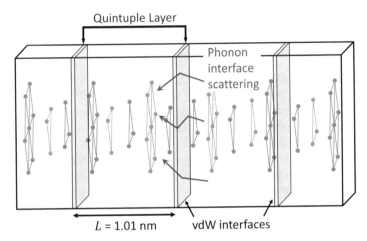

Fig. 1.2 Bi_2Te_3 structure projected in a superlattice structure. Van der Waals (vdW) interfaces exist between quintuple layers and cause an additional anharmonic phonon scattering [1]

Isotropy is fundamental property of Seebeck coefficient S in crystals having Bi_2Te_3 structure with charge carriers of one sign only. Therefore, as follows from Table 1.1 and Formula (1.12), anisotropy of Z of n-type material is high and close to 2, whereas Z of p-type material is almost isotropic. In crystals with anisotropic mobility of charge carriers like Bi_2Te_3, Seebeck coefficient becomes also anisotropic in the presence of charge carriers of opposite sign, especially when contribution of that charge carriers to total electric conductivity becomes tangible.

A.F. Ioffe suggested an idea of improving thermoelectric efficiency of available thermoelectric materials—elemental and binary compound semiconductors by means of producing of the so-called solid solutions (first, binary alloys of elemental semiconductors, binary compound semiconductors and ternary alloys of binary compounds). It opened great opportunities for improving efficiency and for Bi_2Te_3 as well.

Approach to development of improved performance ternary alloys based on Bi_2Te_3 was as follows: in bismuth telluride, the bismuth sublattice forms a negative charge; therefore, electrons should scatter on this sublattice, and, respectively, holes should scatter on tellurium sublattice. Thus, to get improved n-type ternary alloys, the third component (e.g., selenium) should be introduced in the tellurium sublattice, and to get improved p-type ternary alloys, the third component (e.g., antimony) should be introduced in the bismuth sublattice.

As a result, n-type $(Bi_2Te_3)_y(Bi_2Se_3)_{1-y}$ ternary alloys have been developed by S.S. Sinani and G.N. Gordeeva [3] in year 1956 and p-type $(Bi_2Te_3)_x(Sb_2Te_3)_{1-x}$ ternary alloys—by G.I. Shmelev [4] in year 1959. And these materials remain unrivaled in thermoelectric efficiency to nowadays.

By the way, the validity of an idea that alloys can provide higher thermoelectric efficiency has been confirmed later when developing high-temperature thermoelectric materials based on Si - Ge alloy system.

As for p-type $(Bi_2Te_3)_x(Sb_2Te_3)_{1-x}$ ternary alloys, G.I. Shmelev has determined (during the work on PhD thesis) alloy composition value where ratio μ/κ_L was maximal. Note that thermoelectric efficiency of pristine binary compounds Bi_2Te_3 and Sb_2Te_3 was equal to $Z = 1 \times 10^{-3}$ K^{-1}, whereas efficiency of p-type $(Bi_2Te_3)_x(Sb_2Te_3)_{1-x}$ ternary alloy with optimal composition x was equal to $Z = 2.6 \times 10^{-3}$ K^{-1}, which is very close to the best current values $(3.0–3.2) \times 10^{-3}$ K^{-1}.

The increase in thermoelectric efficiency of $(Bi_2Te_3)_x(Sb_2Te_3)_{1-x}$ ternary alloys with growth of Sb_2Te_3 content was due to a decrease in thermal conductivity which reached minimum at content of 70–80 mole % of Sb_2Te_3. In addition, at such content of Sb_2Te_3, ordering of crystal structure of ternary alloy was observed, which caused an increase in mobility of charge carriers. It should be noted that bandgap E_g of $(Bi_2Te_3)_x(Sb_2Te_3)_{1-x}$ ternary alloys increases monotonically with growth of mole fraction of Sb_2Te_3 in alloy, which is very important for thermoelectric generation applications. The bandgap of Bi_2Te_3 is not more than $E_g = 0.13$ eV at 300 K, i.e., $E_g/k_0T = 5$, and with increasing temperature, thermal generation of minority charge carriers degrades sharply thermoelectric efficiency of p-type Bi_2Te_3.

We describe and discuss in the chapter possible techniques of suppressing the negative effect of thermally generated minority charge carriers (electrons) on thermoelectric efficiency of p-type $(Bi_2Te_3)_x(Sb_2Te_3)_{1-x}$ ternary alloys.

1.4 Manufacturing Technology of Thermoelectric Materials

One of the rigorous selection criteria of semiconductor material for thermoelectric generation applications is mechanical strength, since legs made of that material will operate at significant temperature difference and, hence, at high mechanical stress. Furthermore, considering the tendency to decrease consumption of thermoelectric materials, especially, with high content of deficit tellurium, by means of shortening of legs' height, the role of enhanced mechanical strength is growing, since the thermomechanical stresses in thermoelectric leg increase inversely with the square of the height. Therefore, the use of legs made of materials manufactured by traditional crystallization/melting methods (zone melting technique, Czochralski method, Bridgman method) is unrealistic on practice. Materials for practical use in thermoelectric generation applications are now and will always be of polycrystalline or composite nature.

1.4.1 Pressing Technique

As a rule, serial Bi_2Te_3-based bulk materials are produced with powder compaction by pressing that included synthesis of ingots by proper melting technique, grinding of ingots into powder, pressing, and annealing. Technique enables also to do the so-called co-pressing of n- and p-type legs, when, during compaction, n- and p-type legs connect to each other forming ready-made thermocouple. The uniaxial pressing technique provides manufacturing of bulk anisotropic thermoelectric materials with optimal crystallographic texture including ternary alloys of $(Bi_2Te_3)_y(Bi_2Se_3)_{1-y}$ for n-type legs and $(Bi_2Te_3)_x(Sb_2Te_3)_{1-x}$ for p-type legs.

Thermoelectric efficiency Z of Bi_2Te_3 and related alloys is maximal when heat flow direction is parallel to cleavage planes of material (grain). Therefore, disordered orientation of crystalline grains in pressed materials should reduce Z. In addition, the value of Z can degrade due to defects (grain boundaries, oxide films, dislocations, vacancies, microcracks, etc.) that scatter charge carriers. Inherent feature of thermoelectric Bi_2Te_3 and related alloys manufactured by pressing is some self-ordering of grain orientation when cleavage planes are laying mainly perpendicular to the direction of pressing. This is because, during grinding, the ingot of starting material splits along the cleavage planes, and powder particles take the form of plates in which the plane coincides with the cleavage plane. The preferential orientation of the grains causes anisotropy of the properties of pressed materials: σ, κ, and Z have the greatest values in the direction perpendicular to the direction of pressing. The electrical conductivity and thermal conductivity of pressed samples are less than in polycrystals obtained by crystallization from melt.

1.4.2 Spark Plasma Sintering Technique

SPS (spark plasma sintering) technique of powder compaction became recently a frequent practice but in laboratory only. SPS technique allows of compacting powders made of materials that are difficult to compact by standard pressing due to the need to use forces exceeding the strength of press tool materials. In addition, SPS process provides sintering of grains without significant heating of whole load of powder, which is valuable for compacting not entirely stable systems, e.g., nanostructured powders. High-throughput conveyor-type SPS machines are recently available on the market, which is evidence of transfer of SPS technology for wide use in industry.

1.4.3 Extrusion Technique

Ideally, powder compaction technique of Bi_2Te_3 and related alloys should enable obtaining polycrystalline (textured) ingots (rods) with given orientation, since electrical conductivity and thermal conductivity are anisotropic, especially in n-type materials, because of layered structure (Table 1.1). As for p-type materials, at first sight, manufacturing of well-textured polycrystalline materials is not so critical for applications, since Z value at temperatures up to 100 °C is practically isotropic. But, as it will be shown further, that's all wrong, and obtaining oriented polycrystalline (textured) ingots (rods) is actual task for p-type materials.

The presence of evident cleavage planes in Bi_2Te_3 and related alloys provides obtaining of "flakes" during grinding, which, being packed in a mold, enable manufacturing well-textured ingots (rods) during subsequent pressing. This is a decisive point to the use of pressing and SPS techniques for compacting materials based on Bi_2Te_3 and related alloys.

However, materials with even better texture can be manufactured by hot extrusion technique of powder compaction. In this case, cleavage planes of grains align strictly parallel to the axis of extrusion. In addition, plastic deformation of material under high hydrostatic pressure provides effective repairing of structural defects and obtaining polycrystalline ingots (rods) with grain size about 10 μm and density above 96% of single crystal one.

At present time, hot extrusion becomes the main industrial technology for manufacturing high-performance Bi_2Te_3 and related alloy materials. Ingots (rods) of extruded materials with a diameter of up to 30 mm and thermoelectric efficiency at 25 °C $Z = (3.1–3.2) \times 10^{-3}$ K^{-1} for p-type material and $Z = 2.9 \times 10^{-3}$ K^{-1} for n-type material are commercially available (data provided by RusTec LLC, www.rustec-msk.com).

1.5 Options for Improving Efficiency of Thermoelectric Materials

Growing interest in "green energy" and requirement to reduce atmospheric greenhouse gas emissions under Kyoto Protocol stimulated formation of state R&D programs in the field of prospective thermoelectric materials and generators in several countries. Inflow of investments led to involvement of many new researchers in solving different thermoelectric issues, and now research work in thermoelectricity is one of the most active. Much research work is being done for improving thermoelectric efficiency Z of known and novel thermoelectric materials. Different approaches are being studied for that. However, there are not yet tangible results concerned improving Z value of well-known high-performance low-temperature materials Bi_2Te_3 and related alloys. Improving Z has only been shown in materials with initial low Z value. For example, significant progress in improving efficiency of mid-temperature thermoelectric materials—skutterudites—led to increase in Z values which are now close to real use in industry. Skutterudite materials were known for a long time, e.g., $CoSb_3$ was discovered by L.D. Dudkin in year 1956 and described in his PhD work [5].

There is no progress in improving thermoelectric efficiency of low-temperature materials for more than 60 years. The value of $Z = 3 \times 10^{-3}$ K^{-1} remains an unbeaten record, and new approaches and solutions are needed to overcome it.

1.5.1 Nanotechnology and Low-Dimensional Structures

Progress in grinding equipment is evident. High-speed planetary ball mills are commercially available now. Idea to decreasing thermal conductivity and, hence, increasing Z by reducing size of grains with proper increase in the number of boundaries which scatter phonons has become actual again. Let's consider this option more detailed. To provide a marked increase in scattering of phonons, grain sizes should be comparable to mean free path of phonons which equals to about 15 nm in materials like $(Bi_2Te_3)_x(Sb_2Te_3)_{1-x}$. At the same time, crystal structure of Bi_2Te_3 is rhombohedral cell with lattice constant c equals to approximately 5.08 nm (Sect. 1.3). It seems to be difficult to fabricate materials like Bi_2Te_3 with grain sizes that are only three times of the size of crystal unit cell using ordinary grinding of an already crystallized material. Likely, approach could be realized for densely packed cubic materials such as PbTe. And, although researchers have already published reports about successes of increase in Z of nanostructured $(Bi_2Te_3)_x(Sb_2Te_3)_{1-x}$, manufacturers of thermoelectric devices have not yet presented clear evidences that it's so.

It should be noted that the idea of increasing in thermoelectric efficiency by using structures with higher number of grain boundaries for phonon scattering and with barrier effect that provides increase in average energy of charge carriers in flow has always been in sight of researchers. It was found that increasing in thermoelectric efficiency was observed mainly for p-type $(Bi_2Te_3)_x(Sb_2Te_3)_{1-x}$ ternary alloys and no noticeable increase was observed in n-type $(Bi_2Te_3)_y(Bi_2Se_3)_{1-y}$ ternary alloys, and it was concluded that grain boundaries scatter electrons more strongly than holes. Gain in thermoelectric efficiency due to effects associated with grain boundaries did not exceed a few percent.

In contrast to methods of fabricating nanoparticles by grinding of a crystallized material with already formed chemical bonds, M.V. Kovalenko and colleagues [6–9] proposed original approach to building nanoparticles of thermoelectric materials from atoms by liquid-phase synthesis without melting. In proposed method, synthesis of nanoparticles occurs in aqueous solutions through the proper precursors. This approach allows to configure internal structure of nanoparticles, i.e., technology could produce structured materials both with effective scattering of charge carriers on phonons and with transformation of energy spectrum, including changes of effective mass of density of states (DOS) which is one of the key parameters for thermoelectricity. Apparent advantages of this technology in comparison with traditional vapor-phase technology are flexibility, ease in scaling, and low cost of technological processes.

Obviously, manufacturing of high-performance nanomaterials by compaction of nanoparticles will require solving of difficult problems including building a perfect texture of anisotropic materials like Bi_2Te_3, as well as ensuring reliable encapsulation of nanoparticles to prevent oxidation, aggregation, and crystallization at high temperatures in running thermoelectric generators.

Theoretical work published in year 1993 by L.D. Hicks and M.S. Dresselhaus [10] can be considered as kickoff for nanoengineering in thermoelectricity. Under certain assumptions made, it was shown that the reduction in body size in one or a few dimensions to nanometer level can cause due to quantum effects a change of energy spectrum. In transition to low-dimensional structures, the density of states (DOS) changes very strongly. Two key parameters for thermoelectricity—effective mass of charge carriers and Z—depend directly on DOS. Low-dimensional structures are called "quantum well" (2D), "quantum wire" (1D), and "quantum dot" (0D).

Estimations of low-dimensional effects for Bi_2Te_3 done in framework of L.D. Hicks and M.S. Dresselhaus model showed a strong increase in Z when the characteristic size of material grain decreased to 50 Å.

Unfortunately, theory has not yet tested, but development of nanotechnology in thermoelectricity will clear up the matters.

1.6   Novel Thermocouple Made of "Old" Materials

When we speak about options of increasing in efficiency Z, we should keep in mind, first, thermoelectric material operates in generator application over a wide range of temperatures and, second, material which is optimal, e.g., for use in low-temperature range, may be inefficient at higher temperatures. One option to provide optimal (the highest possible) efficiency of thermoelectric leg in wide temperature range is using a leg composed of a few parts (sub-legs) made of several materials with optimal efficiency in specified temperature range. Another option of optimization should also be considered, namely, leg composed of sub-legs made of same material with different crystalline orientation.

Traditionally, legs made of low-temperature materials like Bi_2Te_3 and related alloys are diced out so that cleavage planes in the legs are oriented parallel to heat flux direction. This approach is fully justified for legs made of n-type material, where anisotropy of electrical conductivity is close to 2, since this orientation provides maximum efficiency. However, for p-type legs, this condition is optional. Indeed, when cleavage planes in p-type legs are oriented parallel to heat flux direction, then at room temperature, thermoelectric efficiency of p-type legs is only somewhat (~ 10%) higher than transverse. The rise in temperature induces thermal generation of minority charge carriers (electrons) with high mobility. Thermally generated electrons produce an electromotive force of opposite sign that reduces the overall Seebeck coefficient S in accordance with formula:

$$S = \frac{S_n \sigma_n + S_p \sigma_p}{\sigma_n + \sigma_p}, \qquad (1.13)$$

where indexes n and p describe parameters determined by electrons and holes, respectively.

Table 1.1 shows that electron mobility is higher than hole mobility and electron concentration increases exponentially with temperature, which leads to a sharp decrease in Seebeck coefficient S and even greater decrease in Z, because Z is proportional to the square of S.

Note that isotropy is a fundamental property of Seebeck coefficient in crystal of any kind of structure with charge carriers of same sign only. As it follows from Formula (1.13), in crystals with anisotropy of charge carrier mobility, Seebeck coefficient becomes also anisotropic in the presence of charge carriers of opposite sign. Therefore, negative effect of minority charge carriers can be minimized by using crystal with orientation parallel to axis with minimal mobility of minority charge carriers. Electron mobility in Bi_2Te_3 and related alloys is minimal in direction transverse to cleavage planes. So, to build highly effective thermocouple operating in the temperature range where intrinsic conductivity began, we should use p-type leg with orientation transverse to cleavage planes. Certainly, n-type legs are diced out always with orientation parallel to cleavage planes, since anisotropy of Z in n-type material is high and close to 2.

However, traditionally, thermoelectric generation modules are assembled of n-type and p-type legs diced out with the same orientation parallel to cleavage planes. Old tradition is not always good tradition.

Temperature dependence of the anisotropy of Z in p-type $Bi_{0.5}Sb_{1.5}Te$ ternary alloy has been studied on two samples with sizes $(L \times W \times H)$ 18 mm $\times$ 10 mm $\times$ 10 mm. Ingot of material has been manufactured by hot extrusion technique (see Sect. 1.4.3). Samples were diced out parallel (i.e., with orientation parallel to cleavage planes) and transverse (i.e., with orientation transverse to cleavage planes) to extrusion axis from an ingot of material with $\sigma = 1850$ Ohm$^{-1} \times$ cm^{-1} and $S = 157$ μV/K at room temperature. Specific electric conductivity σ, Seebeck coefficient S, and specific thermal conductivity κ have been measured in temperature range of 50–380 °C. Figures 1.3, 1.4, and 1.5 show obtained temperature dependences, where indexes ($\parallel$) and ($\perp$) indicate the kinetic coefficients measured on samples diced out parallel and transverse to the extrusion axis, i.e., parallel and transverse to cleavage planes. Figure 1.6 shows calculated value of Z. Figures 1.7 and 1.8 show temperature dependences of the anisotropy of Seebeck coefficient $S_\perp/S_\parallel$ and thermoelectric efficiency $Z_\perp/Z_\parallel$. Obtained results confirm concept that variation of S and Z with growth of temperature is noticeably different in samples with working axis along and transverse to cleavage planes. Impressive gain due to change of the working axis of p-type material from traditional

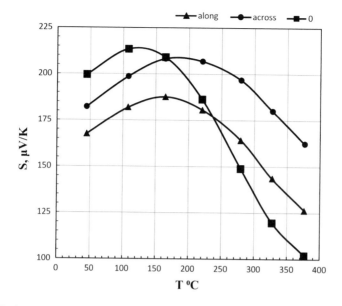

Fig. 1.3   Seebeck coefficient as a function of temperature

Fig. 1.4   Electrical conductivity as a function of temperature

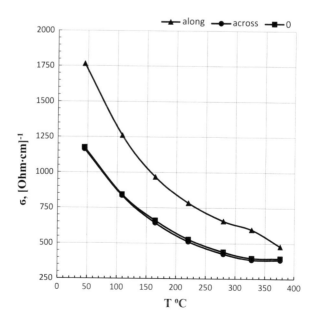

Z. Dashevsky and S. Skipidarov

Fig. 1.5 Thermal
conductivity as a function of
temperature

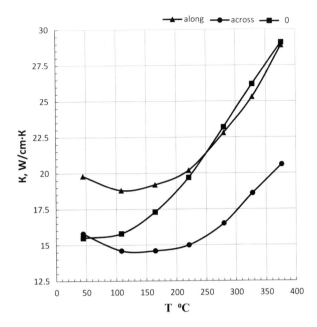

Fig. 1.6 Figure of merit as a function of temperature

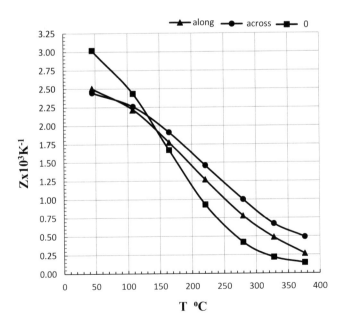

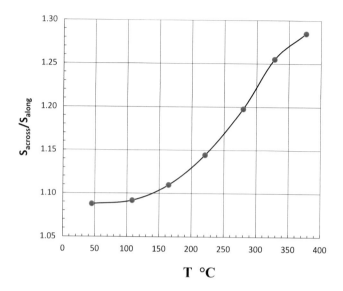

Fig. 1.7 Anisotropy of Seebeck coefficient $S_\perp/S_\parallel$ as a function of temperature

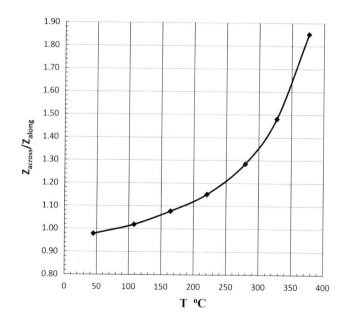

Fig. 1.8 Anisotropy of figure of merit $Z_\perp/Z_\parallel$ as a function of temperature

Hot side

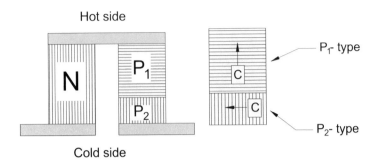

Cold side

Fig. 1.9 Schematic view of thermocouple with combined p-type leg

to alternative reaches at 350 °C up to 50%. That gain in thermoelectric efficiency can easily be obtained on already known materials.

Figures 1.3, 1.4, 1.5, and 1.6 show also temperature dependences (marked "0") of the same kinetic coefficients measured on sample of p-type $Bi_{0.5}Sb_{1.5}Te$ material but with lower level of doping and characterized by $\sigma = 1270$ $Ohm^{-1} \times cm^{-1}$ and $S = 183$ $\mu V/K$ at room temperature. This material with lower level of doping has been optimized for operating temperatures from room temperature and above.

Figure 1.9 suggests optimal design of advanced thermocouple with p-type leg composed of sub-legs with mutually perpendicular orientations of crystalline axes. Bottom sub-leg (cold part P_2) for operation in temperature range up to 130–140 °C should be made of moderately doped material ($\sigma \sim 1200$ $Ohm^{-1} \times cm^{-1}$) with the orientation of working axis parallel to cleavage planes, and top sub-leg (hot part P_1) for higher temperature range should be made of heavily doped material ($\sigma \sim 1800$– 1900 $Ohm^{-1} \times cm^{-1}$) with working direction transverse to cleavage planes.

Figure 1.9 shows a sketch of novel thermocouple.

1.7   Conclusions

Clearly expressed anisotropy of Seebeck coefficient S in p-type $(Bi_2Te_3)_x(Sb_2Te_3)_{1-x}$ ternary alloys at T > 100 °C has been observed for the first time.

Anisotropy of Seebeck coefficient in p-type samples with working axis parallel ($\parallel$) and transverse ($\perp$) to cleavage planes arises from anisotropy of mobility of minority charge carriers (electrons) along different crystalline axes $\mu_\parallel > \mu_\perp$.

At temperatures T > 100 °C, efficiency $Z_\perp$ of extruded samples diced out with working axis transverse to cleavage planes is higher than $Z_\parallel$ of samples diced out with working axis along cleavage planes. At temperature T = 350 °C, efficiency gain equals to ~1.6.

References

1. K.H. Park, M. Mohamed, Z. Aksamija, U. Ravaioli, Phonon scattering due to van der Waals forces in the lattice thermal conductivity of Bi_2Te_3 thin films. J. Appl. Phys. **117**, 015103 (2015). https://doi.org/10.1063/1.4905294
2. B.M. Goltsman, B.A. Kudinov, I.A. Smirnov, *Thermoelectric semiconductor materials based on Bi_2Te_3* (Nauka, Moscow, 1972)
3. S.S. Sinani, G.N. Gordeeva, Bi_2Te_3—Bi_2Se_3 solid solutions as materials for thermoelements. Tech Phys (Sov.). **26**, 2398–2399 (1956)
4. G.L. Shmelev, Materials for thermoelements based on three-component intermetallic alloys. Solid State Phys (Sov.). **1**, 63–75 (1959)
5. L.D. Dudkin, N.K.H. Abrikosov, Physico-chemical investigations of cobalt antimony. J Inorg Chem (Sov.) **1**, 2096–2101 (1956)
6. M.V. Kovalenko, B. Spokoyny, J.-S. Lee, M. Scheele, A. Weber, S. Perera, D. Landry, D.V. Talapin, Semiconductor nanocrystals functionalized with antimony telluride Zintl ions for nanostructured thermoelectrics. J. Am. Chem. Soc. **32**, 6686–6695 (2010). https://doi.org/10.1021/ja909591x
7. M.V. Kovalenko, L. Manna, A. Cabot, Z. Hens, D.V. Talapin, C.R. Kagan, V.I. Klimov, A.I. Rogach, P. Reiss, D.J. Milliron, P. Guyot-Sionnnest, G. Konstantatos, W.J. Parak, T. Hyeon, B.A. Korgel, C.B. Murray, W. Heiss, Prospects of nanoscience with nanocrystals. ACS Nano **9**, 1012–1057 (2015). https://doi.org/10.1021/nn506223h
8. S. Ortega, M. Ibáñez, Y. Liu, Y. Zhang, M.V. Kovalenko, D. Cadavid, A. Cabot, Bottom-up engineering of thermoelectric nanomaterials and devices from solution processed nanoparticle building blocks. Chem. Soc. Rev. **46**, 3510–3528 (2017). https://doi.org/10.1039/c6cs00567e
9. M. Ibáñez, R. Hasler, Y. Liu, O. Dobrozhan, O. Nazarenko, D. Cadavid, A. Cabot, M.V. Kovalenko, Tuning p-type transport in bottom-up-engineered nanocrystalline Pb chalcogenides using alkali metal chalcogenides as capping ligands. ACS Chem Mater **29**(17), 7093–7097 (2017). https://doi.org/10.1021/acs.chemmater.7b02967
10. L.D. Hicks, M.S. Dresselhaus, Effect of quantum-well structures on the thermoelectric figure of merit. Phys. Rev. B **47**, 12727–12731 (1993)

Chapter 2
SnSe: Breakthrough or Not Breakthrough?

Christophe Candolfi, Dorra Ibrahim, Jean-Baptiste Vaney, Selma Sassi, Philippe Masschelein, Anne Dauscher, and Bertrand Lenoir

Abstract The simple binary semiconductor SnSe has not been for long considered as a promising thermoelectric material compared to PbSe. For this reason, its thermoelectric properties have been largely overlooked over the past decades. This compound came back to the forefront of research in thermoelectricity due to very high thermoelectric figure of merit ZT values reported recently in single crystals. This announcement has been the starting point of renewed interest leading to a wealth of experimental and theoretical studies in the past few years with the aim of better understanding its physical properties and optimizing its thermoelectric performances in both *p*- and *n*-type samples. Here, we review the progress of research on transport properties of both single-crystalline and polycrystalline SnSe, highlighting in particular the controversy regarding thermal properties of single crystals and important points that remain to be investigated to determine whether or not the (re) discovery of SnSe may be considered as a breakthrough in thermoelectricity.

2.1 Introduction

The design of highly efficient thermoelectric materials represents a formidable challenge. Such compound should indeed strike subtle balance between electronic and thermal properties [1–3]. High thermopower α and low electrical resistivity ρ typical of heavily doped semiconductors should be reconciled with low total thermal conductivity κ, usually seen in glassy systems, in order to maximize dimensionless thermoelectric figure of merit $\mathrm{ZT} = \frac{\alpha^2 T}{\rho \kappa} = \frac{PT}{\kappa}$ at absolute temperature T [1–3]. Adding another degree of complexity, these mutually incompatible properties cannot be optimized independently due to interdependence through concentration of charge carriers. Most of the best thermoelectric materials possess

C. Candolfi (✉) · D. Ibrahim · J.-B. Vaney · S. Sassi · P. Masschelein · A. Dauscher · B. Lenoir
Institut Jean Lamour, UMR 7198 CNRS—Université de Lorraine, Nancy, France
e-mail: christophe.candolfi@univ-lorraine.fr; dorra.ibrahim@univ-lorraine.fr;
jean-baptiste.vaney@nims.go.jp; selma.sassi@univ-lorraine.fr;
philippe.masschelein@univ-lorraine.fr; anne.dauscher@univ-lorraine.fr;
bertrand.lenoir@univ-lorraine.fr

© Springer Nature Switzerland AG 2019
S. Skipidarov, M. Nikitin (eds.), *Novel Thermoelectric Materials and Device Design Concepts*, https://doi.org/10.1007/978-3-030-12057-3_2

favourable energy band structure with multiple valleys that help to achieve high power factors P [1–3]. Concomitantly, these thermoelectric materials harbour also specific crystallographic features that strongly limit the ability of acoustic phonons to propagate, for instance, localized lattice vibrations ("rattling modes") or chemical bonds that vibrate anharmonically [4–8]. Additional geometric architecting such as nanostructuring or inducing nanoscale domains through controlled synthetic conditions represents another strategy to limit heat transport and, thus, achieve better thermoelectric performances [9–12].

For decades, the family of lead telluride (PbTe) compounds and its selenide analogue PbSe have been thermoelectric materials of choice for operating at medium temperatures (typically near 500 K) [1–3]. These semiconductors gather many important characteristics to be a prospective area of research for achieving high thermoelectric performances [1–3]. Despite the simple cubic rock-salt crystal structure, lattice thermal conductivity is very low due to significant anharmonicity in the lattice that originates from strong anharmonic coupling between transverse optic modes and longitudinal acoustic modes, as shown by neutron spectroscopy techniques [13–15]. The electronic band structure with non-parabolic bands and multiple valleys is the key ingredient that explains good thermoelectric properties achieved by proper doping. In comparison to other chalcogenide semiconductors, PbTe-based compounds have received significant attention in the 1960s and 1970s and are still the subject of numerous studies nowadays [1–3].

Tin selenide SnSe is another representative of vast family of chalcogenide semiconductors which basic crystallographic and physical properties have also been investigated in these decades [16–18]. Although chemically similar to PbTe, SnSe crystallizes with distorted lattice described in orthorhombic space group Pnma [19–21]. In retrospective, it is perhaps not so surprising that this compound has received less attention in this period. Its orthorhombic crystal structure gives rise to anisotropic transport properties, even in polycrystalline samples. In addition, the number of extrinsic dopants that can be introduced in SnSe seemed more restricted than in PbTe-based compounds. For these reasons, significant efforts were devoted to optimizing thermoelectric properties of PbTe and its derivatives.

In 2014, this forgotten "ugly-duckling" compound made remarkable comeback to the forefront of research in thermoelectricity with the announcement of record-breaking ZT values of up to 2.6 at 900 K in single crystals [22, 23]. What was even more surprising was the fact that such high values have been achieved in pristine single crystals, while high ZT values are usually reached once a careful optimization of electronic and thermal properties has been realized. Another striking outcome of this report was very low lattice thermal conductivity measured at high temperatures ranking this compound among the best thermal insulators known to date [22]. Such extremely rare combination of transport properties raised a number of issues: Can one optimize even further thermoelectric performances of single-crystalline SnSe through doping? Is it possible to achieve similar ZT values in polycrystalline samples more suitable for integration in thermoelectric generators? What microscopic mechanisms do impede so efficiently the propagation of heat-carrying acoustic phonons in such simple crystal structure? In addition to these important questions of both practical and fundamental interest, the difference between thermal conductivity measured in this study [22] and that reported in the 1960s [16–18] has rapidly become another source of interrogation and debate within the thermoelectric community. Although consensual answers to some of these questions are yet to be obtained, this announcement stirred up strong interest to revisit in detail physical properties of this compound by modern techniques.

In this chapter, we will provide a brief review of accumulated knowledge acquired on this compound based on experimental and theoretical results obtained over the last few years. We will focus more particularly on low-temperature allotrope of SnSe referred to as α − SnSe which shows the best thermoelectric performances. Our goal is to highlight important structural and chemical aspects that influence its transport properties and on the strategies employed to optimize its thermoelectric performances. As excellent reviews on this compound and, more generally, on tin chalcogenides already exist [24–26], we will only focus on the main aspects and issues that recently emerged and remain to be elucidated in future studies.

2.2 Crystal Structure, Synthesis and Defect Chemistry

2.2.1 Crystal Structure

At room temperature, SnSe adopts an orthorhombic crystal structure (space group Pnma, No. 62; $a = 11.37$ Å, $b = 4.19$ Å and $c = 4.44$ Å at 300 K) which can be thought of as three-dimensional distortion of rock-salt cubic structure of PbTe (Fig. 2.1) [19–21]. The basic building block is composed of two-atom-thick SnSe slabs that run along $b − c$ plane. These slabs are bound by weak Sn − Se bonding along a axis and are corrugated, creating zig-zag chains when the structure is projected along b direction (see Fig. 2.1). Interestingly, this creased honeycomb structure is similar to that of black phosphorus; sheets of black phosphorus called phosphorene are currently attracting attention for fast-response electronic and photonic devices [27].

Upon warming, SnSe undergoes displacive structural transition near 800 K which leaves orthorhombic symmetry of the unit cell unchanged [22, 28]. High-temperature allotrope, referred to as β − SnSe, is described in the space group Cmcm (No. 63) with room-temperature lattice parameters $a = 4.31$ Å, $b = 11.71$ Å and $c = 4.32$ Å [22, 28]. Other allotropes have been identified at high pressures with structural modifications above 27 GPa that give rise to a topologically nontrivial semimetallic state and to superconductivity at 39 GPa with critical temperature of 3.2 K [29–37]. Transition α to β, which has been directly imaged by high-resolution transmission electron microscopy [22], leaves clear signature on both transport and thermodynamic properties. As we will see below, anisotropic crystal structure of SnSe results in significant anisotropy in transport properties, should the samples be in single-crystalline or polycrystalline form.

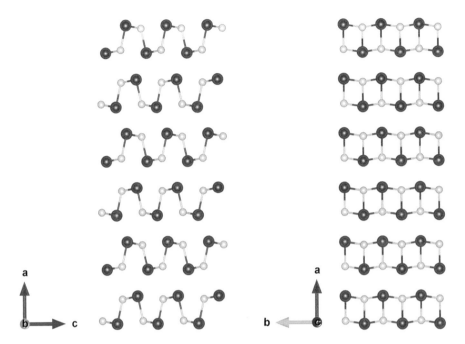

Fig. 2.1 Projection of SnSe crystal structure (space group Pnma, No. 62) along *b* and *c* axis. Atoms of Sn and Se are in red and green, respectively

2.2.2 Synthesis of Polycrystalline and Single-Crystalline SnSe

Since elemental Sn and Se are commercially available with very low impurity levels, both highly pure polycrystalline and single-crystalline samples can be easily synthesized. Large polycrystalline specimens can be obtained by direct melting the powders or shots of the elements at high temperatures in sealed silica ampoules. Unlike other thermoelectric materials for which such technique does not necessarily yield homogeneous ingots because of inherent defect chemistry, polycrystalline SnSe samples appear to be well homogeneous.

Large single crystals can be synthesized by more complex growth techniques. While early studies on SnSe used vapour phase techniques in sealed silica tubes [16–18], another elegant technique is Bridgman technique. This method implies moving the ampoule with carefully shaped end through large temperature gradient with the highest temperature above the melting point of the compound. The crystal then solidifies from a seed at shaped end of the ampoule and progressively forms along its length. This process has been carried out successfully in both horizontal and vertical configurations to obtain large, centimetre-sized single crystals of SnSe. However, one major issue regarding the growth of single-crystalline SnSe is related to phase transition which undergone near 800 K, that is, below the melting point of this compound (1134 K). Together with high vapour pressure of Se, the volume increase across β-to-α transition (around 2.5%) can lead to breakage of ampoule during the process. This problem can be mitigated by sealing the first silica ampoule into the second one to prevent loss of Se vapour and oxidation of the elements. The single crystals obtained by such techniques have typically centimetre-size lateral dimensions and can be easily cleaved along a and c axes [22, 38–54]. Of note, other techniques related to hydro- and solvothermal syntheses have been also successfully used to obtain single-crystalline plates or nanobelts [55, 56]. Obtaining large crystals is a necessary prerequisite to study in detail the components of transport properties tensors and allows for more sophisticated spectroscopic techniques to be employed.

2.2.3   Defect Chemistry

As in many thermoelectric materials [1–3], defect chemistry plays an important role in determining transport properties of SnSe. Although this predisposition to defects is not as critical in SnSe as, for instance, in well-known Bi_2Te_3-based alloys, nevertheless, presence of defects can result in varying physical properties along the growth direction in single crystals. In particular, such variations can be problematic when combining transport properties measured on samples taken at different positions in the crystal to calculate ZT values.

Pioneering works on SnSe single crystals grown by vapour techniques have probed the full range of concentrations of charge carriers showing that only p-type crystals can be obtained with hole concentration varying between 3×10^{15} and 2×10^{18} cm^{-3} at 300 K [16–18]. Hole concentration strongly depends on the details of synthetic procedure and possible subsequent annealing step employed (a process referred to as saturation annealing). Although this p-type nature was supposed to be due to an excess of Se atoms [16–18], it was not clear at that time whether Sn vacancies, anti-site defects or both types of defects were responsible for this behaviour. Furthermore, laboratory structural and chemical probes such as X-ray diffraction or energy-dispersive spectroscopy are not sensitive enough to discriminate between different amounts and types of defects. The determination of the main type of defects has been recently addressed by detailed electronic band structure calculations [57]. These theoretical studies are consistent with native p-type doping primarily due to Sn vacancies which are energetically favoured over anti-site defects under both Sn- and Se-rich conditions. Another possible source of extrinsic p-type doping has been recently advanced based on the presence of $SnSe_2$ microdomains in single crystals grown under various experimental conditions [58]. These inhomogeneities are believed to yield interfacial charge transfer to occur from SnSe to $SnSe_2$ resulting in overall hole-doped SnSe matrix. Yet, as mentioned by the authors, high cooling rate employed during the growth process is likely the key parameter that leads to nucleation of this secondary phase. Several other recent studies performed on single crystals have indeed not reported the presence of such domains while these pristine crystals remain intrinsically hole-doped.

2.3 Electronic Band Structure and Electronic Properties

2.3.1 Electronic Band Structure of α − SnSe

Significant efforts have been devoted to determining both theoretically and experimentally the main features of electronic band structure of SnSe that govern its transport properties. While some dispersion in computed band gap exists in literature due to the details of the calculations and the method used, most of studies indicate band gap on the order of 0.6–0.9 eV [22, 43, 59–72]. This value is consistent with experimental band gap values of 0.86–0.92 eV obtained by optical absorption spectroscopy [22, 54].

The band gap is indirect with valence band maximum (VBM) lying along Γ − Z direction, while conduction band minimum (CBM) is located along Γ − Y direction (Fig. 2.2) [60]. The overall structure of valence and conduction bands is rather complex with several maxima and minima, respectively, close to each other in energy. In particular, the highest VBM shows "pudding-mould-like" shape, that is, a band that shows dispersive part and flatter portion nearby [60]. Such shape has in particular been proposed to be beneficial for achieving good thermoelectric performances [73]. Furthermore, the second valence band maximum lies just below the main valence band maximum.

Indirect nature of band gap implies anisotropic density-of-states (DOS) effective masses that differ significantly between three crystallographic axes. DOS effective mass ellipsoids were calculated to be equal to 0.74 m_0, 0.31 m_0 and 0.16 m_0 (m_0 is

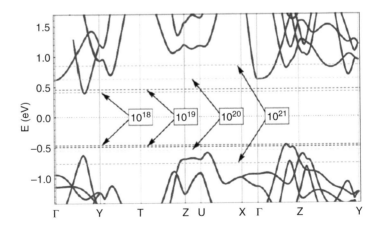

Fig. 2.2 Electronic band structure along high-symmetry directions of low-temperature structure of SnSe (space group Pnma, No 62) calculated by Korringa-Kohn-Rostoker method [60]. Dashed horizontal lines correspond to various hole and electron concentrations used to illustrate the bands that participate to conduction at these concentrations. Pudding-mould-like band is located along Γ − Z direction. Reproduced with permission from Kutorasinski et al. [60]. Copyright 2018 by the American Physical Society. https://doi.org/10.1103/PhysRevB.91.205201

bare electron mass) for the highest VBM and 2.40 m_0, 0.11 m_0 and 0.15 m_0 for the lowest CBM along a, b and c axes, respectively [59]. The second VBM shows similar values along a and c directions (0.90 m_0 and 0.15 m_0, respectively), while it differs more significantly along b direction (0.12 m_0) [59]. The presence of multiple local extremum near band edges is an important aspect to understand good electronic properties achieved in this compound as we will see below.

Based on calculated band structure, electrical conductivity and thermopower have been computed using Boltzmann transport approach for a wide range of temperatures and p- and n-type doping levels [60]. The results evidenced significant anisotropy in transport properties between different crystallographic axes, as expected for such layered crystal structure. The anisotropy depends on concentration of charge carriers, and while, at relatively high concentrations, thermopower shows only little variations between three directions, more significant differences may appear at concentrations of charge carriers ~10^{17} cm^{-3}. One of the main results of these calculations is related to the prediction that n-type SnSe should potentially exhibit higher thermoelectric performances with respect to p-type analogues [60].

2.3.2 Angle-Resolved Photoemission Spectroscopy

Because the details of electronic band structure can depend on specific computational method used, direct experimental probes of electronic band structure are particularly relevant to unveil its main characteristics in both pristine and hole-doped crystals. The growth of high-quality large single crystals and the ease with which these crystals can be cleaved make the angle-resolved photoemission spectroscopy (ARPES) a powerful tool for such investigations. Experimental investigation of SnSe valence band structure by this technique has been recently addressed by several groups as function of doping level (undoped and Na-doped single crystals) and temperature [27, 58, 74–76]. Probably due to low doping level studied, the presence of Na had little influence on the main characteristics of band structure of SnSe [76]. Regarding pristine SnSe, all these studies consistently evidenced the presence of highly anisotropic valence bands with multiple valleys in agreement with available calculated band structures (Fig. 2.3).

DOS effective masses have been derived from collected ARPES $k - E$ maps along the main directions ($\Gamma - X$, $\Gamma - Y$ and $\Gamma - Z$). The results obtained along $\Gamma - Z$ and $\Gamma - Y$ directions indicated effective masses on the order of 0.2 m_0 along b and c axes for both directions. The values along a axis are slightly higher as predicted theoretically. These results are thus consistent with each other's and in line with predicted values. Nevertheless, among these studies, in [76] have been found values significantly higher than those obtained in other studies by one order of magnitude (effective masses of up to 2.23 m_0) even at the lowest temperature probed. This study also evidenced that effective masses tend to decrease with increasing temperature up to 600 K.

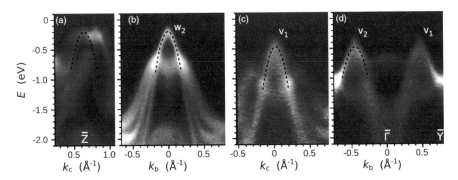

Fig. 2.3 Energy dispersion of the highest pockets in valence bands of SnSe measured by ARPES in the vicinity of Z (panels **a** and **b**) and Y (panels **c** and **d**) points which correspond to two perpendicular directions parallel to the layers. Overlaid dashed black curves represent fits using parabolic dispersions to estimate effective masses. Reproduced with permission from Pletikosić et al. [27]. Copyright 2018 by the American Physical Society. https://doi.org/10.1103/PhysRevLett. 120.156403

These results help to understand the beneficial role of this multivalley band structure on ZT values. Indeed, overall ZT values along c axis of crystal structure are related to band parameters through the following relation [77]:

$$ZT \propto \sum_i \gamma_i \tau_i \sqrt{\frac{m_{ai}^* m_{bi}^*}{m_{ci}^*}}$$

where m_{ai}^*, m_{bi}^* and m_{ci}^* are DOS effective masses along a, b and c axes of band i, τ_i is relaxation time of charge carriers along transport direction for band i and γ_i is degeneracy of band i which is equal to 2 in SnSe. This simple formula explains, for instance, poorer values observed along a axis. In this direction, ZT values are proportional to coefficient $\sqrt{\frac{m_{bi}^* m_{ci}^*}{m_{ai}^*}}$ which is significantly smaller due to the less dispersive valence bands, resulting in higher DOS effective mass m_{ai}^* compared to m_{bi}^* and m_{ci}^*. Hence, these experimental results evidence the beneficial contribution of several valence bands to transport properties and provide an insightful guidance for developing relevant transport models to describe the variations of these bands with hole concentration (in so-called Pisarenko-Ioffe plot).

2.3.3 Electronic Transport Properties of Pristine and Doped Single-Crystalline SnSe

Because of inherent defect chemistry of SnSe, hole concentration of synthesized single crystals can vary from sample to sample. Since the discrimination of such small amounts of defects is very difficult, the most effective metric is the characterization of its electrical properties and, notably, its concentration of charge carriers. Pristine single-crystalline SnSe has been always obtained so far as lightly doped p-type semiconductor [22, 38–54]. This state is reflected by high electrical resistivity and thermopower values measured with hole concentration ~10^{17} cm^{-3} at 300 K. As expected from its orthorhombic crystal structure, transport properties are anisotropic, and those studies in single-crystalline specimens require careful examination along three crystallographic axes. Due to low hole concentrations, electrical resistivity values are typically on the order of few Ohm × m at 300 K and show semiconducting behaviour with increasing temperature. Concomitantly, thermopower can reach values as high as 800 µV × K^{-1} at 300 K [22, 38–54]. While the anisotropy remains moderate in thermopower values, electrical resistivity is more anisotropic following the sequence $\rho_a > \rho_c > \rho_b$ along a, b and c crystallographic axes. However, this is not to say that thermopower cannot show some significant anisotropy as predicted theoretically [60] and demonstrated experimentally [42].

Only few studies on doped single crystals have been undertaken so far [38, 43, 45, 46, 78–80]. While substitutions of Na, Ag and Bi for Sn have been considered, only solid solution SnSe − SnS has been considered as a mean to substitute on Se site. Content of Na as low as 3 at. % is enough to enhance significantly hole concentration with maximum values higher than 10^{19} cm^{-3} at 300 K [38, 43, 45]. Although this increase in hole concentration induces a decrease in thermopower values, the net result of doping is positive with an overall increase in power factor at high temperatures [38, 43, 45]. Introduction of similar amount of Ag acts similarly with analogous trends in electrical properties as a function of Ag content [43, 46]. Intriguingly, the substitution of Sn by 6 at. % of Bi switches electrical conduction to n type over the whole temperature range covered (300–800 K) [78]. With power factors that rival those obtained in p-type samples, these results show that good thermoelectric performances can be equally achieved in both n- and p-type crystals. Without exceptions, for all these dopants, the best thermoelectric performances have been achieved along b axis as in pristine single crystals [38, 43, 45, 46, 78–80]. Because electronic band structure calculations have indicated that better performances could be achieved in n-type SnSe [60], further studies on single crystals with dopants known to induce n-type conduction such as Br (see below) would be worthwhile.

2.3.4 Electronic Transport Properties of Pristine and Doped Polycrystalline SnSe

Although the study of single crystals is illuminating in determining basic physical mechanisms governing transport properties and those possible evolutions as a function of doping level, carefully controlled environments required for growth and weak mechanical properties make these crystals poor candidates for direct integration in thermoelectric generators. The assessment of thermoelectric

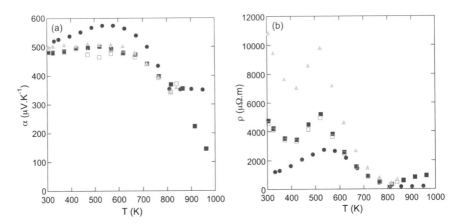

Fig. 2.4 Temperature dependence of (**a**) thermopower and (**b**) electrical resistivity of undoped polycrystalline SnSe measured parallel (triangle symbols) and perpendicular (square symbols) to pressing direction. The measurements have been performed on two different samples showing good reproducibility (distinguished by filled and open symbols). Average values of the data measured on single-crystalline SnSe [22] (filled circle symbols) have been added for comparison. Material reproduced from Sassi et al. [81], Appl. Phys. Lett., 104, 212,105 (2014)

performances in polycrystalline samples thus rapidly became a central topic within the thermoelectric community.

The first reports on these aspects [81, 82] have demonstrated that pristine SnSe behaves as lightly doped p-type semiconductor with values of electrical resistivity and thermopower in line with those achieved in single-crystalline samples (Fig. 2.4). The anisotropy seen in single crystals is not fully suppressed and survives in polycrystals. The transport properties thus need to be measured both parallel and perpendicular to pressing direction to estimate correctly ZT values.

The most surprising result evidenced by these investigations [81, 82], and widely confirmed in the following studies, is higher thermal conductivity values measured with respect to reported single crystals. Even though the values reached at high temperatures are very low for compound with such simple crystal structure (~ 0.5 W $\times$ m^{-1} $\times$ K^{-1} at 800 K [81, 82]), the difference between polycrystalline and single-crystalline samples is particularly unexpected since the presence of grain boundary scattering in the former should in principle result in lower values. The origin of this discrepancy will be specifically addressed in detail below. In pristine polycrystalline samples, maximum ZT values of 0.5 have been achieved at 823 K [81, 82]. Noteworthy, orthorhombic-to-orthorhombic structural transition undergone by SnSe near this temperature leads to significant loss of Se and to irreversible plastic deformation of the samples [81]. This clearly precludes any use of this compound at temperatures above 800 K, a conclusion further corroborated by the rapid degradation at temperatures above 873 K observed in oxidation studies on polycrystalline SnSe [83].

With the aim of further tuning concentration of charge carriers, numerous dopants have been considered resulting in either p-type or n-type electrical conduction. For *p*-type doping, the main elements known to effectively dope PbTe and PbSe have been primarily investigated. These include alkali metals (Na [83–89], K [85, 87, 90], Li [87]) and alkaline-earth elements (Ca, Sr and Ba [91]). In addition, various transition metals (Ag [82, 92–95], Cu [91, 96, 97], Zn [91, 98], Al [97], Tl [99], Ti [100] and In [97, 101]), metalloids (Pb [89, 97], Te [102–104], S [105, 106], Ge [91, 107–109]) as well as some lanthanides (Sm [110] and LaCl$_3$ [111]) have also been investigated. In contrast, halogens Cl [112, 113], Br [114] and I [106] as well as Bi [113] were shown to induce *n*-type electrical conduction. Only few studies have used so far a multiple-doping approach either on single site (*n*-type Sn$_{0.74}$Pb$_{0.20}$Ti$_{0.06}$Se [100]) or on both sites (such as in *p*-type Sn$_{0.99}$Na$_{0.01}$Se$_{0.84}$Te$_{0.16}$ [102]). In most cases, the introduction of foreign element increases in concentration of charge carriers which help to optimize power factor. Due to low doping levels, most of these dopants have little influence on lattice thermal conductivity, giving rise ultimately to enhanced ZT values near 800 K. So far, peak ZT values ranging between 0.8 and 1.3 have been achieved in either *p*-type or *n*-type samples.

The variations of electronic properties with concentration of charge carriers and, more particularly, the evolution of thermopower with hole concentration have been successfully modelled by considering a simple parabolic band model assuming that either one or two valence bands govern transport properties [87]. For this analysis, several sets of available data in literature on both polycrystalline and single-crystalline samples have been considered (Fig. 2.5).

The average DOS effective mass m^*_{DOS} used in transport equations can be calculated from the values along three axes determined theoretically or experimentally by ARPES using the formula $m^*_{DOS} = \left(m^*_{ai} m^*_{bi} m^*_{ci} \right)^{1/3} \gamma^{2/3}$ (with the degeneracy factor γ equal to 2 for SnSe). Each of these models (with one or two valence bands) describes equally well the variations of thermopower with hole concentration at 300 K for concentrations varying between 10^{17} cm^{-3} and 10^{20} cm^{-3} [87]. The best fit value of m^*_{DOS} (1.1 m_0) agrees very well with the value predicted by electronic band structure calculations or by ARPES experiments. These results consistently show that single parabolic band model is enough to describe satisfactorily the evolution of thermopower as function of doping level in *p*-type SnSe.

Further attempts at optimizing thermoelectric properties of SnSe have been considered through the formation of solid solutions with SnS [105], SnTe [115] or PbSe [116, 117]. Solid solutions are usually a convenient way to further control the concentration of charge carriers and to decrease lattice thermal conductivity via enhanced point defect scattering. Although this last mechanism is less pronounced in the present case due to very low lattice thermal conductivity already exhibited by this compound, these solid solutions resulted in slightly enhanced ZT values with peak value of 0.8 achieved at 823 K for SnSe$_{0.8}$S$_{0.2}$ [105]. A notable exception is the system SnSe − PbSe for which very high ZT values of up to 1.7 at 873 K have been reported [116]. In this last study, the introduction of PbSe beyond the solubility limit led to formation of nanoprecipitates, the influence of which on transport properties has been proposed as the main key factor. However, the results reported in another

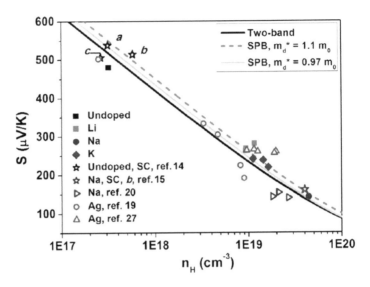

Fig. 2.5 Pisarenko-Ioffe plot (thermopower as function of hole concentration) at 300 K. Data obtained both on single-crystalline (SC) and polycrystalline doped SnSe have been added. Two-band model and two single parabolic band (SPB) models with different DOS effective masses (denoted as m_d^*) have been considered to generate black solid, grey dashed and grey solid curves, respectively; m_d^* value of 0.97 corresponds to that determined theoretically, and value of 1.1 represents the best fit to the data. References [14, 15, 19, 20, 27] correspond to References [22, 38, 82, 84, 94] in the present chapter, respectively. Reprinted (adapted) with permission from Wei et al. [87], J. Am. Chem. Soc. 138, 8875 (2016). Copyright 2016 by the American Chemical Society

study [117] on the same system show much lower values with a maximum twice as low as the above-mentioned value (around 0.85 at 800 K). Further work seems, therefore, necessary to determine the reasons behind such significant discrepancy between two studies.

Finally, some studies on composite samples have been reported recently where incorporation of MoS_2, $MoSe_2$ or graphene into SnSe has been studied [118, 119]. However, none of those have led to significantly higher ZT values compared to that achieved in pristine SnSe.

2.4 Thermal Properties

2.4.1 Specific Heat of SnSe

The specific heat C_p provides valuable information on a given material below and above room temperature. At low temperatures, the specific heat represents energy-averaged response that encodes low-energy features of phonon subsystem associated

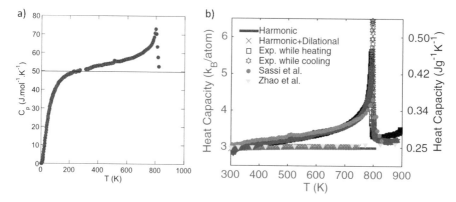

Fig. 2.6 (a) Temperature dependence of the specific heat of SnSe measured at low and high temperatures [120]. (b) High-temperature specific heat of SnSe determined experimentally by differential scanning calorimetry and calculated from phonon density of states measured by inelastic neutron scattering. The data are compared to those shown in panel **a** [120] and measured by Zhao et al. [22] on single crystals. (Panel **a**) Material reproduced from Sassi et al. [120]. Materials Today: Proceedings 2, 690 (2015). (Panel **b**) Reproduced Fig. 12 with permission from Bansal et al. [27]. Copyright 2016 by the American Physical Society. https://doi.org/10.1103/PhysRevB.94. 054307

with optical phonons or zone-boundary excitations. At high temperatures, this thermodynamic property is an essential component to be measured accurately in order to correctly assess the thermal conductivity. The specific heat of SnSe has been studied on both polycrystalline and single-crystalline specimens (Fig. 2.6).

At low temperatures (Fig. 2.6a), the results are consistent with semiconducting ground state observed in pristine SnSe giving rise to null electronic contribution to experimental uncertainty [120, 121]. Near 10 K, specific heat data, when plotted as $\frac{C_p}{T^3}$ vs. T, deviate from conventional Debye behaviour typified by pronounced peak. This peak has been interpreted as either Einstein contributions or as Schottky contribution [120, 121]. In this last model [121], the origin of this peak has been discussed in terms of hard and soft substructures related to short and weaker bonds within the layers, respectively. Thus, these results show that the lattice dynamics of SnSe is dominated by two energy scales, in line with two main regions observed in phonon density of states [44, 66, 122].

Temperature dependence of specific heat at high temperatures steadily increases with increasing in temperatures up to about 600 K (Fig. 2.6a, b). Above this temperature, nearly linear increase in C_p turns into non-linear variation. Near 800 K, clear peak is observed which corresponds to orthorhombic-to-orthorhombic transition [22, 81, 122]. The transition appears as a broad peak between 650 and 800 K in agreement with its continuous nature [22]. In order to determine the specific heat values over this temperature range for calculating total thermal conductivity, it is, therefore, necessary to extrapolate linearly the values measured below 600 K up to 800 K.

2.4.2 Thermal Conductivity of Single-Crystalline and Polycrystalline SnSe

One of the main ingredients that led to the very high ZT values claimed in single-crystalline SnSe is very low lattice thermal conductivity κ_L values measured along different axes of the crystal structure [22]. At 300 K, the values were reported to be 0.46, 0.70 and 0.67 W $\times$ m^{-1} $\times$ K^{-1} which further decrease to reach 0.24, 0.34 and 0.31 W $\times$ m^{-1} $\times$ K^{-1} at 700 K along a, b and c axes, respectively [22]. Of note, due to high electrical resistivity values measured in pristine single crystals, electron contribution to thermal conductivity $\kappa_e = \frac{LT}{\rho}$, where L is Lorenz number, is negligible so that $\kappa(T) \equiv \kappa_L(T)$.

Soon after this report, the question regarding lattice thermal conductivity of single-crystalline SnSe has rapidly become a central issue for several reasons. First, compared to lattice thermal conductivity values measured in the 1960s [16–18], values reported by Zhao et al. [22] are significantly lower. Second, subsequent studies on transport properties of polycrystalline pristine samples have reported higher lattice thermal conductivity values than in single crystals [81, 82]. This result is counterintuitive since the presence of grain boundaries is expected to provide additional phonon scattering channel which should lead to the opposite trend, that is, lower values should be observed in pristine polycrystalline samples. Third, the anisotropy observed between three crystallographic axes was not following the sequence $\kappa_L^a < \kappa_L^c < \kappa_L^b$ predicted by ab initio phonon calculations [123]. All these elements raised two fundamental questions: What could be the origin of such discrepancies between polycrystalline and single-crystalline samples? Were the values measured in single crystals by Zhao et al. [22] underestimated?

A first important point to answer these questions has been raised by Wei et al. [124] and is related to the density of samples used in the study of Zhao et al. [22]. Using thermal conductivity, specific heat and thermal diffusivity values reported, these authors pointed out that the density of samples measured was corresponding to only 88% of theoretical density of SnSe from crystallographic data [124]. This lower density could provide a natural explanation of lower values measured in [22] and would thus point to a significant underestimation of lattice thermal conductivity of single-crystalline SnSe. With higher κ_L values, the contradiction between values measured in polycrystalline and single-polycrystalline samples would then vanish, thereby recovering more conventional trend in κ_L from single to polycrystals.

Beyond this simplest explanation, several mechanisms have been advanced to try to explain the difference between single-crystalline and polycrystalline specimens as well as lower values measured in the former [125]. Notably, air sensitivity under ambient conditions has been evoked as a possible problem. However, several studies have pointed out that SnSe is not air sensitive enough to consistently explain the observed differences in κ_L. Another possible reason discussed is the presence of "vast off-stoichiometry" in single crystals studied by Zhao et al. [22, 45]. Since SnSe inherently possesses defects in the form of Sn vacancies, large amounts of such defects would tend to decrease the density in comparison to more stoichiometric crystals and would act as efficient source of phonon scattering leading to lower κ_L values. However, estimated concentration of Sn vacancies should be as large as 16.5% to reconcile these different aspects [45, 124]. This concentration is much higher than those obtained on single crystals grown under various synthetic conditions in the 1960s [16–18], and it remains unclear whether the structure of SnSe can accommodate such vacancy concentrations or, equivalently, whether SnSe remains stable with such off-stoichiometry.

Another point of concern related to this explanation is rooted in band gap measured on these single crystals. Assuming that such large numbers of defects are indeed possible, we may expect these to have a significant influence on electronic band structure. In particular, significant variation in band gap could be anticipated. However, direct measurements of band gap width by absorption spectroscopy [22, 54] yielded results consistent with band gap of around 0.9 eV expected for stoichiometric SnSe. Both results could perhaps still be reconciled, but this would require a rather surprising set of circumstances, since large number of defects should indeed result in modifications of electronic band structure but with net result of constant band gap. Of note, investigations on the influence of off-stoichiometry in polycrystalline $Sn_{1-x}Se$ (with $x = 0.01, 0.02$ and 0.04) are in agreement with binary $Sn - Se$ phase diagram indicating that initial Sn deficiency is accommodated by formation of secondary phase $SnSe_2$ rather than by creating Sn vacancies [126]. Hence, whether significant Sn deficiency is possible in SnSe remains to be demonstrated and would require further theoretical and experimental investigations. In particular, calculations of electronic band structure with large number of Sn vacancies should provide insights into the expected band structure. Moreover, systematic measurements of band gap by absorption spectroscopy should be performed on single crystals with different concentrations of charge carriers to determine whether varying concentration of defects indeed can lead to modifications of band gap.

Recent studies on pristine single-crystalline SnSe with densities close to theoretical density for stoichiometric SnSe have been reported. These investigations have found lattice thermal conductivity higher than the one reported by Zhao et al. [22]. For instance, Ibrahim et al. [42] measured κ_L values of 1.2, 2.3 and 1.7 W $\times$ m^{-1} $\times$ K^{-1} at 300 K along a, b and c axes (Fig. 2.7).

These values, between two and three times higher than those reported in prior studies, are consistent with early reports in the 1960s and with the sequence predicted by ab initio calculations [16–18, 123]. Upon warming, these values decrease to reach a value of 1 W $\times$ m^{-1} $\times$ K^{-1} at 700 K in b and c directions. As a consistency check, the values obtained by low-temperature measurements match very well with those obtained by laser flash technique at high temperatures [42]. Since these measurements were carried out using different measurement technology, it avoids any bias related to particular technology.

Below 300 K, κ_L exhibits crystalline-like behaviour with pronounced Umklapp peak (also referred to as dielectric maximum) that reaches the values 52, 27 and 20 W $\times$ m^{-1} $\times$ K^{-1} at 15 K along b, c and a directions, respectively (see Fig. 2.7) [42, 127]. Because the magnitude of this peak is sensitive to crystal

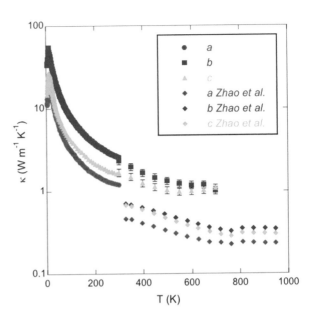

Fig. 2.7 Temperature dependence of lattice thermal conductivity measured on single-crystalline SnSe along a, b and c axes [42]. The data are compared to those reported by Zhao et al. [22]

imperfections, it usually can be considered as effective metric of the quality of grown crystal. The amplitude of this peak follows the sequence seen in κ_L values with the lowest values observed along a axis [42]. This result is not surprising since this direction corresponds to interlayer direction along which phonon mean free path is expected to be the shortest due to stronger phonon scattering at interface between the layers.

Another recent study on Na-doped and Ag-doped single crystals further corroborates all these results [127]. The measurements of thermal conductivity at low temperatures are consistent with the above-mentioned main features. In particular, Umklapp peak reported along b axis for pristine crystal is similar to that measured by Ibrahim et al. [42]. Upon doping, the amplitude of this peak is reduced due to additional point defect scattering induced by doping atoms. What makes these results particularly interesting is that these measurements have been performed on crystals for which very high ZT values had been reported at high temperatures (with maximum ZT of 2 at 800 K [43]). This astonishing difference between both studies has been discussed in terms of inhomogeneous distribution of dopants in the crystals. Such discrepancy clearly warrants further experimental investigations to determine whether ZT values of up to 2 reported [43] are really achievable in SnSe or are rather due to an erroneous combination of transport properties measured on samples with different hole concentrations due to inhomogeneous distribution of doping species.

In polycrystalline SnSe, measurements need to be performed both parallel and perpendicular to pressing direction due to anisotropic crystal structure [81, 82]. Compared to single crystals, published data on polycrystals exhibit less dispersion, and most of these studies agree on the values in pristine SnSe. At 300 K, measured κ_L values are 0.9 and 1.4 W $\times$ m^{-1} $\times$ K^{-1} in parallel and perpendicular direction,

Fig. 2.8 Lattice thermal conductivity as function of temperature measured on pristine polycrystalline SnSe parallel (circle symbol) and perpendicular (triangle symbol) to pressing direction [120]. The difference observed near room temperature is due to thermal radiations present in low-temperature measurements. Material reproduced from Sassi et al. [120], Materials Today: Proceedings 2, 690 (2015)

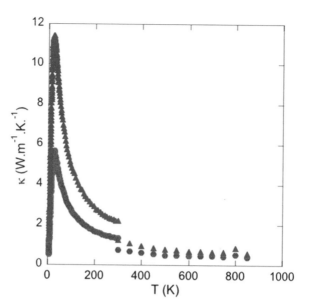

respectively [81, 82]. High amplitude Umklapp peak observed in single crystals survive in polycrystals with respective peak values of 12 and 6 W × m^{-1} × K^{-1} near 25 K [120]. Upon increasing temperature, κ_L values decrease to reach very low values that are close to 0.5 W × m^{-1} × K^{-1} in both directions (Fig. 2.8).

One important aspect when estimating lattice thermal conductivity of doped polycrystalline samples is related to determination of Lorenz number L which values sensitively depend on temperature and on scattering mechanism. Thus, L values can significantly deviate at high temperatures from conventional degenerate limit of free electron or hole gas (2.45 × 10^{-8} V^2 × K^{-2}). As mentioned earlier, single parabolic band model is enough to explain the trends in electronic properties with concentration of charge carriers. This model can be also used to compute the temperature dependence of Lorenz number in doped samples. This approach gives typically in the result Lorenz numbers (1.5–2.0) × 10^{-8} V^2 × K^{-2} at high temperatures [87, 127]. When subtracting electronic contribution with these L values, similar κ_L values are then observed in doped compounds regardless of the element used for doping. Nevertheless, some differences can be observed since solid solutions or heavy doping can provide an additional phonon scattering source [105, 116, 117].

2.4.3 Microscopic Origin of Low Lattice Thermal Conductivity of SnSe

Thanks to the possibility to grow large, high-quality single crystals of SnSe, advanced spectroscopic tools can be used to pinpoint the microscopic origin of very low lattice thermal conductivity of this compound. To this end, inelastic neutron scattering is the experiment of choice since this allows to directly probe acoustic and optical phonon dispersions. Such experiment carried out on SnSe has revealed several striking features that make it distinct from other rock-salt semi-conductors [44, 122]. These results have shown that continuous structural phase transition near 800 K is associated with condensation of phonon mode that strongly softens upon warming (Fig. 2.9).

Intriguingly, this mode remains strongly anharmonic below temperature of phase transition over wide temperature range. This anharmonicity, beyond that associated with conventional thermal expansion of the unit cell, gives rise to anharmonic phonon-phonon interactions which effectively scatter heat-carrying acoustic pho-nons and naturally leads to very low lattice thermal conductivity values measured. This type of lattice-instability-driven suppression of heat transport is reminiscent to those seen in the related rock-salt semiconductors PbTe, SnTe and GeTe [13–15]. However, some important differences exist between these three compounds and SnSe. PbTe is close to ferroelectric transition, but its crystal structure remains stable because softening of transverse optic mode is only partial [13–15]. The cubic crystal structure of both SnTe and GeTe exhibits displacive distortion giving rise to ferroelectric rhombohedral structure, while SnSe is not ferroelectric in its low-temperature orthorhombic phase. Thus, SnSe represents model system to study in more detail the role of this type of lattice instability and its related anharmonicity on thermal transport.

2.5 Conclusion

The simple orthorhombic semiconductor SnSe has focused attention in recent years owing to record-breaking ZT values at high temperatures reported in Bridgman-grown single crystals. This announcement has revitalized interest for this compound to gain in-depth understanding of its structural, chemical and physical properties. Thanks to multivalley structure of valence bands revealed by various electronic band structure calculations and confirmed by ARPES experiments, SnSe exhibits good electronic transport properties near 800 K. Combined with very low lattice thermal conductivity due to some peculiarities in its bonding scheme, this results in thermo-electric performances that rival those achieved in PbTe-based thermoelectric materials.

Several extrinsic dopants have been successfully used to tailor its thermoelectric properties revealing that SnSe is in fact as chemically flexible as other rock-salt semiconductors. These investigations have raised ZT values from 0.5 at 800 K in pristine samples up to around 1.0 at 800 K in properly optimized polycrystalline samples. These lower values with respect to those announced in single crystals come from significant difference in lattice thermal conductivity measured. The fact that higher values have been measured in polycrystals challenges usual trend that single crystals should possess higher lattice thermal conductivity due to absence of grain boundary scattering. This counterintuitive finding has raised concern about true

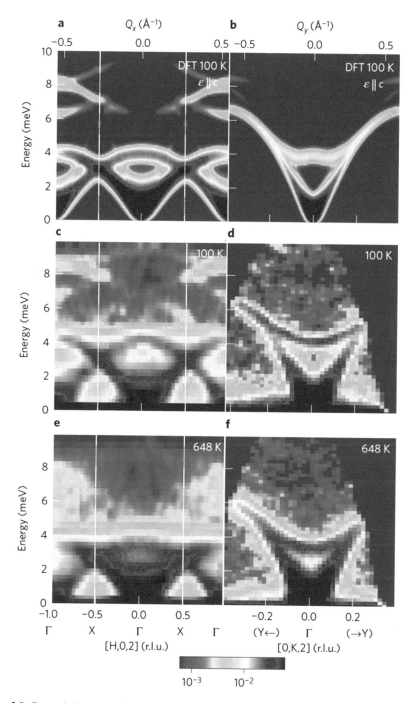

Fig. 2.9 Dynamical structure factors calculated by DFT at 100 K (panels **a** and **b**) and measured experimentally on single-crystalline SnSe at 100 K (panels **c** and **d**) and at 648 K (panels **e** and **f**) along $\Gamma - X$ and $\Gamma - Y$ directions. These measurements show strong softening experienced by low-energy optic phonons upon warming. Data reproduced from Li et al. [44] Nat. Phys. **11**, 1063 (2015)

lattice thermal conductivity values of single-crystalline SnSe, an issue which is still debated at the time of writing this review.

As a concluding remark, does the (re)discovery of SnSe be considered as a breakthrough? On one hand, announcement of record-breaking ZT values has led to plethora of theoretical and experimental investigations of its physical properties, should it be in bulk form or as thin film down to only one single layer. In this sense, our knowledge on this so far largely overlooked material has been significantly advanced over the last years thanks to the modern techniques employed to probe its basic properties. On the other hand, lattice thermal conductivity values required to reach these high ZT values seem to have been significantly underestimated, casting serious doubt on the possibility to experimentally achieve such high thermoelectric performances. Thus, honorific title "best thermoelectric material ever discovered" that SnSe is sometimes dubbed is likely overemphasized. Although very intriguing on its own, this simple binary compound probably does not represent a real break-through in thermoelectricity. Yet, this compound and its Te- and S-based analogues remain a fascinating area of research to explore in years to come.

References

1. H.J. Goldsmid, *Thermoelectric refrigeration* (Temple Press Books Ltd, London, 1964). https://doi.org/10.1007/978-1-4899-5723-8
2. D. M. Rowe (ed.), *Thermoelectrics and its Energy Harvesting* (CRC Press, Boca Raton, FL, 2012)
3. G. Tan, L.-D. Zhao, M.G. Kanatzidis, Chem. Rev. **116**, 12123 (2016)
4. T. Takabatake, K. Suekuni, T. Nakayama, E. Kaneshita, Rev. Mod. Phys. **86**, 669 (2014)
5. M.M. Koza, H. Mutka, Y. Okamoto, J. Yamaura, Z. Hiroi, Phys. Chem. Chem. Phys. **17**, 24837 (2015)
6. D.J. Safarik, T. Klimczuk, A. Llobet, D.D. Byler, J.C. Lashley, J.R. O'Brien, N.R. Dilley, Phys. Rev. B **85**, 014103 (2012)
7. M.D. Nielsen, V. Ozolins, J.P. Heremans, Energy Environ. Sci. **6**, 570 (2013)
8. Y. Bouyrie, C. Candolfi, S. Pailhès, M.M. Koza, B. Malaman, A. Dauscher, J. Tobola, O. Boisron, L. Saviot, B. Lenoir, Phys. Chem. Chem. Phys. **17**, 19751 (2015)
9. J. Androulakis, I. Todorov, J. He, D.Y. Chung, V. Dravid, M.G. Kanatzidis, J. Am. Chem. Soc. **133**, 10920 (2011)
10. L.-D. Zhao, V.P. Dravid, M.G. Kanatzidis, Energy Environ. Sci. **7**, 251 (2014)
11. Z.-G. Chen, G. Han, L. Yang, L. Cheng, J. Zou, Prog. Nat. Sci.: Mater. Int. **22**, 535 (2012)
12. J.-F. Li, W.-S. Liu, L.-D. Zhao, M. Zhou, NPG Asia Mater. **2**, 152 (2010)
13. O. Delaire, J. Ma, K. Marty, A.F. May, M.A. McGuire, M.-H. Du, D.J. Singh, A. Podlesnyak, G. Ehlers, M.D. Lumsden, B.C. Sales, Nat. Mater. **10**, 614 (2011)
14. C.W. Li, J. Ma, H.B. Cao, A.F. May, D.L. Abernathy, G. Ehlers, C. Hoffmann, X. Wang, T. Hong, A. Huq, O. Gourdon, O. Delaire, Phys. Rev. B **90**, 214303 (2014)
15. C.W. Li, O. Hellman, J. Ma, A.F. May, H.B. Cao, X. Chen, A.D. Christianson, G. Ehlers, D.J. Singh, B.C. Sales, O. Delaire, Phys. Rev. Lett. **112**, 175501 (2014)
16. J.D. Wasscher, W. Albers, C. Haas, Solid State Electron. **6**, 261 (1963)
17. H. Maier, D.R. Daniel, J. Electron. Mater. **6**, 693 (1977)
18. W. Albers, C. Haas, H. Ober, G.R. Schodder, J.D. Wasscher, J. Phys. Chem. Solids **23**, 215 (1962)

19. M.J. Peters, L.E. McNeil, Phys. Rev. B **41**, 5893 (1990)
20. R. Sharma, Y. Chang, J. Phase Equilib. **7**, 68 (1986)
21. H. Okamoto, J. Phase Equilib. **19**, 293 (1998)
22. L.-D. Zhao, S.-H. Lo, Y. Zhang, H. Sun, G. Tan, C. Uher, C. Wolverton, V.P. Dravid, M.G. Kanatzidis, Nature **508**, 373 (2014)
23. J.P. Heremans, Nature **508**, 327 (2014)
24. Z.-G. Chen, X. Shi, L.-D. Zhao, J. Zou, Prog. Mater. Sci. **97**, 283 (2018)
25. R. Moshwan, L. Yang, J. Zou, Z.-G. Chen, Adv. Funct. Mater. **27**, 1703278 (2017)
26. W. Li, Y. Wu, S. Lin, Z. Chen, J. Li, X. Zhang, L. Zheng, Y. Pei, ACS Energy Lett. **2**, 2349 (2017)
27. I. Pletikosić, F. von Rohr, P. Pervan, P.K. Das, I. Vobornik, R.J. Cava, T. Valla, Phys. Rev. Lett. **120**, 156403 (2018)
28. M. Sist, J. Zhang, B. Brummerstedt Iversen, Acta Crystallogr. B **72**, 310 (2016)
29. J.J. Yan, F. Ke, C.L. Liu, L. Wang, Q.L. Wang, J.K. Zhang, G.H. Li, Y.H. Han, Y.Z. Ma, C.X. Gao, Phys. Chem. Chem. Phys. **18**, 5012 (2016)
30. T. Chattopadhyay, A. Werner, H.G. von Schnering, J. Pannetier, Rev. Phys. Appl. (Paris) **19**, 807 (1984)
31. I. Loa, R.J. Husband, R.A. Downie, S.R. Popuri, J.-W.G. Bos, J. Phys. Condens. Matter **27**, 072202 (2015)
32. J. Zhang, H.Y. Zhu, X.X. Wu, H. Cui, D.M. Li, J.R. Jiang, C.X. Gao, Q.S. Wang, Q.L. Cui, Nanoscale **7**, 10807 (2015)
33. S.M. de Souza, H.O. da Frota, D.M. Trichês, A. Ghosh, P. Chaudhuri, M.S. dos Santos Gusmao, A.F.F. de Figueiredo Pereira, M.C. Siqueira, K.D. Machado, J.C. de Lima, J. Appl. Crystallogr. **49**, 213 (2016)
34. S. Alptekin, J. Mol. Model. **17**, 2989 (2011)
35. L. Makinistian, E.A. Albanesi, Comput.Mater. Sci. **50**, 2872 (2011)
36. Y.A. Timofeev, B.V. Vinogradov, V.B. Begoulev, Phys. Solid State **39**, 207 (1997)
37. X. Chen, P. Lu, X. Wang, Y. Zhou, C. An, Y. Zhou, C. Xian, H. Gao, Z. Guo, C. Park, B. Hou, K. Peng, X. Zhou, J. Sun, Y. Xiong, Z. Yang, D. Xing, Y. Zhang, Phys. Rev. B **96**, 165123 (2017)
38. L.-D. Zhao, G. Tan, S. Hao, J. He, Y. Pei, H. Chi, H. Wang, S. Gong, H. Xu, V.P. Dravid, C. Uher, G.J. Snyder, C. Wolverton, M.G. Kanatzidis, Science **351**, 141 (2016)
39. C. Julien, M. Eddrief, I. Samaras, M. Balkanski, Mater. Sci. Eng. B-Adv. **15**, 70 (1992)
40. A. Elkorashy, J. Phys. Chem. Solids **51**, 289 (1990)
41. A. Elkorashy, J. Phys. Chem. Solids **47**, 497 (1986)
42. D. Ibrahim, J.-B. Vaney, S. Sassi, C. Candolfi, V. Ohorodniichuk, P. Levinsky, C. Semprimoschnig, A. Dauscher, B. Lenoir, Appl. Phys. Lett. **110**, 032103 (2017)
43. K. Peng, X. Lu, H. Zhan, S. Hui, X. Tang, G. Wang, J. Dai, C. Uher, G. Wang, X. Zhou, Energy Environ. Sci. **9**, 454 (2016)
44. C.W. Li, J. Hong, A.F. May, D. Bansal, S. Chi, T. Hong, G. Ehlers, O. Delaire, Nat. Phys. **11**, 1063 (2015)
45. D. Wu, L. Wu, D. He, L.D. Zhao, W. Li, M. Wu, M. Jin, J. Xu, J. Jiang, L. Huang, Y. Zhu, M.G. Kanatzidis, J. He, Nano Energy **35**, 321 (2017)
46. M. Jin, H. Shao, H. Hu, D. Li, J. Xu, G. Liu, J. Cryst. Growth **460**, 112 (2017)
47. M. Jin, H. Shao, H. Hu, D. Li, H. Shen, J. Xu, J. Alloys Compd. **712**, 857 (2017)
48. M. Nassary, Turk. J. Phys. **33**, 201 (2009)
49. V. Bhatt, K. Gireesan, G. Pandya, J. Cryst. Growth **96**, 649 (1989)
50. B. Nariya, A. Dasadia, M. Bhayani, A. Patel, A. Jani, Chalcogenide Lett. **6**, 549 (2009)
51. A. Agarwal, S.H. Chaki, D. Lakshminarayana, Mater. Lett. **61**, 5188 (2007)
52. A. Agarwal, M. Vashi, D. Lakshminarayana, N. Batra, J. Mater. Sci-Mater. El **11**, 67 (2000)
53. A. Agarwal, P. Patel, D. Lakshminarayana, J. Cryst. Growth **142**, 344 (1994)
54. J.G. Yu, A. Yue, O. Stafsudd, J. Cryst. Growth **54**, 248 (1981)
55. M. Hong, Z.-G. Chen, L. Yang, T.C. Chasapis, S.D. Kang, Y. Zou, J. Mater. Chem. A **5**, 10713 (2017)

56. J. Guo, J. Jian, J. Liu, B. Cao, R. Lei, Z. Zhang, Nano Energy **38**, 569 (2017)
57. Y. Huang, C. Wang, X. Chen, D. Zhou, J. Du, S. Wang, L. Ning, RSC Adv. **7**, 27612 (2017)
58. Z. Wang, C. Fan, Z. Shen, C. Hua, Y. Hu, F. Sheng, Y. Lu, H. Fang, Z. Qiu, J. Lu, Z. Liu, W. Liu, Y. Huang, Z.-A. Xu, D.W. Shen, Y. Zheng, Nat. Commun. **9**, 47 (2018)
59. G. Shi, E. Kioupakis, J. Appl. Phys. **117**, 065103 (2015)
60. K. Kutorasinski, B. Wiendlocha, S. Kaprzyk, J. Tobola, Phys. Rev. B **91**, 205201 (2015)
61. Y. Suzuki, H. Nakamura, Phys. Chem. Chem. Phys. **17**, 29647 (2015)
62. S. Chen, K. Cai, W. Zhao, Phys. Rev. B **407**, 4154 (2012)
63. R.L. González-Romero, A. Antonelli, J.J. Meléndez, Phys. Chem. Chem. Phys. **19**, 12804 (2017)
64. Y. Huang, L. Li, Y.-H. Lin, C.-W. Nan, J. Phys. Chem. C **121**, 17530 (2017)
65. A. Shafique, Y.-H. Shin, Sci. Rep. **7**, 506 (2017)
66. R. Guo, X. Wang, Y. Kuang, B. Huang, Phys. Rev. B **92**, 115202 (2015)
67. X. Guan, P. Lu, L. Wu, L. Han, G. Liu, Y. Song, S. Wang, J. Alloys Compd. **643**, 116 (2015)
68. D.D. Cuong, S.H. Rhim, J.-H. Lee, S.C. Hong, AIP Adv. **5**, 117147 (2015)
69. K. Tyagi, B. Gahtori, S. Bathula, N.K. Singh, S. Bishnoi, S. Auluck, A.K. Srivastava, A. Dhar, RSC Adv. **6**, 11562 (2016)
70. J. Yang, G. Zhang, G. Yang, C. Wang, Y.X. Wang, J. Alloys Compd. **644**, 615 (2015)
71. G. Ding, G. Gao, K. Yao, Sci. Rep. **5**, 9567 (2015)
72. A.J. Hong, L. Li, H.X. Zhu, Z.B. Yan, J.M. Liu, Z.F. Ren, J. Mater. Chem. A **3**, 13365 (2015)
73. K. Kuroki, R. Arita, J. Phys. Soc. Jpn. **76**, 08370 (2007)
74. K. Zhang, K. Deng, J. Li, H. Zhang, W. Yao, J. Denlinger, Y. Wu, W. Duan, S. Zhou, Phys. Rev. Mater. **2**, 054603 (2018)
75. V. Tayari, B.V. Senkovskiy, D. Rybkovskiy, N. Ehlen, A. Fedorov, C.-Y. Chen, J. Avila, M. Asensio, A. Perucchi, P. di Pietro, L. Yashina, I. Fakih, N. Hemsworth, M. Petrescu, G. Gervais, A. Grüneis, T. Szkopek, Phys. Rev. B **97**, 045424 (2018)
76. Q. Lu, M. Wu, D. Wu, C. Chang, Y.-P. Guo, C.-S. Zhou, W. Li, X.-M. Ma, G. Wang, L.-D. Zhao, L. Huang, C. Liu, J. He, Phys. Rev. Lett. **119**, 116401 (2017)
77. D. M. Rowe (ed.), *CRC Handbook of Thermoelectrics* (CRC Press, London, 1995)
78. A.T. Duong, G.D. Van Quang Nguyen, S.K. Van Thiet Duong, J.Y. Song, J.K. Lee, J.E. Lee, S. Park, T. Min, J. Lee, J. Kim, S. Cho, Nat. Commun. **7**, 13713 (2016)
79. T.M.H. Nguyen, A.T. Duong, G. Duvjir, T.L. Trinh, V.Q. Nguyen, J. Kim, S. Cho, Bull. Am. Phys. Soc. **62** (2017)
80. T.T. Ly, G. Duvjir, T. Min, J. Buyn, T. Kim, M.M. Saad, N.T.M. Hai, S. Cho, J. Lee, J. Kim, Phys. Chem. Chem. Phys. **19**, 21648 (2017)
81. S. Sassi, C. Candolfi, J.-B. Vaney, V. Ohorodniichuk, P. Masschelein, A. Dauscher, B. Lenoir, Appl. Phys. Lett. **104**, 212105 (2014)
82. C.-L. Chen, H. Wang, Y.-Y. Chen, T. Day, G.J. Snyder, J. Mater. Chem. A **2**, 11171 (2014)
83. Y. Li, J.P. Heremans, J.-C. Zhao, J. Alloys Compnd. **669**, 224 (2016)
84. E.K. Chere, Q. Zhang, K. Dahal, F. Cao, J. Mao, Z. Ren, J. Mater. Chem. A **4**, 1848 (2016)
85. Z.-H. Ge, D. Song, X. Chong, F. Zheng, L. Jin, X. Qian, L. Zheng, R.E. Dunin-Borkowski, P. Qin, J. Feng, L.-D. Zhao, J. Am. Chem. Soc. **139**, 9714 (2017)
86. H.-Q. Leng, M. Zhou, J. Zhao, Y.-M. Han, L.-F. Li, RSC Adv. **6**, 9112 (2016)
87. T.R. Wei, G. Tan, X. Zhang, C.F. Wu, J.F. Li, V.P. Dravid, G.J. Snyder, M.G. Kanatzidis, J. Am. Chem. Soc. **138**, 8875 (2016)
88. B. Cai, J. Li, H. Sun, P. Zhao, F. Yu, L. Zhang, D. Yu, Y. Tian, B. Xu, J Alloys Compd **727**, 1014 (2017)
89. Y.K. Lee, K. Ahn, J. Cha, C. Zhou, H.S. Kim, G. Choi, S.I. Chae, J.-H. Park, S.-P. Cho, S.H. Park, Y.-E. Sung, W.B. Lee, T. Hyeon, I. Chung, J. Am. Chem. Soc. **139**, 10887 (2017)
90. Y.-X. Chen, Z.-H. Ge, M. Yin, D. Feng, X.-Q. Huang, W. Zhao, J. He, Adv. Funct. Mater. **26**, 6836 (2016)
91. B.R. Ortiz, H. Peng, A. Lopez, P.A. Parilla, S. Lany, E.S. Toberer, Phys. Chem. Chem. Phys. **17**, 19410 (2015)

92. L. Zhang, J. Wang, Q. Sun, P. Qin, Z. Cheng, Z. Ge, Z. Li, S. Dou, Adv. Energy Mater. **7**, 1700573 (2017)
93. H. Guo, H. Xin, X. Qin, J. Zhang, D. Li, Y. Li, C. Song, C. Li, J Alloys Compd. **689**, 87 (2016)
94. H. Leng, M. Zhou, J. Zhao, Y. Han, L. Li, J. Electron. Mater. **45**, 527 (2015)
95. C.-H. Chien, C.-C. Chang, C.-L. Chen, C.-M. Tseng, Y.-R. Wu, M.-K. Wu, C.-H. Lee, Y.-Y. Chen, RSC Adv. **7**, 34300 (2017)
96. N.K. Singh, S. Bathula, B. Gahtori, K. Tyagi, D. Haranath, A. Dhar, J. Alloys Compd. **668**, 152 (2016)
97. J. Gao, G. Xu, Intermetallics **89**, 40 (2017)
98. J.C. Li, D. Li, X.Y. Qin, J. Zhang, Scripta Mater. **126**, 6 (2017)
99. V. Kucek, T. Plechacek, P. Janicek, P. Ruleova, L. Benes, J. Navratil, C. Drasar, J. Electron. Mater. **45**, 2943 (2016)
100. F. Li, W. Wang, X. Qiu, Z.-H. Zheng, P. Fan, J. Luo, B. Li, Inorg. Chem. Front. **4**, 1721 (2017)
101. J.H. Kim, S. Oh, Y.M. Kim, H.S. So, H. Lee, J.-S. Rhyee, S.-D. Park, S.-J. Kim, J. Alloys Compd. **682**, 785 (2016)
102. T.-R. Wei, C.-F. Wu, X. Zhang, Q. Tan, L. Sun, Y. Pan, J.-F. Li, Phys. Chem. Chem. Phys. **17**, 30102 (2015)
103. M. Hong, Z.-G. Chen, L. Yang, T.C. Chasapis, S.D. Kang, Y. Zou, G.J. Auchterlonie, M.G. Kanatzidis, G.J. Snyder, J. Zou, J. Mater. Chem. A **5**, 10713 (2017)
104. S. Chen, K. Cai, W. Zhao, Phys. B: Condens. Matter **407**, 4154 (2012)
105. Y.-M. Han, J. Zhao, M. Zhou, X.-X. Jiang, H.-Q. Leng, L.-F. Li, J. Mater. Chem. A **3**, 4555 (2015)
106. Q. Zhang, E.K. Chere, J. Sun, F. Cao, K. Dahal, S. Chen, G. Chen, Z. Ren, Adv. Energy Mater. **5**, 1500360 (2015)
107. M. Gharsalla, F. Serrano-Sanchez, N.M. Nemes, F.J. Mompean, J.L. Martinez, M.T. Fernandez-Diaz, F. Elhalouani, J.A. Alonso, Sci. Rep. **6**, 26774 (2016)
108. Y. Fu, J. Xu, G.-Q. Liu, X. Tan, Z. Liu, X. Wang, H. Shao, H. Jiang, B. Liang, J. Electron. Mater. **46**, 3182 (2017)
109. J. Gao, Z. Shao, G. Xu, Int. J. Appl. Ceram. Tec. **14**, 963 (2017)
110. J. Gao, H. Zhu, T. Mao, L. Zhang, J. Di, G. Xu, Mater. Res. Bull. **93**, 366 (2017)
111. F. Li, W. Wang, Z.-H. Ge, Z. Zheng, J. Luo, P. Fan, B. Li, Materials **11**, 203 (2018)
112. G. Han, S.R. Popuri, H.F. Greer, L.F. Llin, J.W.G. Bos, W. Zhou, D.J. Paul, H. Ménard, A.R. Knox, A. Montecucco, J. Siviter, E.A. Man, W. Li, M.C. Paul, M. Gao, T. Sweet, R. Freer, F. Azough, H. Baig, T.K. Mallick, D.H. Gregory, Adv. Energy Mater. **7**, 1602328 (2017)
113. X. Wang, J. Xu, G. Liu, Y. Fu, Z. Liu, X. Tan, H. Shao, H. Jiang, T. Tan, J. Jiang, Appl. Phys. Lett. **108**, 083902 (2016)
114. D. Li, X. Tan, J. Xu, G. Liu, M. Jin, H. Shao, H. Huang, J. Zhang, J. Jiang, RSC Adv. **7**, 17906 (2017)
115. S. Siol, A. Holder, B.R. Ortiz, P.A. Parilla, E.S. Toberer, S. Lany, A. Zakutayev, RSC Adv. **7**, 24747 (2017)
116. G. Tang, W. Wei, J. Zhang, Y. Li, X. Wang, G. Xu, C. Chang, Z. Wang, Y. Du, L.-D. Zhao, J. Am. Chem. Soc. **138**, 13647 (2016)
117. T.-R. Wei, G. Tan, C.-F. Wu, C. Chang, L.-D. Zhao, J.-F. Li, Appl. Phys. Lett. **110**, 053901 (2017)
118. S. Yang, J. Si, Q. Su, H. Wu, Mater. Lett. **193**, 146 (2017)
119. X.-Q. Huang, Y.-X. Chen, M. Yin, D. Feng, J. He, Nanotechnology **28**, 105708 (2017)
120. S. Sassi, C. Candolfi, J.-B. Vaney, V. Ohorodniichuk, P. Masschelein, A. Dauscher, B. Lenoir, Mater. Today: Proc **2**, 690 (2015)
121. S.R. Popuri, M. Pollet, R. Decourt, M. Viciu, J.-W.G. Bos, Appl. Phys. Lett. **110**, 253903 (2017)
122. D. Bansal, J.W. Hong, C.W. Li, A.F. May, W. Porter, M.Y. Hu, D.L. Abernathy, O. Delaire, Phys. Rev. B **94**, 054307 (2016)

123. J. Carrete, N. Mingo, S. Curtarolo, Appl. Phys. Lett. **105**, 101907 (2014)
124. P.-C. Wei, S. Bhattacharya, J. He, S. Neeleshwar, R. Podila, Y.Y. Chen, A.M. Rao, Nature **539**, E1–E2 (2016)
125. L.-D. Zhao, C. Chang, G. Tan, M.G. Kanatzidis, Energy Environ. Sci. **9**, 3044 (2016)
126. S.T. Lee, M.J. Kim, G.-G. Lee, S.G. Kim, S. Lee, W.-S. Seo, Y.S. Lim, Curr. Appl. Phys. **17**, 732 (2017)
127. S. Wang, S. Hui, K. Peng, T.P. Bailey, W. Liu, Y. Yan, X. Zhou, X. Tang, C. Uher, Appl. Phys. Lett. **112**, 142102 (2018)

Chapter 3
Tin Sulfide: A New Nontoxic Earth-Abundant Thermoelectric Material

Hong Wu, Xu Lu, Xiaodong Han, and Xiaoyuan Zhou

Abstract Lead-free tin sulfide (SnS) consisting of earth-abundant and nontoxic elements has intrinsically low thermal conductivity and large Seebeck coefficient, which indicates that it has a potential application in the area of waste heat recovery. In this chapter, we elucidate the origin of low thermal conductivity by analysis of the crystal structure and concentration of charge carriers and unveil the reason for large Seebeck coefficient found in p-type doped samples based on band edge morphology examination. Finally, we achieve the highest value of dimensionless figure of merit ZT ~ 1.1 at 870 K for p-type highly doped SnS single crystal samples due to intrinsic high mobility of charge carriers in SnS single crystal, which holds a potential for application in the area of waste heat recovery.

3.1 Introduction

Seebeck effect in thermoelectric (TE) materials enables harvesting electric energy directly from thermal energy, including waste heat, by means of TE generators, which can play influential role in alleviating worldwide energy and environmental issues [1, 2]. In general, dimensionless figure of merit ZT determines conversion efficiency of TE device [3], $ZT = \frac{S^2 T}{\rho \kappa}$, where S, ρ, T, and κ are Seebeck coefficient, electrical resistivity, absolute temperature, and total thermal conductivity, respectively. The basic math in the definition of figure of merit requires good TE materials which possess high S, low ρ, and low κ. However, the underlying physics dictates that S is proportional to ρ. In other words, high S and low ρ cannot coexist in the same material. In addition, total thermal conductivity (κ) consists of lattice part (κ_L) and electronic part (κ_e) which are governed by lattice vibration and electrical

H. Wu · X. Lu · X. Zhou (✉)
College of Physics, Chongqing University, Chongqing, P. R. China
e-mail: xdhan@bjut.edu.cn

X. Han (✉)
Beijing Key Laboratory of Microstructure and Property of Advanced Materials,
Beijing University of Technology, Beijing, P. R. China
e-mail: xiaoyuan2013@cqu.edu.cn

© Springer Nature Switzerland AG 2019
S. Skipidarov, M. Nikitin (eds.), *Novel Thermoelectric Materials and Device Design Concepts*, https://doi.org/10.1007/978-3-030-12057-3_3

conductivity, respectively. Thus, these parameters (S, κ_e, and ρ) are strongly coupled with concentration of charge carriers, which hinder greatly the efforts to obtain high ZT. Fortunately, κ_L is a relatively independent parameter, and thus reducing κ_L (enhancing phonons scattering) is most frequently employed as an effective method to improve TE performance, such as nanostructuring [4, 5], point defects [6], and strong anharmonicity [7]. Other strategies to enhance power factor (S^2/ρ) include optimization of charge density [8] and band structure engineering [9].

Recently, bulk SnSe single crystal [10] exhibited ultrahigh ZT ~ 2.6 at 923 K along crystalline b-axis, which is attributed to strong lattice anharmonicity. Tin sulfide (SnS), possessing an analogous crystal structure with SnSe, is also expected to have large anharmonicity, which would give rise to intrinsically low lattice thermal conductivity. As theoretical calculations predicted [11], SnS-based materials have high S and low κ. However, most polycrystalline SnS samples reach the highest ZT ~ 0.65 at 850 K only [12], since mobility of charge carriers is inevitably deteriorated due to the presence of many grain boundaries and disorder orientation. To further improve TE performance of SnS materials, bulk SnS-based single crystals are grown by modified Bridgman methods [13]. Facilitated by ordered orientation and lack of grain boundary nature, ZT in a single crystal SnS doped with 2 at. % Na reaches the highest value ~1.1 along b-axis direction. In this chapter, we have reviewed the detailed reasons for intrinsically low thermal conductivity and large Seebeck coefficient for SnS-based materials. Additionally, the influence of different doping elements substituting on Sn sites on electrical transport properties in SnS-based materials is discussed.

3.2 Crystal Structure

Figure 3.1a displays that SnS crystal possesses 2D layered crystal structure (orthorhombic P_{nma} space group) at room temperature, which has zigzagged atomic chains along c-axis and two-atom-thick SnS slabs. In that structure, each Sn atom links to adjacent three S atoms, consisting of two S atoms in $b - c$ plane and one S atom along a-axis. The distance of Sn $-$ S bond of 2.690 Å in $b - c$ plane is larger than that of 2.612 Å along a-axis, which indicates strong Sn $-$ S bond along a-axis within this slab. Meanwhile, the layers stacked along a-axis are bonded by van der Waals force. Figure 3.1c shows the crystal structure containing coordination polyhedra of SnS$_7$ including three short bonds Sn $-$ S and four long bonds that give rise to anharmonicity to reduce thermal conductivity.

To further comprehend crystal structural information, the investigations involving scanning transmission electron microscopy high-angle annular dark field (STEM-HAADF) and energy-dispersive X-ray spectrum (EDS) mapping are performed on Sn$_{0.98}$Na$_{0.02}$S single crystal along [001] direction. Figure 3.2c shows explicitly that SnS crystal is a layered structure, implying that it is easy to cleave SnS single crystals (both doped and undoped) along the (100) planes. It is clearly seen

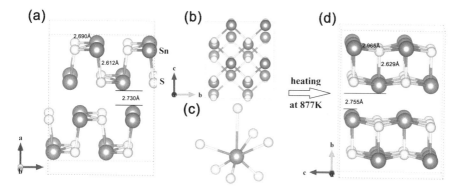

Fig. 3.1 SnS crystal structures: (**a**) P_{nma} phase. (**b**) P_{nma} phase along crystalline a-axis. (**c**) The resonant bonding in P_{nma} space group. (**d**) C_{mcm} phase of high temperature

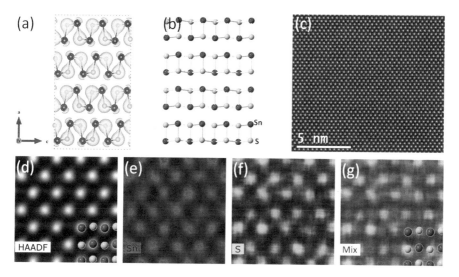

Fig. 3.2 (**a**) The charge density of pristine SnS. (**b**) The model of atomic arrangement projection of SnS (P_{nma}) and (**c**) HAADF image of $Sn_{0.98}Na_{0.02}S$ sample along the [001] direction. (**d–f**) STEM-EDS mapping of $Sn_{0.98}Na_{0.02}S$ crystal and (**g**) the mixture mapping [13]

that Sn and S atoms are located on different columns of [001] projection from Fig. 3.2d–g.

He et al. [14] reported HAADF image and EDS mappings of SnS single crystal along [010] zone axis. Similar to SnSe, SnS crystal structure undergoes second-order displacive phase transition from threefold coordinated P_{nma} to fivefold coordinated C_{mcm} phase with higher symmetry at 877 K. Figure 3.1d exhibits the layers piled up along b-axis and two-atom slab similar to that in P_{nma} phase. Additionally, each Sn atom is now bound to four neighboring S atoms with equal distance (2.965 Å) in

$a - c$ plane and one additional atom with short distance (2.629 Å) along b-axis (i.e., 4 + 1 coordination). Therefore, it is crucial to understand the difference in structure between P_{nma} and C_{mcm} phases for studying this system.

3.3 Electronic Band Structure and BoltzTrap Calculation of SnS

To shed light on the physical mechanism of charge carriers transport properties, calculations of electronic structure of low-temperature phase (P_{nma} space group) of SnS have been carried out by several groups, and the results were similar. As displayed in Fig. 3.3a, calculated results exhibit SnS indirect bandgap $E_g \sim 0.89$ eV, which is smaller than that obtained from experimentally optical measurements (~1.1 eV) shown in Fig. 3.3c. In general, it is reasonable assuming that slight underestimation (~10%) for bandgap has negligible influence on the morphology of band edges, which directly determines electronic transport properties. The morphology of near band edge will be investigated in the following section.

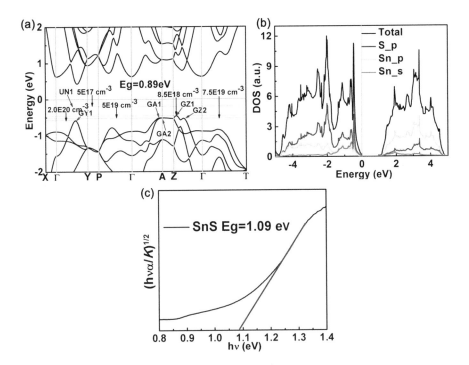

Fig. 3.3 (**a**) Calculated electronic band structure of SnS and density of charge carriers as function of Fermi level. (**b**) Total and partial density of states (DOS) for pristine SnS [13]. (**c**) Optical bandgap for pristine SnS crystal

Table 3.1 Effective masses of SnS obtained by fitting the band structure

Valence band tops	VBM (eV)	$\frac{m^*_{kx}}{m_0}$	$\frac{m^*_{ky}}{m_0}$	$\frac{m^*_{kz}}{m_0}$	$\frac{m^*_c}{m_0}$	$\frac{m^*_b}{m_0}$	$\frac{m^*_d}{m_0}$
GZ1	0.0	3.051	0.338	0.222	0.385	0.612	0.971
GA1	0.055	6.781	0.740	0.358	0.698	1.215	3.063
GA2	0.059	20.01	6.270	0.415	1.145	3.734	9.409
GZ2	0.071	1.278	0.798	0.233	0.474	0.619	0.983
GY1	0.123	0.319	0.291	0.169	0.240	0.251	0.397

The extremum position of the top valence band and the offset of each extremum from the first valence band maximum (VBM) are also listed in Table [13]

Valence band features: As shown in Fig. 3.3a, valence band maximum GZ1 (0.0, 0.0, 0.448) situates along $\Gamma - Z$ direction, and the second valence band maximum GA1 (0.466, 0.0, 0.466) lies in $\Gamma - A$ direction. Energy difference between these two valence band extrema is ~0.055 eV (see Table 3.1). Band effective masses of these extremum points (GZ1, GA1, GA2, GZ2, and GY1) are listed in Table 3.1. All extremum points exhibit large anisotropy along three principle directions, and calculated effective mass along k_x direction is larger than that along other two directions (k_y and k_z) owing to layered crystal structure of SnS. Hao et al. [15] reported the following effective masses: for the first band $m^*_{kx} = 2.63m_0$, $m^*_{ky} = 0.32m_0$, and $m^*_{kz} = 0.32m_0$ and for the second band $m^*_{kx} = 8.48m_0$, $m^*_{ky} = 0.29m_0$, and $m^*_{kz} = 0.54m_0$, where m_0 is the free electron mass. These results are confirmed by experimental access to valence band dispersion of SnS via angle-resolved photoemission spectroscopy (ARPES) [14]. The density of states (DOS) effective mass for the second band is about ~$3.063m_0$, which is three times to that of the first extremum (~$0.971m_0$). Considering Seebeck coefficient is directly proportional to DOS effective mass, enlarged Seebeck coefficient is readily linked to the existence of such heavy bands. Based on BoltzTrap calculation, it is found that when hole concentration reaches ~8.5×10^{18} cm^{-3}, Fermi level drops down to top valence band (GZ1) at 300 K. Significantly, when hole concentration reaches ~5.0×10^{19} cm^{-3}, Fermi level is lowered down to the second heavy band (GA1). In such case, multiple valence bands will participate in transportation of holes, which is beneficial for the enhancement of S^2/ρ. Therefore, optimization of concentration of charge carriers in SnS-based materials is crucial as proved by combination of calculated DOS effective mass and BoltzTrap results. Specifically, Fig. 3.3b shows DOS of pure SnS, in which total DOS near the top of valence band are mainly contributed by p orbitals of S atoms.

Conduction band features: Profile of conduction band edge for *n*-type SnS was also investigated [15]. Conduction band minimum is located along $\Gamma - Y$ direction, and the second extreme point lies at Γ point. Energy difference between two conduction band extrema is about 0.18 eV, which is much larger than that of valence bands. Large energy difference impedes the contribution from the second conduction band to electrons transport. Fermi level positions calculated with varied electron concentration show that Fermi level can reach the second conduction band edge only when electron concentration is over ~2.0×10^{20} cm^{-3}. In addition, band effective masses of the first conduction band minimum ($m_{kx}^* = 0.23m_0$, $m_{ky}^* = 0.13m_0$ and $m_{kz}^* = 0.17m_0$) are much smaller than that of the valence band. As shown in Fig. 3.3b, total DOS near bottom of conduction band mainly stems from p orbitals of Sn atoms. These results give the physical origin of differences in charge carriers transport properties of n-type and p-type SnS.

3.4 Origin of Intrinsic Low Thermal Conductivity of SnS

As mentioned above, low thermal conductivity is critical for materials to attain high TE performance. In practice, low lattice thermal conductivity can be reached for multiple reasons including reasons aside from extrinsic strategies [16], such as solid solutions, reducing grain size, or creating dislocations. However, it is more favorable to search for materials with intrinsically low lattice thermal conductivity driven by strong lattice anharmonicity, low speed of sound, or less contribution to heat capacity by acoustic phonons [16].

Large anharmonicity: harmonicity represents situation when force enforcing deviation of atom from its equilibrium position is proportional to atom displacement, while anharmonicity can be considered as situation when displacement depends not proportionally on force and even in high extent beyond proportionality. Universally, crystal structures with bond hierarchy [17], anisotropic bond [18], or a lone pair of electrons [19] tend to possess a stronger anharmonicity which is characterized through average Grüneisen parameter. If the main thermal energy is transported by acoustic phonons and intrinsic scattering (phonon-phonon) is a dominant mechanism to hinder thermal conduction process, then κ_L can be expressed as [16]:

$$\kappa_L = A \frac{\theta_D^3 \delta^{1/3} \overline{M}}{\gamma^2 N^{2/3} T}$$

where A is physical constant, θ_D is Debye temperature, δ is average volume per atom, $\overline{M}$ is average atomic mass, N is atomic number in primitive unit cell, T is absolute temperature, and γ is Grüneisen parameter. The equation shows that large Grüneisen parameter γ which indicates strong anharmonicity gives rise to intrinsically low lattice thermal conductivity. Based on previous theory, Morelli et al. [19] reported that low κ_L of Cu_3SbSe_3 is attributed to strong anharmonicity due to existent lone pair electrons. As displayed in Fig. 3.2a, strongly localized charge density near Sn atoms, which is contributed by $5s$ electron of Sn atom, is observed along layers bonded by van der Waals force in SnS crystal structure. Such lone pair electrons lead to strong anharmonicity induced by asymmetric chemical bonds in SnS, for the same reason as in SnSe [20]. Calculated average Grüneisen parameters of SnS samples are 3.9, 2.1, and 2.3 along a-axis, b-axis, and c-axis, respectively [12], which are comparable with SnSe ($\overline{\gamma}_a = 4.1$, $\overline{\gamma}_b = 2.1$, and $\overline{\gamma}_c = 2.3$)

[15]. Therefore, it is concluded that SnS-based materials should have low κ_L as indicated by large Grüneisen parameter.

3.5 Synthesis and Thermoelectric Performance of Polycrystalline SnS-Based Materials

3.5.1 Synthesis Methods

Tan et al. [21] synthesized SnS-based polycrystalline by means of mechanical alloying (MA) at 450 rpm for 15 h in argon atmosphere (>99.5%) combined with spark plasma sintering (SPS) at 933 K in vacuum for 5 min under pressure of 50 MPa. Zhou et al. [12] reported about Na-doped polycrystalline SnS synthesized by traditional solid-state reaction followed by hot pressing. Wang et al. [22] reported successful synthesis of SnS polycrystalline samples via wet chemistry method.

3.5.2 Thermoelectric Performance of Undoped and p-Type doped Polycrystalline SnS

Experimentally, Tan et al. [21] first systematically investigated p-type doping effect of Ag impurity atoms on Sn sites for polycrystalline SnS. It was found that bandgap of SnS (1.21 eV) shifts to lower value (1.18 eV) with 0.5 at. % Ag doping. Hole concentration of undoped SnS ($\sim 10^{14}$ cm^{-3}) is enlarged to $\sim 3.6 \times 10^{18}$ cm^{-3} by 1 at. % Ag doping at 300 K. However, hole mobility decreases from 6 cm$^2 \times$ V$^{-1} \times$ s^{-1} for pure SnS to 2 cm$^2 \times$ V$^{-1} \times$ s^{-1} for 1 at. % Ag-doped SnS, which results in low electrical conductivity for Ag-doped polycrystalline SnS samples. In addition, low κ_L (0.40 W $\times$ m$^{-1} \times$ K^{-1} at 873 K) of polycrystalline Ag$_{0.005}$Sn$_{0.995}$S is obtained as expected owing to intrinsic strong anharmonicity assisted by point defects and grain boundary (see Fig. 3.4).

As a result, polycrystalline SnS sample with 0.5 at. % Ag doping reaches ZT of 0.6 at 923 K. It is obvious that relatively low hole concentration ($\sim 3.6 \times 10^{18}$ cm^{-3}) for Ag-doped polycrystalline SnS is not optimized, which leads to low TE performance. Recently, Zhou et al. [12] reported TE properties of Na-doped polycrystalline SnS samples. The highest resultant hole concentration reached $\sim 2 \times 10^{19}$ cm^{-3} for 2 at. % Na-doped SnS. As is the same case for SnSe, Na is a much more efficient dopant than Ag in SnS as well. Unfortunately, due to the degradation of hole mobility for polycrystalline SnS, the highest ZT of Na-doped polycrystalline SnS is ~ 0.65. In addition, Wang et al. [22] reported that the highest ZT value of 0.14 for SnS synthesized by chemical precipitation methods is obtained at 850 K owing to low total thermal conductivity (~ 0.29 W $\times$ m$^{-1} \times$ K^{-1}). As described above, all SnS samples have intrinsically low thermal conductivity attributed to strong

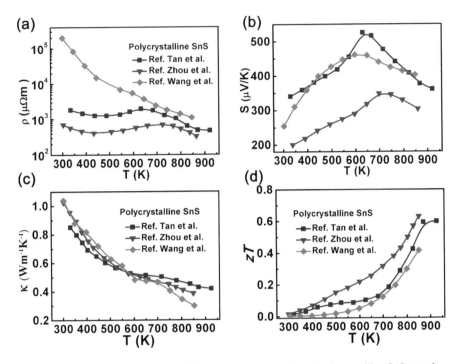

Fig. 3.4 Temperature dependences of thermoelectric properties of polycrystalline SnS samples. Zhou et al. [12], Tan et al. [21], Wang et al. [22]

anharmonicity. As a result, further enhancement in TE performance of SnS relies on increasing in mobility and concentration of charge carriers.

3.6 Synthesis and Thermoelectric Performance of Pristine and *p*-Type Doped SnS Single Crystals

3.6.1 Synthesis

High-purity elemental Na chunks (99.9%), S powder (99.99%), and Sn granules (99.999%) were weighted in the stoichiometry of $Sn_{1-x}Na_xS$ ($x = 0$, 0.1 at. %, 1 at. %, 2 at. %, 3 at. %, 4 at. %) and loaded into conical tubes with argon atmosphere. The tubes were evacuated under ~5×10^{-4} Pa and sealed and then placed in other (larger diameter) quartz tubes, which were evacuated (~5×10^{-4} Pa) and flame-sealed. The outer tube was used to protect samples from oxidation because the inner quartz tube breaks easily owing to phase transition from C_{mcm} space group (high-temperature phase) to P_{nma} space group (low-temperature phase). The tube was put in vertical furnace (temperature gradient) and then was slowly heated up to 1273 K over 20 h and soaked at temperature for 15 h. Subsequently, the furnace was cooled to 1100 K and then maintained at that temperature as the tube was moved down with a rate of 1.5 mm × h^{-1}. Finally, Sn$_{1-x}$Na$_x$S single crystal ingots with diameter of 14 mm and height of 45 mm were obtained. Similar synthesis processes were also reported by Nassary et al. [23] using the furnace with dual temperature zones and moving quartz tube and He et al. [14] utilizing vertical furnace with controllable temperature gradient cooling rate.

3.6.2   Thermoelectric Performance of Pristine SnS Single Crystal

Figure 3.5 shows temperature-dependent thermoelectric properties of pure SnS single crystal along three crystalline directions (*a*-axis, *b*-axis, and *c*-axis) which can be confirmed by electron backscattered diffraction (EBSD) analysis using

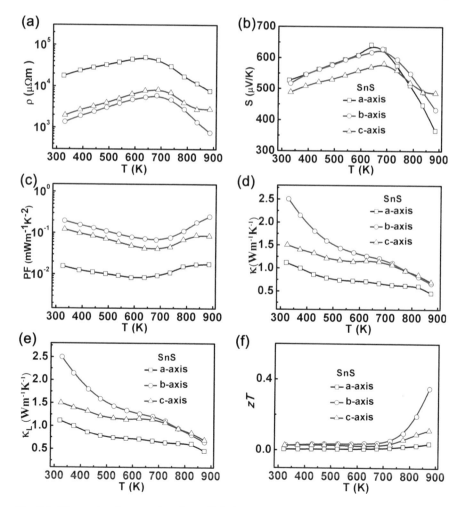

Fig. 3.5 Thermoelectric properties of pure SnS single crystals

scanning electron microscopy (SEM). As shown in Fig. 3.5, layered crystal structure of SnS results in strong textured character, which indicates SnS-based materials have strong anisotropy of thermoelectric properties. It is noted that total thermal conductivity for SnS crystal along a-axis (0.46 W $\times$ m^{-1} $\times$ K^{-1} at 873 K) is smaller than that along two in-plane directions (0.67 W $\times$ m^{-1} $\times$ K^{-1} along b-axis and 0.70 W $\times$ m^{-1} $\times$ K^{-1} along c-axis at 873 K) due to the presence of van der Waals interface. However, electrical conductivity is also limited (131.7 S $\times$ m^{-1} at 884 K) along out of plane direction due to small hole mobility. Finally, the highest ZT ~ 0.4 along crystalline b-axis is obtained at 873 K. He et al. [14] obtained the highest ZT ~ 1 along b-axis. However, average ZT that directly determines the conversion efficiency for SnS single crystal is lower than that for other mainstream TE materials, which is attributed to low hole concentration ($\sim 2.0 \times 10^{17}$ cm^{-3}).

3.6.3 Thermoelectric Performance of p-Type Doped SnS Single Crystals

Electrical Transport Properties

In order to further enhance electrical properties of SnS crystal, sodium (Na) doping on Sn sites was performed to increase in concentration of charge carriers. TE properties of Na-doped samples were investigated from room temperature to 873 K. Figure 3.6 shows temperature-dependent electrophysical and TE properties of SnS-based single crystals along b-axis because this direction exhibits better TE performance than other two directions. As shown distinctly in Fig. 3.6a, electrical resistivity (ρ) of sample doped with 2 at. % Na is two orders of magnitude lower than that of pristine SnS single crystals along b-axis at 300 K. As shown in Fig. 3.6b, c, outstanding electrical properties result from increased hole concentration due to Na doping. As displayed in Fig. 3.6b, the highest hole concentration reaches $\sim 2.3 \times 10^{19}$ cm^{-3} for 2 at. % Na-doped SnS crystal, while the pristine SnS sample has concentration of charge carriers only of $\sim 2.0 \times 10^{17}$ cm^{-3} caused by intrinsic Sn atom vacancy. Hall mobility of charge carriers $\mu_H = (\rho e n_H)^{-1}$ is depicted in Fig. 3.6c, where ρ is electrical resistivity, n_H is Hall concentration of charge carriers, and e is electron charge value. It is noted that Hall mobility of p-type doped SnS samples is slightly lower than pure SnS sample due to impurity effect. Most importantly, Hall mobility of SnS-based single crystals is much larger than that of polycrystalline SnS-based materials over whole temperature range owing to the absence of grain boundary scattering and perfect orientation. Additionally, Fig. 3.6d displays that Seebeck coefficient (S) decreases with increasing concentration of charge carriers in accordance with Mott formula [16]. However, it is readily found that S of SnS-based materials is larger than that of SnSe-based materials for the same Hall concentration of charge carriers [24], which could be attributed to possess gradual valence band edge structure for SnS-based materials. As shown in Fig. 3.6e, the highest power factor (~ 2.0 mW $\times$ m^{-1} $\times$ K^{-2}) for 2 at. % Na-doped

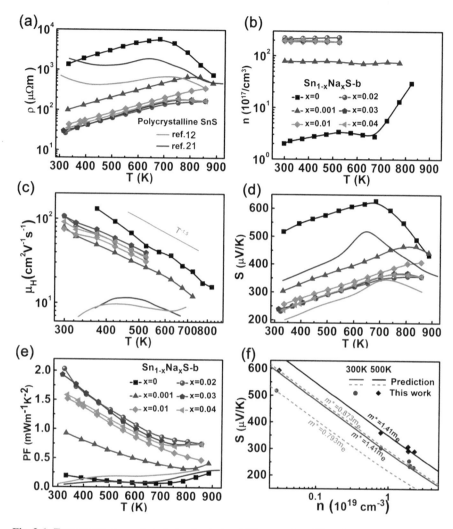

Fig. 3.6 Temperature-dependent electrophysical and TE properties of $Sn_{1-x}Na_xS$ ($x = 0, 0.1$ at. %, 1 at. %, 2 at. %, 3 at. %, 4 at. %) single crystals along crystallographic b-axis [13]

SnS single crystal along b-axis is achieved at room temperature and is maintained at ~0.75 mW $\times$ m^{-1} $\times$ K^{-2} at 873 K. The power factor of Na-doped SnS single crystals is much larger than that of undoped SnS crystal along b-axis and is much higher than all polycrystalline samples in whole temperature range. The reason is that largely increased hole concentration of SnS single crystals using Na doping lowers down Fermi level and makes the second heavy hole band participating in charge carriers transport, which results in enhanced Seebeck coefficient. As shown in Fig. 3.6f, the relation between S and n_H of p-type doped samples at 300 K could be fitted using large DOS effective mass ($1.41m_e$), while that of pristine SnS single crystal could be fitted using DOS effective mass of $0.793m_e$. This is the sign for multiple bands taking part in charge carriers transport, which is beneficial to enhance power factor. It was also reported that Hall concentration for p-type doped SnS crystal can reach $\sim 3.5 \times 10^{19}$ cm^{-3}, which results in a power factor (~ 3.0 mW $\times$ m^{-1} $\times$ K^{-2}) along crystalline b-axis at 300 K [14], suggesting further enhancement in TE performance of SnS-based single crystals is possible if higher concentration of charge carriers can be reached.

Consolidated data related to room temperature concentration of charge carriers obtained by different methods [13] are given in Table 3.2.

Thermal Properties

Figure 3.7 shows temperature-dependent total thermal conductivity (κ_{total}) and lattice part (κ_L) along crystalline b-axis direction for all samples. The minimum value of $\kappa_{\text{total}} \sim 0.61$ W $\times$ m^{-1} $\times$ K^{-1} for all single crystals along b-axis is achieved at 873 K. As shown in Fig. 3.7b, lattice part (κ_L), calculated by $\kappa_L = \kappa_{\text{total}} - \kappa_e$, where electronic part (κ_e) is calculated via Wiedemann-Franz law ($\kappa_e = LT\rho^{-1}$), decreases

Table 3.2 Room temperature concentration of charge carriers obtained by different methods [13]

Fermi points	Fermi level (eV)	Calculation (cm^{-3})	Samples Sn$_{1-x}$Na$_x$S	SPB model (cm^{-3})	Hall concentration of charge carriers (cm^{-3})
UN1	−0.26	5.0E17	$x = 0$	3.7E17	2.0E17
GZ1	0.0	8.5E18	$x = 0.001$	8.1E18	7.8E18
			$x = 0.01$	2.0E19	1.9E19
			$x = 0.02$	2.6E19	2.3E19
			$x = 0.03$	2.4E19	2.0E19
			$x = 0.04$	2.5E19	1.9E19
GA1	0.055	5.0E19			

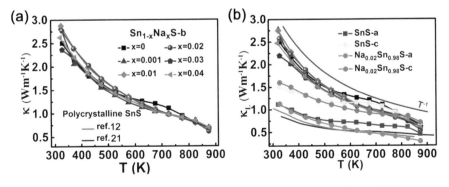

Fig. 3.7 (a) The total thermal conductivity of all single crystal samples along b-axis. (b) The lattice thermal conductivity [13]

rapidly with increasing temperature and follows T^{-1} law, suggesting dominance of phonon-phonon scattering processes. Finally, the minimal value of $\kappa_{total} \sim 0.54$ W $\times$ m^{-1} $\times$ K^{-1} for 2 at. % Na-doped SnS single crystal is obtained at 873 K, which is ascribed to intrinsic strong anharmonicity as mentioned earlier. Additionally, Fig. 3.7b shows that the lattice thermal conductivity (κ_L) for polycrystalline SnS is smaller than that of single crystals along crystalline three directions at low temperature due to grain boundaries and disordered orientation.

Figure of Merit

As displayed in Fig. 3.8b, the dimensionless figure of merit (ZT) for all single crystals along b-axis increases with concentration of charge carriers, and the highest value reaches ~1.1 for 2 at. % Na-doped SnS single crystal at 870 K. This value was higher than that of PbS-based materials with the same Na doping [25]. As shown in Fig. 3.8c, we summarize the highest ZT values and average ZT value of several representative SnS-based materials as reported previously [12–14, 21, 22] over temperature range from 300 K to 870 K. It is obviously found that 2 at. % Na-doped SnS single crystal is among the best TE sulfides in terms of both peak ZT value and average ZT value. Significantly, average ZT ~ 0.54, improved by

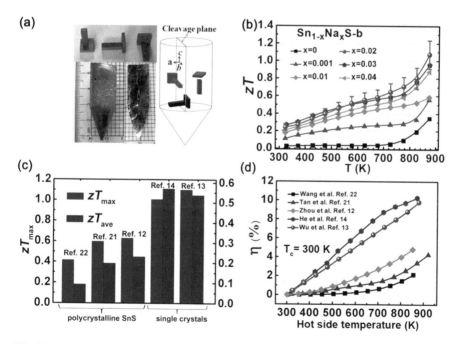

Fig. 3.8 (**a**) The photos of as-prepared single crystals, cleaved samples, and square-shaped samples for TE properties measurement. (**b**) ZT values of Sn$_{1-x}$Na$_x$S ($x = 0$, 0.1 at. %, 1 at. %, 2 at. %, 3 at. %, 4 at. %) single crystals along crystallographic b-axis. (**c**) The highest ZT value and average ZT value of SnS-based materials. (**d**) The calculated efficiency (η) of SnS-based materials with the cold side temperature (300 K)

~180% as compared to SnS polycrystalline materials in temperature range from 300 K to 870 K, is attained, and calculated efficiency (η) of single crystal samples is superior to that of polycrystalline SnS-based materials in Fig. 3.8d, which suggests Na-doped SnS-based single crystals are more plausible for practical TE applications.

3.7 Summary and Outlook

In this chapter, we have unveiled that the unique crystal structure of SnS-based materials gives rise to low lattice thermal conductivity and demonstrated the effect of different doping elements substituting on Sn sites on charge carriers transport properties of SnS-based materials. The profile of valence band edge of pure SnS was investigated by DFT calculations, which is related to high power factor of p-type doped SnS samples. Furthermore, the detailed synthesis method of SnS single crystals was described, and high TE performance of Na-doped samples was experimentally identified, which is ascribed to increased hole concentration and promoted hole mobility through eliminating grain boundaries.

It should be noted that concentration of charge carriers for all SnS-based materials so far has not reached optimized value due to the low doping efficiency. Thus, improving the doping efficiency for SnS-based materials is imperative. Furthermore, thermoelectric device consists of parts made of p-type and n-type materials, and it is, therefore, important to investigate TE performance of n-type SnS-based materials. If these problems can be solved successfully, SnS-based materials composed of earth-abundant, low-cost, and nontoxic chemical elements would hold the potential for application in the area of waste heat recovery.

References

1. L.E. Bell, Cooling, heating, generating power, and recovering waste heat with thermoelectric systems. Science **321**, 1457–1461 (2008)
2. M.S. Dresselhaus, G. Chen, M.Y. Tang, et al., New directions for low-dimensional thermoelectric materials. Adv. Mater. **19**, 1043–1053 (2007)
3. G.J. Snyder, E.S. Toberer, Complex thermoelectric materials. Nat. Mater. **7**, 105–114 (2008)
4. K.F. Hsu, S. Loo, F. Guo, W. Chen, et al., Cubic $AgPb_mSbTe_{2+m}$: bulk thermoelectric materials with high figure of merit. Science **303**, 818–821 (2004)
5. J.P. Heremans, C.M. Thrush, D.T. Morelli, Thermopower enhancement in lead telluride nanostructures. Phys. Rev. B **70**, 115334 (2004)
6. Y. Pei, L. Zheng, W. Li, S. Lin, et al., Interstitial point defect scattering contributing to high thermoelectric performance in SnTe. Adv. Electron Mater. **2**, 1600019 (2016)
7. M.D. Nielsen, V. Ozolins, J.P. Heremans, Lone pair electrons minimize lattice thermal conductivity. Energy Environ. Sci. **6**, 570–578 (2013)
8. Y.Z. Pei, A.D. LaLonde, N.A. Heinz, et al., Stabilizing the optimal carrier concentration for high thermoelectric efficiency. Adv. Mater. **23**, 5674–5678 (2011)

9. J. Yang et al., On the tuning of electrical and thermal transport in thermoelectrics: an integrated theory-experiment perspective. npj Comp. Mater. **2**, 15015 (2016)

10. L.D. Zhao, S.H. Lo, Y. Zhang, et al., Ultralow thermal conductivity and high thermoelectric figure of merit in SnSe crystals. Nature **508**, 373–377 (2014)

11. R. Guo, X. Wang, Y. Kuang, B. Huang, First-principles study of anisotropic thermoelectric transport properties of IV-VI semiconductor compounds SnSe and SnS. Phys. Rev. B **92**, 115202 (2015)

12. B. Zhou, S. Li, W. Li, J. Li, et al., Thermoelectric properties of SnS with Na-doping. ACS Appl. Mater. Interfaces **9**, 34033–34041 (2017)

13. H. Wu, X. Lu, G.Y. Wang, K.L. Peng, et al., Sodium-doped tin sulfide single crystal: a nontoxic earth-abundant material with high thermoelectric performance. Adv. Energy Mater. **8**(20), 1800087 (2018). https://doi.org/10.1002/aenm.201800087

14. W.K. He, D.Y. Wang, J.F. Dong, et al., Remarkable electron and phonon band structures lead to a high thermoelectric performance $ZT > 1$ in earth-abundant and eco-friendly SnS crystals. J. Mater. Chem. A **6**, 10048–10056 (2018)

15. S.Q. Hao, V.P. Dravid, M.G. Kanatzidis, C. Wolverton, Prediction of high figure of merit plateau in SnS and solid solution of (Pb,Sn)S. APL Mater. **4**, 104505–104501 (2016)

16. X.Y. Zhou, Y.C. Yan, X. Lu, et al., Routes for high performance thermoelectric materials. Mater. Today **21**(9), 974–988 (2018). https://doi.org/10.1016/j.mattod.2018.03.039

17. W.J. Qiu, L.L. Xi, P. Wei, X.Z. Ke, J.H. Yang, W.Q. Zhang, Part crystalline part liquid state and rattling like thermal damping in materials with chemical bond hierarchy. Proc. Natl. Acad. Sci. U. S. A. **111**, 15031–15035 (2014)

18. S. Wang, J. Yang, L.H. Wu, P. Wei, J.H. Yang, W.Q. Zhang, Y. Grin, Anisotropic multicenter bonding and high thermoelectric performance in electron poor CdSb. Chem. Mater. **27**, 1071–1081 (2015)

19. D.T. Morelli, V. Jovovic, J.P. Heremans, Instrinsically minimal thermal conductivity in cubic I-V-VI2 semiconductors. Phys. Rev. Lett. **101**, 035901–035905 (2008)

20. K. Peng, X. Lu, H. Zhan, S. Hui, et al., Broad temperature plateau for high ZTs in heavily doped *p*-type SnSe single crystals. Energy Environ. Sci. **9**, 454–460 (2016)

21. Q. Tan, L.D. Zhao, J.F. Li, et al., Thermoelectrics with earth abundant elements: low thermal conductivity and high thermopower in doped SnS. J. Mater. Chem. A **2**, 17302–17306 (2014)

22. C. Wang, Y.D. Chen, J. Jiang, et al., Improved thermoelectric properties of SnS synthesized by chemical precipitation. RSC Adv. **7**, 16795 (2017)

23. M.M. Nassary, Temperature dependence of the electrical conductivity, hall effect and thermo-electric power of SnS single crystals. J. Alloys Compd. **398**, 21–25 (2005)

24. K.L. Peng et al., Grain size optimization for high-performance polycrystalline SnSe thermo-electrics. J. Mater. Chem. A **5**, 14053–14060 (2017)

25. L.D. Zhao, J. He, S. Hao, et al., Raising the thermoelectric performance of p-type PbS with endotaxial nanostructuring and valence-band offset engineering using CdS and ZnS. J. Am. Chem. Soc. **134**, 16327–16336 (2012)

Chapter 4
SnTe-Based Thermoelectrics

Wen Li, Jing Tang, Xinyue Zhang, and Yanzhong Pei

Abstract Semiconducting PbTe has historically led to majority of the advancements in thermoelectrics. The realized thermoelectric figure of merit, ZT, has shown to be one of the highest. However, the concern on toxicity of Pb hinders its large-scale terrestrial applications, which motivates numerous efforts recently to be put on advancing its alternative analogue SnTe. To be as thermoelectrically efficient as that of p-type PbTe, concepts for ZT enhancement in SnTe are focused on the reductions of concentration of charge carriers, valence band offset, and lattice thermal conductivity. The synergy of these strategies, including band structure and defect engineering, significantly increases ZT from 0.4 up to 1.8 for SnTe thermoelectrics. Importantly, many of these high ZT materials remain a nontoxic composition. In this chapter, compositional designs for optimizing concentration of charge carriers, band structure/defect engineering, and solubility manipulation are introduced correspondingly. The strategies discussed here lead not only to promote SnTe as a highly efficient and eco-friendly alternative for p-type PbTe but open similar possibilities for advancing other thermoelectrics as well.

4.1 Introduction

Direct conversion of heat into electricity without any emissions and moving parts due to Seebeck effect in thermoelectric materials attracts increasing attentions as a sustainable and green energy technology. However, relatively low conversion efficiency of thermoelectric materials and, hence, generators hinders large-scale applications. The conversion efficiency depends on dimensionless thermoelectric figure of merit of thermoelectric material $ZT = \frac{S^2 T}{\rho(\kappa_E + \kappa_L)}$, where ρ, S, T, κ_E, and κ_L are electrical resistivity, Seebeck coefficient, absolute temperature, electronic, and lattice components of thermal conductivity, respectively. Therefore, high S but low ρ and $\kappa = \kappa_E + \kappa_L$ are desired for superior ZT.

W. Li · J. Tang · X. Zhang · Y. Pei (✉)
School of Materials Science and Engineering, Tongji University, Shanghai, China
e-mail: liwen@tongji.edu.cn; yanzhong@tongji.edu.cn

© Springer Nature Switzerland AG 2019
S. Skipidarov, M. Nikitin (eds.), *Novel Thermoelectric Materials and Device Design Concepts*, https://doi.org/10.1007/978-3-030-12057-3_4

Parameters S, ρ, and κ_E are strongly coupled to each other via concentration of charge carriers, band structure, and scattering of charge carriers [1, 2]. Early strategy for electronic performance enhancements in thermoelectrics relies largely on optimizing concentration of charge carriers for maximizing power factor S^2/ρ and then ZT enhancements. This is realized typically by chemical doping.

Parameter κ_L is the only independent parameter determining ZT; therefore, minimizing κ_L plays a key role for enhancing ZT [3]. The reduction of κ_L has been achieved by introduction of point defects [4–8], dislocations [9–11], and using nanostructures [12–16] for phonon scattering, leading to a significant ZT enhancement. However, κ_L in many thermoelectric materials has approached to amorphous limit [13], which motivates the numerous efforts on the development of electronic strategies for further ZT enhancement.

Very recently, newly developed strategies of band engineering [17], including band convergence [18]/nestification [19] and resonant states [20, 21], successfully decouple the correlation among the electronic parameters at some degree. An improvement in S^2/ρ by increasing band degeneracy N_v has been demonstrated in various materials [19, 22–33], and then significant ZT enhancements have been realized.

PbTe is a historical thermoelectric material that has led to most of the advancements in the field using the strategies mentioned above. Value ZT of PbTe, particularly, p-type conduction, has been enhanced to be one of the highest among known thermoelectrics. The band convergence approach is particularly interesting; the existence of L and Σ valence bands with a small energy offset ($\Delta E_{L-\Sigma} \sim 0.15$ eV) [34–40] enables an energy alignment of these valence bands (Fig. 4.1). Both L and Σ valence bands have high N_v (4 for L and 12 for Σ), which can be engineered to converge within a few $k_B T$ to provide effective N_v of 12~16 and, therefore, improved electronic performance [9, 18, 22, 23, 41–43]. Alternatively, phonon scattering through nanostructures [13, 14], dislocations [9], and alloying defects [6] for effective reduction of κ_L has also led to significant enhancement in ZT [9]. Unfortunately,

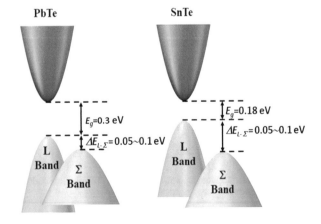

Fig. 4.1 A schematic diagram of band structures for PbTe and SnTe at room temperature

toxicity of Pb limits large-scale terrestrial applications of PbTe, which motivates strongly the development of high-performance alternatives without toxic chemicals.

Due to the same crystal structure and very similar band structure (Fig. 4.1) as compared to PbTe, SnTe attracts increasing attentions recently, as potentially eco-friendly alternative for PbTe. This is also driven by its similar chemical properties with those of PbTe, for a compatible device fabrication technique. However, ZT in pristine SnTe is only 0.6 [44], even with optimized concentration of charge carriers, which is much lower than ZT of pristine p-PbTe [45] with an optimal p. This can be understood to be a result of the inferiorities in SnTe, including (1) significantly high intrinsic concentration of charge carriers due to the existence of Sn vacancies, much higher than that needed for thermoelectrics [44]; (2) large energy offset ($\Delta E_{L-\Sigma} \sim 0.3$ eV [46]) between L and Σ valence bands, leading the high-valley degeneracy band (Σ band with N_v of 12) barely contributing to electronic transport; and (3) higher κ_L of ~3 W $\times$ m^{-1} $\times$ K^{-1} presumably resulting from its lighter constituent atomic mass, as compared to that of PbTe (~2 W $\times$ m^{-1} $\times$ K^{-1}) [45] at room temperature. Fortunately, the strategies for ZT enhancements successfully realized in p-PbTe pave the way for similar ZT enhancements in SnTe [47–49]. Numerous efforts have been devoted on optimizing concentration of charge carriers [44, 50], valence band offset reduction [25, 51–67], resonant states [24, 60, 68–77], and lattice thermal conductivity reductions [6, 69, 78–83], all of which have realized a significant enhancement in thermoelectric SnTe.

This chapter introduces successful strategies of optimizing concentration of charge carriers, band engineering, and defect engineering for ZT enhancements in SnTe. In more details, (1) concentration of charge carriers has been effectively reduced through Sn vacancy compensation by excess of Sn and chemical doping by donors such as trivalent cations or halogens anions, respectively; (2) the energy offset between valence bands has been reduced through substitution of Sn by divalent cations; (3) the resonant states have been introduced by In doping; and (4) the lattice thermal conductivity reduction has been achieved through additional phonon scattering by introducing point defects, nanostructures, and lattice softening. Utilization of these strategies successfully enhance ZT of SnTe to be as high as 1.8 [84]. This promotes SnTe as highly efficient thermoelectric material with eco-friendly composition as a top alternative to PbTe thermoelectrics. Moreover, new strategies of crystal structure engineering and dislocation are discussed for possible further advancements in SnTe thermoelectrics. All the strategies discussed in this chapter are believed to be equally capable of enhancing the performance in both existing and new thermoelectrics.

4.2   Principles for Advancing Thermoelectrics

On the basis of parabolic band model with acoustic scattering, which is capable to approximate electronic transport for most thermoelectrics [17], the maximum ZT of a given material can be determined by thermoelectric quality factor B [1]:

$$B = T \times \frac{2k_B^2\hbar}{3\pi} \times \frac{C_l N_v}{m_I^* E_{def}^2 \kappa_L}$$

where m_I^* is the inertial effective mass, E_{def} is deformation potential coefficient, C_l is combined elastic moduli, k_B is Boltzmann constant, $\hbar$ is reduced Planck constant, and T is absolute temperature.

A high ZT relies on high B; therefore, high N_v but low m_I^*, E_{def}, and κ_L are expected. Although low m_I^* and E_{def} have been demonstrated to be beneficial to ZT in PbTe [85] and PbSe [86], respectively, an effective reduction on both still remains challenging in a given material. Therefore, the existing efforts for enhanced B factor mainly focus on band engineering for increasing N_v and defect engineering for reducing κ_L. It should be noted that the highest possible ZT strongly depends on concentration of charge carriers even with high B factor. Therefore, optimizing concentration of charge carriers is essential to realize maximal ZT in a given thermoelectric material.

4.3 Proven Strategies for Advancing SnTe Thermoelectrics

4.3.1 Optimization of Concentration of Charge Carriers

Due to the existence of high concentration of cation vacancies, pristine SnTe shows Hall concentration of charge carriers $n_H \sim 10^{20}$ cm^{-3} or higher, which deviates much from that needed ($\sim 10^{18}$–10^{19} cm^{-3}) for maximizing ZT (Fig. 4.2b) [44]. In order to realize the maximum ZT in SnTe, n_H has to be reduced, and successful strategies include both impurity and self-doping. In more details, doping with trivalent Sb/Bi

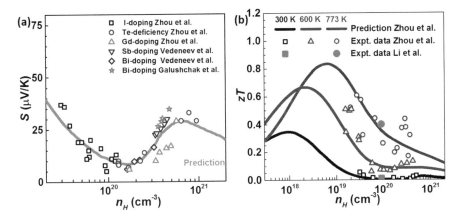

Fig. 4.2 Dependences on Hall concentration of charge carriers (n_H): (**a**) of Seebeck coefficient at 300 K and (**b**) of ZT at 300 K, 600 K, and 773 K for SnTe with doping for varying concentration of charge carriers only [27, 44, 50, 87]. The predictions are based on a two-band model [44]

on Sn site or negative monovalent iodine I on Te site enables an effective reduction in n_H, as shown in Fig. 4.2a [44, 50, 87]. Alternatively, excess of Sn effectively compensates cation vacancies and thus reduces hole concentration as well [56]. Synergy of both Sn excess and I doping is found to be even more effective [26, 27].

Limited by solubility of these dopants, the lowest n_H obtained so far in SnTe and its solid solutions are about $\sim 3 \times 10^{19}$ cm^{-3}, which is realized through I doping. Such concentration of charge carriers approaches to optimal concentration (n_{opt}) at $T > 773$ K according to $n_{opt} \sim \left(m_d^* T\right)^{1.5}$, where m_d^* is the density of states mass (DOS) [88] (Fig. 4.2b). Such optimization of n_H enables maximum ZT ~ 0.6 at 773 K [44] in pristine SnTe, which is inferior to that of pristine PbTe or PbSe. It should be noted that compositional complexity required for band and microstructure manipulations to realize a higher ZT in SnTe would presumably reduce the solubility of dopants. Therefore, the development of effective dopants for varying n_H in a broad range remains important for advancing SnTe thermoelectrics.

With doping with Gd on Sn site, n_H can be reduced to be as low as 10^{19} cm^{-3}, while n_H can be as high as $\sim 10^{21}$ cm^{-3} once Sn deficiency is applied. Such a broad range of n_H enables well understanding of electronic transport properties using a two-band model (Kane band approximation for L band and parabolic band approximation for Σ band) [44]. The model-predicted dependences of Seebeck coefficient and ZT (solid curves in Fig. 4.2a) on n_H fit well with the experimental measurements. Peak Seebeck coefficient is observed at $n_H \sim 6 \times 10^{20}$ cm^3, which is attributed to the contribution of low-lying heavy Σ band, experimentally indicating two-valence band structure in SnTe. This offers possibilities for electronic performance enhancements through engineering of two-valence bands to be aligned/converged with offset $\Delta E_{L-\Sigma}$ in a few $k_B T$ [11, 18].

4.3.2 Band Engineering

The principle of large number of transporting bands for electronic performance enhancement can be understood simply as an increase in electrical conductivity without explicit reduction in Seebeck coefficient [18]. Although $n_H > 2 \times 10^{20}$ cm^{-3} could involve the contribution from low-lying Σ band to the transport properties at room temperature in pristine SnTe (Fig. 4.2a), this n_H are much higher than that needed for optimized ZT, and, therefore, optimization of n_H can realize only ZT < 0.8 in pristine SnTe (Fig. 4.2b).

Band convergence, an effective approach to have both contributions of all converged bands to electronic transport properties, has often been demonstrated in PbTe thermoelectrics with a significantly improved ZT [9, 18, 22, 23, 41, 42]. Two-valence band structure with energy offset in SnTe is very similar to that of PbTe, which leads to an expectation of similar enhancements in ZT through band convergence. However, larger energy offset in SnTe compared to that of PbTe indicates that higher solubility and more effectiveness solvent are required for well-converging these two-valence bands in SnTe.

Sharing with PbTe the similarities in chemical properties and band structure, where Pb substitutions by divalent Mg [41], Mn [23], Cd [22], Eu [9], and Yb [42] are proven to be effective for reducing $L - \Sigma$ valence band offset, SnTe is expected to have a similar band convergence effect through substituting Sn with alkaline earths and transition metals. Typical band structures for pristine SnTe and $Sn_{26}MTe_{27}$ (M = Mn, Mg, Cd) alloys [27, 89] calculated on the base of density functional theory (DFT) are shown in Fig. 4.3a–d. And calculated energy change of $L - \Sigma$ valence band offset for SnTe alloys with various substitutions of Sn by monovalent and divalent impurities [27, 51–53, 56, 89, 90] is surveyed in Fig. 4.4.

Calculated band structures for SnTe with Mn, Mg, and Cd substitution on Sn site at a given concentration of 1/27 (Fig. 4.3b–d) show an effective reduction in $\Delta E_{L-\Sigma}$, meaning effective convergence of L and Σ valence bands. These have been demonstrated experimentally as well [27, 53, 56]. Similar effects are also observed in SnTe with Ca, Ag, Zn, and Hg substitution on Sn site through both theoretical calculations [52, 54, 89] (Fig. 4.4) and experimental measurements [51, 52, 54, 62]. Therefore, sufficient convergence of two valence bands is expectable in SnTe through cation substitutions as well, where these substitutions have either a high solubility (~10 mol. % or higher, such as Ca, Mg, Mn, and Ag) or a strong effect on converging valence bands (such as Cd, Zn, and Hg). Although calculated results show a certain reduction in $\Delta E_{L-\Sigma}$ for SnTe with Eu, Sr, Yb, or Ba substitution on Sn site, there are still no clear evidences of experimental realization. In many cases, cation substitutions increase in bandgap (E_g) of SnTe simultaneously, which could weaken the influence of thermally excited minority charge carriers at high temperatures [53, 54, 56].

The sufficient reduction of $\Delta E_{L-\Sigma}$ leads to redistribution of many holes from light L to heavy Σ valence band, resulting in an increase in Seebeck coefficient at given concentration of charge carriers, as compared to that of pristine SnTe (Fig. 4.5). As a result, significant ZT enhancements are experimentally observed in SnTe through heavy substitution of Sn with Ca [51], Mg [53], Mn [27], Cd [56], Hg [54], and Ag [52] (Fig. 4.6a). Although calculated band structure indicates Zn as very effective substitution for converging valence bands, the degree of increase in ZT for SnTe with Zn substitution on Sn site is very slight (Fig. 4.6a), which is mainly due to low solubility of ZnTe in SnTe.

Similar enhancements in electronic performance could also be achieved by introduction of resonant states. This is demonstrated in Tl-doped PbTe [20], where locally distorted band structure leads to a significant increase in Seebeck coefficient. Such effect is also observed in Ga- [95, 96] and In-doped [97–99] PbTe. It is, therefore, expected that similar enhancements in electronic performance can be obtained in SnTe as well, once resonant states can be introduced to locate close to Fermi level.

Density of states (DOS) for $Sn_{0.963}Te$ and $Sn_{0.926}M_{0.037}Te$ (M = Ga, In, Tl) [76] is calculated and shown in Fig. 4.3e. A hump at valence band edge, which can be regarded as a typical signal of the resonant states [20], is observed in SnTe with In doping only. The absence of the resonant states in Ga- and Tl-doped SnTe is attributed to weak bonding between s orbit of Ga and p orbit of Te and downward

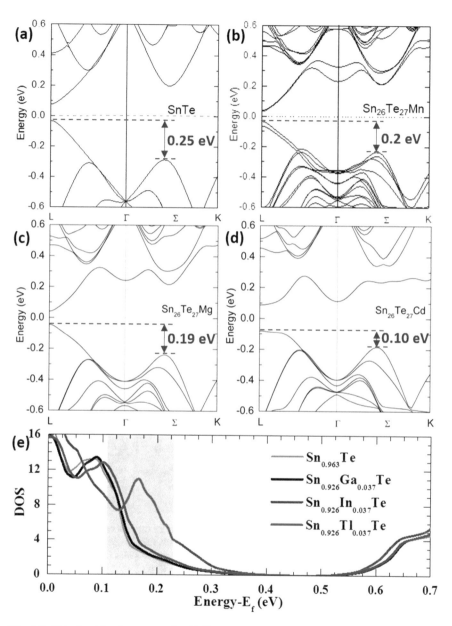

Fig. 4.3 Calculated band structures for (**a**) $Sn_{27}Te_{27}$ [89], (**b**) $Sn_{26}MnTe_{27}$ [27], (**c**) $Sn_{26}MgTe_{27}$ [89], and (**d**) $Sn_{26}CdTe_{27}$ [89]. (**e**) Density of states (DOS) for $Sn_{0.963}Te$ and $Sn_{0.926}M_{0.037}Te$ (M = Ga, In, Tl) [76]. Reproduced with permission, Copyright 2018, Royal Society of Chemistry

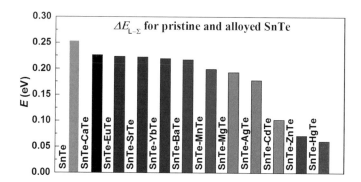

Fig. 4.4 Calculated $L - \Sigma$ valence band offset ($\Delta E_{L-\Sigma}$) for SnTe, $Sn_{16}AgTe_{17}$ [52], and $Sn_{26}MTe_{27}$ (M = Ca, Sr, Ba, Eu, Yb, Mg, Mn, Cd, Zn, Hg) alloys [27, 51, 53, 56, 89, 90]

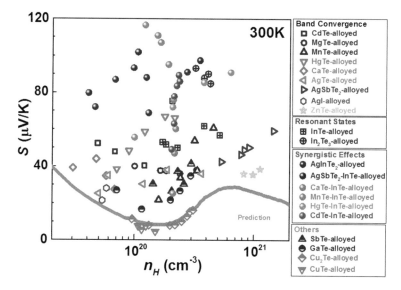

Fig. 4.5 Room temperature dependences of Seebeck coefficient on Hall concentration of charge carriers for various SnTe alloys with converged valence bands [27, 51–54, 56, 61–63], resonant states [68, 69], synergistic effects of band convergence and resonant states [52, 60, 71, 73–75, 91], or other effects [6, 79, 92, 93] with comparison to pristine SnTe [44]

shift of $6s$ orbit of Tl [76], respectively. The experimentally measured Seebeck coefficient in both SnTe $-$ InTe [68] and SnTe $-$ In_2Te_3 [69] alloys, locates well above of Pisarenko curve (gray line, Fig. 4.5). This indicates the existence of resonant states through In substitution on Sn site. As a result, significant ZT enhancement (Fig. 4.6b) is realized.

Based on calculated band structures (Fig. 4.3) and experimental observation on thermoelectric properties, both band convergence and resonant states are indeed beneficial for improving ZT in SnTe materials but act in different ways. This

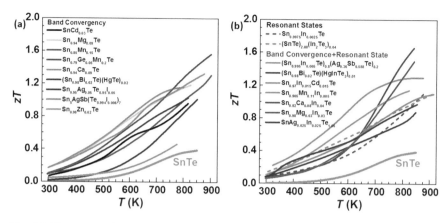

Fig. 4.6 Temperature dependences of ZT for (**a**) SnTe-based alloys with converged valence bands [27, 51, 53, 54, 56, 61–63, 84], (**b**) resonant states [68, 69], and synergistic effects [60, 70, 71, 73, 74, 91, 94]

suggests possible further enhancement in electronic performance and then ZT through a synergy of both. This motivated the research on thermoelectric properties of SnTe with co-doping, where one of dopants takes mainly the responsibility for valence band convergence and the other for creating resonant states. This is typified by co-doped SnTe alloys, where dopants can be In and Mn, In and Hg, In and Mg, In and Cd, and In and Ag. All these materials show further increased room temperature Seebeck coefficient at a given Hall concentration of charge carriers (Fig. 4.5). By this way, further increased ZT can be obtained (Fig. 4.6b).

As discussed above, the limitation of solubility of dopants in SnTe might hinder the further convergence of valence bands and, therefore, a further ZT enhancement. Therefore, a further increase in solubility of some species, particularly those offering a strong effect on converging the valence bands, is expected to open possibilities for further enhancing ZT of SnTe.

Among the abovementioned high-performance SnTe thermoelectric alloys, $Sn_{1-x}Mn_xTe$ shows the most promising ZT due to the highest solubility of MnTe (~15 mol. %) [27]. However, there is no indication that ZT stops to increase in $Sn_{1-x}Mn_xTe$ alloys even when MnTe solubility is engineered to be higher than (15 mol. %). Recently, efforts have been put on increasing in solubility of MnTe for further improving ZT of SnTe thermoelectrics. Considering much higher solubility (~50 mol. %) of MnTe in GeTe [102, 103] as well as the capability of forming full compositional range solid solution between GeTe and SnTe [104, 105], significantly increased MnTe solubility in SnTe can be expected with the help of GeTe [106]. This is realized, and the solubility of MnTe in SnTe successfully increases from ~15 mol. % [27] to ~25 mol. % with the help of 5 mol. % GeTe (Fig. 4.7a). The composition-dependent lattice parameters for $Sn_{0.95-x}Ge_{0.05}Mn_xTe$ [84] confirm the above expectation (Fig. 4.7b). Importantly, such significant increased solubility successfully enables well-converged valence bands at MnTe concentration of ~20

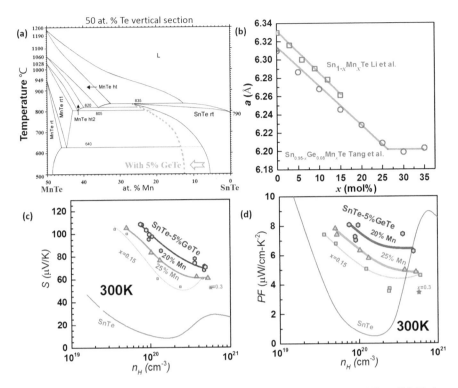

Fig. 4.7 (**a**) Phase diagram of SnTe — MnTe [100, 101] with an increase in solubility of MnTe due to the existence of 5 mol. % GeTe, (**b**) composition-dependent lattice parameters for $Sn_{0.95-x}Ge_{0.05}Mn_xTe$ [84] and $Sn_{1-x}Mn_xTe$ [27], (**c**) dependences of Seebeck coefficient on Hall concentration of charge carriers, and (**d**) power factor (PF) for SnTe-based materials [27, 84]

mol. % for maximizing electronic performance (Fig. 4.7c, d). This opens successfully further possibilities for dopants with low solubility or insoluble ones but strongly converges valence bands of SnTe, to be reconsidered for advancing SnTe thermoelectrics through solubility engineering approach.

Relying much on electronic performance enhancements by band convergence, resonant states, and synergistic effects of both, a peak ZT as high as ~1.6 [27, 94] is achieved in single-phased SnTe alloys (Fig. 4.6), which are much higher than that of pristine SnTe (ZT ~ 0.4).

4.3.3 Defect Engineering

Pristine SnTe shows intrinsic lattice thermal conductivity $\kappa_L \sim 3\ W \times m^{-1} \times K^{-1}$, which is high as compared to that of other IV–VI thermoelectrics, such as PbTe [45], PbSe [107], PbS [108], and SnSe [109]. Therefore, numerous efforts have also been

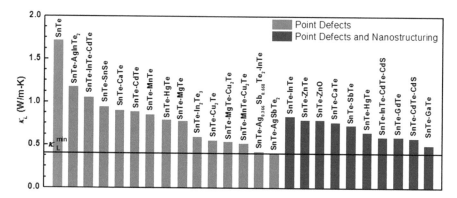

Fig. 4.8 Comparison of measured lowest lattice thermal conductivity κ_L for SnTe solid solutions [6, 25–27, 51–56, 60–62, 69, 110] and composites [51, 55, 56, 68, 69, 78, 79, 92, 111, 112] at high temperatures. The black line shows the amorphous limit of SnTe estimated according to Debye-Cahill model [113]

put on reducing κ_L for ZT enhancement in SnTe thermoelectrics, through additional phonon scattering by point defects (including substitutions [6, 25–27, 51–56, 60–62, 69, 110], interstitials [6, 25, 26], and vacancies [60, 61, 69]) and nanostructures [51, 55, 56, 68, 69, 78, 79, 92, 111, 112]. Both strategies have been widely demonstrated in IV–VI and many other thermoelectrics, and the resultant reduction in κ_L for SnTe thermoelectrics at high temperatures is surveyed in Fig. 4.8.

In reality, most of ZT enhancements achieved in SnTe-based alloys mentioned above partially come from contribution of reduced κ_L due to additional phonon scattering introduced by point defects. This type of phonon scattering originates from mass and strain fluctuations between the host and the guest atoms [114]. Reduction of κ_L achieved in SnTe-based alloys, either with tellurides or selenide having cation-to-anion ratio of 1, is mainly due to the existence of substitutional point defects. The minimal κ_L of 0.8 W $\times$ m^{-1} $\times$ K^{-1} is obtained in alloy consisting of 91 mol. % SnTe and 9 mol. % MgTe [53], because of large mass and size contrasts introduced and high concentration of point defects.

The fluctuations on mass and strain can be maximized by interstitial defects and vacancies; therefore, an effective phonon scattering for reducing κ_L can be expected in materials with these types of point defects as compared to that of substitutions [8]. Indeed, κ_L as low as 0.5 W $\times$ m^{-1} $\times$ K^{-1} is obtained in SnTe $-$ Cu$_2$Te alloys (Fig. 4.8), which approaches to its amorphous limit of 0.4 W $\times$ m^{-1} $\times$ K^{-1} [6] according to Debye-Cahill model [113]. This dramatic reduction in κ_L comes from strong phonon scattering by coexisting interstitial and substitutional Cu defects [6], which is evidenced by the comparably low formation energies for these two types of defects according to DFT calculations [6]. Debye-Callaway thermal model, taking these two types of point defects for phonon scattering, predicts successfully the experimentally observed κ_L reduction, supporting that interstitial Cu defects act as the dominant source for phonon scattering [6]. Importantly, alloying SnTe with

Cu$_2$Te shows a negligible effect on electronic transport properties (Fig. 4.5), leading to a pure thermal strategy for enhancing ZT up to unity in SnTe in the form of a solid solution [6]. This enables the introduction of Cu interstitials as effective approach for reducing κ_L without downgrading electronic performance.

Alternatively, vacancies could in principle offer a similar reduction in κ_L as interstitials. In SnTe $-$ In$_2$Te$_3$ alloys, both vacancy and substitutional defects are simultaneously introduced [69], leading to significantly reduced κ_L of 0.6 W $\times$ m^{-1} $\times$ K^{-1} (Fig. 4.8). Relatively higher κ_L, as compared to that in SnTe $-$ Cu$_2$Te alloy, can be understood by lower solubility of In$_2$Te$_3$ in SnTe and thus lower vacancy concentration. Recently, a surprisingly low κ_L of ~0.4 W $\times$ m^{-1} $\times$ K^{-1} has been realized in SnTe $-$ AgSbTe$_2$ system [60, 61]. Alloying SnTe with AgSbTe$_2$ enables dual effects of point defects and lattice softening (reduced sound velocity) on reducing κ_L.

In addition to point defects for phonon scattering, both nanoprecipitates and nanostructuring are widely used to further reduce κ_L of SnTe, through introducing high-density boundary interfaces as shown in Fig. 4.8. Successful reduction of κ_L has been achieved in SnTe $-$ InTe alloys synthesized by ball milling [68], SnTe $-$ InTe $-$ CdTe $-$ CdS [55]/SnTe $-$ CdTe $-$ CdS [56]/SnTe $-$ ZnO [112] by adding insoluble solvents, and SnTe $-$ HgTe [17]/SnTe $-$ GaTe [79]/SnTe $-$ SbTe [92]/SnTe $-$ GdTe [111] by precipitation. Taking SnTe $-$ SbTe [92] as an example, nanoprecipitates are introduced by over-solubility of SbTe ($\geq$4 mol. %), which is found to be effective sources for a strong phonon scattering by boundary interfaces, as shown in Fig. 4.9. The lowest reported value of $\kappa_L \sim 0.5$ W $\times$ m^{-1} $\times$ K^{-1} in SnTe composites is obtained in Sn$_{0.93}$Ga$_{0.1}$Te, approaching to amorphous limit (0.4 W $\times$ m^{-1} $\times$ K^{-1}) as well.

Fig. 4.9 TEM images and corresponding electron diffraction pattern for Sn$_{0.96}$Sb$_{0.04}$Te [92]. Reproduced with permission, Copyright 2016, Royal Society of Chemistry

Although low value of κ_L close to amorphous limit has been achieved in SnTe $-$ AgSbTe$_2$ solid solutions and Sn$_{0.93}$Ga$_{0.1}$Te composites, majority of high-ZT SnTe materials still show much higher κ_L than the amorphous limit. This open up space for further ZT enhancement in SnTe through further reduction in κ_L.

4.3.4 Synergy of Both Band and Defect Engineering

In principle, electronic strategies of band engineering and optimizing concentration of charge carriers are independent to thermal strategy of defect engineering for enhancing of overall ZT. Therefore, the synergy of multiple effects of these strategies would lead to greater enhancement in ZT of SnTe thermoelectrics.

This is demonstrated in many cases. Alloying with MgTe introduces the synergistic effects of band convergence and point defect scattering simultaneously, leading to peak ZT of 1.2 (Fig. 4.6a) [53]. Further optimizing concentration of charge carriers by I doping successfully realizes peak ZT up to 1.3 (Fig. 4.6a) [27]. Similar effects result in increased ZT > 1 in SnTe $-$ AgI system (Fig. 4.6a) [63]. The synergistic effects of AgSbTe$_2$ alloying (for band convergence, point defect scattering, and lattice softening) and I doping (for optimizing concentration of charge carriers) realize increased peak ZT of 1.2 (Fig. 4.6a) [61]. Optimizing concentration of charge carriers by using an excess of Sn, Sn$_{1.03}$Te $-$ CdTe system with converged bands and substitutional defects realizes an enhancement of ZT up to unity [56]. Introducing insoluble CdS/ZnS nanoprecipitates leads to additional reduction in κ_L (Fig. 4.8) and thus a higher ZT $\sim$ 1.3 in Sn$_{1.03}$Te $-$ CdTe $-$ CdS/ZnS composites (Fig. 4.10) [56]. Similarly, simultaneous optimizing concentration of

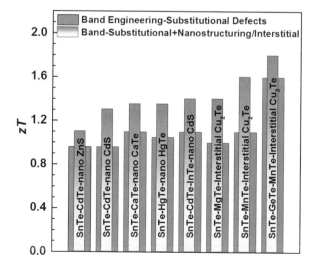

Fig. 4.10 Peak ZT for SnTe-based alloys and composites [25, 26, 51, 54–56, 84]

charge carriers by Bi doping, valence band convergence and nanostructuring by over-solubility HgTe alloying contribute to an enhanced peak ZT up to 1.35 (Fig. 4.10) [54]. Similar synergic effects realize a very comparable peak ZT in SnTe $-$ CaTe system as well (Fig. 4.10) [51].

In substitution on Sn site is found to introduce resonant states for enhancing the electronic performance of SnTe, where the solvents are either InTe [68] or In_2Te_3 [52, 69]. In addition to phonon scattering due to alloy defects in the matrix phase, a further κ_L reduction can be realized by nanostructures through ball milling in SnTe $-$ InTe (Fig. 4.8) [68]. As a result, the synergic effects of both resonant states and phonon scattering by multidimensional defects lead to peak ZT up to 1.1 (Fig. 4.6b) [68]. Moreover, valence band convergence by CdTe alloying, resonant states by InTe doping, and CdS nanoprecipitates synergistically increase in peak ZT up to 1.4 and also an average ZT up to 0.8 in $Sn_{0.97}Cd_{0.015}In_{0.015}Te - 3$ mol. % CdS (Fig. 4.10) [55]. The synergistic effects of band convergence, resonant states, and substitutions for enhancing ZT have been realized in many SnTe-based thermoelectrics co-doped with In and Hg [71], In and Mg [94], In and Mn [74], In and Ag [52], and In and Ca [91].

Alloying with MnTe/MgTe enables effective valence band convergence due to high solubility and results in a significantly increased ZT in SnTe solid solutions. However, κ_L value remains much higher than amorphous limit (Fig. 4.8). Note that alloying with Cu_2Te offers purely thermal effect on reducing κ_L without any detrimental effects on electronic performance [6]. This motivates introduction of both interstitial point defects in SnTe $-$ MnTe/MgTe alloys for further reducing κ_L and for realizing superior ZT [25, 26]. Eventually, synergy of effectively converged $L - \Sigma$ valence bands and low $\kappa_L \sim 0.5\ W \times m^{-1} \times K^{-1}$ (Fig. 4.8) leads to peak ZT as high as 1.8 and 1.4 in Cu_2Te alloyed SnTe $-$ MnTe [84] and SnTe $-$ MgTe [25] solid solutions, respectively (Fig. 4.10). This high ZT realized in nontoxic composition promote SnTe as a top eco-friendly alternative for conventional p-PbTe thermoelectrics.

4.4 New Possibilities for Advancing SnTe-Based Thermoelectrics

4.4.1 Crystal Structure Engineering

Since the location of band extrema in Brillouin zone can be significantly different and these different locations have significantly different symmetrical operations in Brillouin zone, these lead to different band degeneracies. This indicates that symmetry engineering would in principle offer new degrees of freedom for band structure and thus transport property manipulations for ZT enhancements. This concept has been recently proposed and demonstrated in GeTe [28], an analogue to SnTe. Note that both SnTe and GeTe undergo phase transition from NaCl-type cubic structure to rhombohedral structure as temperature decreases, while the difference relies on different phase transition temperatures (~700 K and ~100 K for GeTe [115] and SnTe [116], respectively). This rhombohedral structure can be regarded as the distortion along [110] direction of cubic crystallographic structure [28]. This directional lattice distortion leads to the reduction in energy of L valence band, and, eventually, Σ valence band becomes the dominant transporting band in rhombohedral GeTe [28]. Additionally, reduced symmetry (from cubic to rhombohedral) splits 4 L pockets into 3 L + 1 Z and 12 Σ pockets into 6 Σ + 6 η, which increases degree of freedom of band degeneracy [117]. Bi doping on Ge site is found to control effectively the distortion degree from the cubic structure and engineer the split of low-symmetry bands to be converged [28]. In addition, symmetry reduction would simultaneously lead to reduction in κ_L. Eventually, both high effective N_v and extraordinary ZT of 2.4 are achieved at ~600 K in low-symmetry rhombohedral GeTe but close to its cubic structure [28]. This suggests a possibility for ZT enhancement in SnTe as well by a similar crystal structure engineering approach, once the crystal structure of SnTe is stabilized in a rhombohedral but close to cubic structure at working temperatures.

4.4.2 Dislocations

Phonons, which transport heat, span in a broad range of frequencies. Demonstrated strategies of point defects through alloying and boundary interfaces through nanostructuring for reducing κ_L in SnTe rely mainly on scattering of phonons with high and low frequencies, respectively. However, mid-frequency phonons contribute largely to κ_L as well. However, these phonons are not explicitly scattered by existing strategies for reducing κ_L of many thermoelectrics. Introduction of dense dislocations would offer a great potential for further advancing SnTe thermoelectrics, because dislocations have the capability for strongly scattering mid-frequency phonons. This strategy has recently been demonstrated in SnTe analogue compounds PbTe [9] and PbSe [11], where both extremely low κ_L and significantly increased ZT are realized. Dense in-grain dislocations can be introduced by a collapse of vacancy clusters [11] and by a multiplication [9]. Therefore, proper synthesis procedure for creating dense dislocations would in principle offer possibilities for further reducing κ_L and thus for increasing ZT of SnTe.

4.5 Conclusion

Successful strategies of optimizing concentration of charge carriers and band engineering for electronic performance enhancements and defect engineering for reducing the lattice thermal conductivity are summarized here for SnTe-based thermoelectrics. All these approaches contribute to great enhancements in thermoelectric performance achieved in SnTe. Peak ZT as high as 1.8 has been realized in SnTe with nontoxic composition, which is realized by synergy of both thermal and electronic effects. This clearly demonstrates SnTe as high-efficient and eco-friendly alternative for *p*-PbTe thermoelectrics. The strategies surveyed here for advancing SnTe-based thermoelectrics are believed in principle to offer equal opportunities for improving other thermoelectric materials.

References

1. H.J. Goldsmid, *Introduction to Thermoelectricity* (Springer, Heidelberg, 2009)
2. X.Y. Zhang, Y.Z. Pei, npj Quantum Mater. **2**, 68 (2017)
3. Z. Chen, X. Zhang, Y. Pei, Adv. Mater. **30**, 1705617 (2018)
4. W. Li, S. Lin, X. Zhang, Z. Chen, X. Xu, Y. Pei, Chem. Mater. **28**, 6227–6232 (2016)
5. L. Hu, T. Zhu, X. Liu, X. Zhao, Adv. Funct. Mater. **24**, 5211–5218 (2014)
6. Y. Pei, L. Zheng, W. Li, S. Lin, Z. Chen, Y. Wang, X. Xu, H. Yu, Y. Chen, B. Ge, Adv. Electron. Mater. **2**, 1600019 (2016)
7. J. Shen, X. Zhang, Z. Chen, S. Lin, J. Lin, W. Li, S. Lin, Y. Chen, Y. Pei, J. Mater. Chem. A **5**, 5314–5320 (2017)
8. J. Shen, X. Zhang, S. Lin, J. Li, Z. Chen, W. Li, Y. Pei, J. Mater. Chem. A **4**, 15464–15470 (2016)
9. Z. Chen, Z. Jian, W. Li, Y. Chang, B. Ge, R. Hanus, J. Yang, Y. Chen, M. Huang, G.J. Snyder, Y. Pei, Adv. Mater. **29**, 1606768 (2017)
10. S.I. Kim, K.H. Lee, H.A. Mun, H.S. Kim, S.W. Hwang, J.W. Roh, D.J. Yang, W.H. Shin, X.S. Li, Y.H. Lee, Science **348**, 109–114 (2015)
11. Z. Chen, B. Ge, W. Li, S. Lin, J. Shen, Y. Chang, R. Hanus, G.J. Snyder, Y. Pei, Nat. Commun. **8**, 13828 (2017)
12. B. Poudel, Q. Hao, Y. Ma, Y.C. Lan, A. Minnich, B. Yu, X.A. Yan, D.Z. Wang, A. Muto, D. Vashaee, X.Y. Chen, J.M. Liu, M.S. Dresselhaus, G. Chen, Z.F. Ren, Science **320**, 634–638 (2008)
13. Y. Pei, J. Lensch-Falk, E.S. Toberer, D.L. Medlin, G.J. Snyder, Adv. Funct. Mater. **21**, 241–249 (2011)
14. K. Biswas, J. He, I.D. Blum, C.-I. Wu, T.P. Hogan, D.N. Seidman, V.P. Dravid, M.G. Kanatzidis, Nature **489**, 414–418 (2012)
15. K. Biswas, J. He, Q. Zhang, G. Wang, C. Uher, V.P. Dravid, M.G. Kanatzidis, Nat. Chem. **3**, 160–166 (2011)
16. K. Biswas, J. He, G. Wang, S.-H. Lo, C. Uher, V.P. Dravid, M.G. Kanatzidis, Energy Environ. Sci. **4**, 4675–4684 (2011)
17. Y. Pei, H. Wang, G.J. Snyder, Adv. Mater. **24**, 6125–6135 (2012)
18. Y. Pei, X. Shi, A. LaLonde, H. Wang, L. Chen, G.J. Snyder, Nature **473**, 66–69 (2011)
19. S. Lin, W. Li, Z. Chen, J. Shen, B. Ge, Y. Pei, Nat. Commun. **7**, 10287 (2016)
20. J.P. Heremans, B. Wiendlocha, A.M. Chamoire, Energy Environ. Sci. **5**, 5510–5530 (2012)
21. J.P. Heremans, V. Jovovic, E.S. Toberer, A. Saramat, K. Kurosaki, A. Charoenphakdee, S. Yamanaka, G.J. Snyder, Science **321**, 554–557 (2008)
22. Y. Pei, A.D. LaLonde, N.A. Heinz, G.J. Snyder, Adv. Energy Mater. **2**, 670–675 (2012)
23. Y. Pei, H. Wang, Z.M. Gibbs, A.D. LaLonde, G.J. Snyder, NPG Asia Mater. **4**, e28 (2012)
24. L. Zhang, Q. Peng, C. Han, J. Wang, Z.-H. Ge, Q. Sun, Z.X. Cheng, Z. Li, S.X. Dou, J. Mater. Chem. A **6**, 2507–2516 (2018)
25. L. Zheng, W. Li, S. Lin, J. Li, Z. Chen, Y. Pei, ACS Energy Lett. **2**, 563–568 (2017)

26. W. Li, L. Zheng, B. Ge, S. Lin, X. Zhang, Z. Chen, Y. Chang, Y. Pei, Adv. Mater. **29**, 1605887 (2017)
27. W. Li, Z. Chen, S. Lin, Y. Chang, B. Ge, Y. Chen, Y. Pei, J. Mater. **1**, 307–315 (2015)
28. J. Li, X. Zhang, Z. Chen, S. Lin, W. Li, J. Shen, I.T. Witting, A. Faghaninia, Y. Chen, A. Jain, L. Chen, G.J. Snyder, Y. Pei, Joule **2**, 976–987 (2018)
29. J. Li, Z. Chen, X. Zhang, Y. Sun, J. Yang, Y. Pei, NPG Asia Mater. **9**, e353 (2017)
30. X. Liu, T. Zhu, H. Wang, L. Hu, H. Xie, G. Jiang, G.J. Snyder, X. Zhao, Adv. Energy Mater. **3**, 1238–1244 (2013)
31. W. Liu, X. Tan, K. Yin, H. Liu, X. Tang, J. Shi, Q. Zhang, C. Uher, Phys. Rev. Lett. **108**, 166601 (2012)
32. C.G. Fu, S.Q. Bai, Y.T. Liu, Y.S. Tang, L.D. Chen, X.B. Zhao, T.J. Zhu, Nat. Commun. **6**, 7 (2015)
33. C.G. Fu, T.J. Zhu, Y.Z. Pei, H.H. Xie, H. Wang, G.J. Snyder, Y. Liu, Y.T. Liu, X.B. Zhao, Adv. Energy Mater. **4**, 1400600 (2014)
34. Y.I. Ravich, B.A. Efimova, I.A. Smirnov, *Semiconducting Lead Chalcogenides* (Plenum Press, New York, 1970)
35. G. Nimtz, B. Schlicht, *Springer Tracts in Modern Physics*, vol 98 (1983), pp. 1–117
36. F. Herman, R.L. Kortum, I.B. Ortenburger, J.P. Van Dyke, J. Phys. Colloq. **29**, 62–77 (1968)
37. Y.I. Ravich, in *Lead Chalcogenides: Physics and Applications*, ed. by D. Khokhlov, (Taylor & Fransics Group, New York, 2003), pp. 1–34
38. R.S. Allgaier, J. Appl. Phys. **32**, 2185–2189 (1961)
39. A.J. Crocker, L.M. Rogers, J. Phys. Colloq. **29**(C4), 129–132 (1968)
40. A.A. Andreev, V.N. Radionov, Sov. Phys. Semicond. **1**, 145–148 (1967)
41. Y. Pei, A.D. LaLonde, N.A. Heinz, X. Shi, S. Iwanaga, H. Wang, L. Chen, G.J. Snyder, Adv. Mater. **23**, 5674–5678 (2011)
42. Z. Jian, Z. Chen, W. Li, J. Yang, W. Zhang, Y. Pei, J. Mater. Chem. C **3**, 12410–12417 (2015)
43. G. Tan, F. Shi, S. Hao, L.-D. Zhao, H. Chi, X. Zhang, C. Uher, C. Wolverton, V.P. Dravid, M.G. Kanatzidis, Nat. Commun. **7**, 12167 (2016)
44. M. Zhou, Z.M. Gibbs, H. Wang, Y.M. Han, C.N. Xin, L.F. Li, G.J. Snyder, Phys. Chem. Chem. Phys. **16**, 20741–20748 (2014)
45. Y. Pei, A. LaLonde, S. Iwanaga, G.J. Snyder, Energy Environ. Sci. **4**, 2085–2089 (2011)
46. L. Rogers, J. Phys. D. Appl. Phys. **1**, 845 (1968)
47. S. Li, X.F. Li, Z.F. Ren, Q. Zhang, J. Mater. Chem. A **6**, 2432–2448 (2018)
48. W. Li, Y. Wu, S. Lin, Z. Chen, J. Li, X. Zhang, L. Zheng, Y. Pei, ACS Energy Lett. **2**, 2349–2355 (2017)
49. R. Moshwan, L. Yang, J. Zou, Z.G. Chen, Adv. Funct. Mater. **27**, 18 (2017)
50. V.P. Vedeneev, S.P. Krivoruchko, E.P. Sabo, Semiconductors **32**, 241–244 (1998)
51. R. Al Rahal Al Orabi, N.A. Mecholsky, J. Hwang, W. Kim, J.-S. Rhyee, D. Wee, M. Fornari, Chem. Mater. **28**, 376–384 (2016)
52. A. Banik, U.S. Shenoy, S. Saha, U.V. Waghmare, K. Biswas, J. Am. Chem. Soc. **138**, 13068–13075 (2016)
53. A. Banik, U.S. Shenoy, S. Anand, U.V. Waghmare, K. Biswas, Chem. Mater. **27**, 581–587 (2015)
54. G.J. Tan, F.Y. Shi, J.W. Doak, H. Sun, L.D. Zhao, P.L. Wang, C. Uher, C. Wolverton, V.P. Dravid, M.G. Kanatzidis, Energy Environ. Sci. **8**, 267–277 (2015)
55. G.J. Tan, F.Y. Shi, S.Q. Hao, H. Chi, L.D. Zhao, C. Uher, C. Wolverton, V.P. Dravid, M.G. Kanatzidis, J. Am. Chem. Soc. **137**, 5100–5112 (2015)
56. G.J. Tan, L.D. Zhao, F.Y. Shi, J.W. Doak, S.H. Lo, H. Sun, C. Wolverton, V.P. Dravid, C. Uher, M.G. Kanatzidis, J. Am. Chem. Soc. **136**, 7006–7017 (2014)
57. J. He, X. Tan, J. Xu, G.-Q. Liu, H. Shao, Y. Fu, X. Wang, Z. Liu, J. Xu, H. Jiang, J. Jiang, J. Mater. Chem. A **3**, 19974–19979 (2015)
58. G. Tan, F. Shi, S. Hao, H. Chi, T.P. Bailey, L.-D. Zhao, C. Uher, C. Wolverton, V.P. Dravid, M.G. Kanatzidis, J. Am. Chem. Soc. **137**, 11507–11516 (2015)

59. W. Haijun, C. Chang, D. Feng, Y. Xiao, X. Zhang, Y. Pei, L. Zheng, W. Di, S. Gong, Y. Chen, J. He, M.G. Kanatzidis, L.-D. Zhao, Energy Environ. Sci. **8**, 3298–3312 (2015)
60. M.K. He, D. Feng, D. Wu, Y.D. Guan, J.Q. He, Appl. Phys. Lett. **112**(6), 063902 (2018). https://doi.org/10.1063/1.5018477
61. G.J. Tan, S.Q. Hao, R.C. Hanus, X.M. Zhang, S. Anand, T.P. Bailey, A.J.E. Rettie, X.L. Su, C. Uher, V.P. Dravid, G.J. Snyder, C. Wolverton, M.G. Kanatzidis, ACS Energy Lett. **3**, 705–712 (2018)
62. Z.Y. Chen, R.F. Wang, G.Y. Wang, X.Y. Zhou, Z.S. Wang, C. Yin, Q. Hu, B.Q. Zhou, J. Tang, R. Ang, Chin. Phys B **27**, 5 (2018)
63. A. Banik, K. Biswas, J. Solid-State Electron **242**, 43–49 (2016)
64. H.C. Wang, J. Hwang, C. Zhang, T. Wang, W.B. Su, H. Kim, J. Kim, J.Z. Zhai, X. Wang, H. Park, W. Kim, C.L. Wang, J. Mater. Chem. A **5**, 14165–14173 (2017)
65. M.H. Lee, D.-G. Byeon, J.-S. Rhyee, B. Ryu, J. Mater. Chem. A **5**, 2235 (2017)
66. S. Acharya, J. Pandey, A. Soni, Appl. Phys. Lett. **109**, 133904 (2016)
67. J. He, J. Xu, G.-Q. Liu, H. Shao, X. Tan, Z. Liu, J. Xu, H. Jiang, J. Jiang, RSC Adv. **6**, 32189 (2016)
68. Q. Zhang, B.L. Liao, Y.C. Lan, K. Lukas, W.S. Liu, K. Esfarjani, C. Opeil, D. Broido, G. Chen, Z.F. Ren, Proc. Nat. Acad. Sci. U S A. **110**, 13261–13266 (2013)
69. G. Tan, W.G. Zeier, F. Shi, P. Wang, G.J. Snyder, V.P. Dravid, M.G. Kanatzidis, Chem. Mater. **27**, 7801–7811 (2015)
70. J.Q. Li, N. Yang, S.M. Li, Y. Li, F.S. Liu, W.Q. Ao, J. Electron. Mater. **47**, 205–211 (2018)
71. X. Tan, G. Liu, J. Xu, X. Tan, H. Shao, H. Hu, H. Jiang, Y. Lu, J. Jiang, J. Mater. **4**, 62–67 (2018)
72. Y.M. Zhou, H.J. Wu, Y.L. Pei, C. Chang, Y. Xiao, X. Zhang, S.K. Gong, J.Q. He, L.-D. Zhao, Acta Mater. **125**, 542–549 (2017)
73. X.F. Tan, X.J. Tan, G.Q. Liu, J.T. Xu, H.Z. Shao, H.Y. Hu, M. Jin, H.C. Jiang, J. Jiang, J. Mater. Chem. C **5**, 7504–7509 (2017)
74. L. Wang, X.J. Tan, G.Q. Liu, J.T. Xu, H.Z. Shao, B. Yu, H.C. Jiang, S. Yue, J. Jiang, ACS Energy Lett. **2**, 1203–1207 (2017)
75. D.K. Bhat, U.S. Shenoy, J. Phys. Chem. C **121**, 7123–7130 (2017)
76. X.J. Tan, G.Q. Liu, J.T. Xu, H.Z. Shao, J. Jiang, H.C. Jiang, Phys. Chem. Chem. Phys. **18**, 20635–20639 (2016)
77. M. Zhou, Z.M. Gibbs, H. Wang, Y. Han, L. Li, G.J. Snyder, Appl. Phys. Lett. **109**, 042102 (2016)
78. L.-D. Zhao, X. Zhang, H. Wu, G. Tan, Y. Pei, Y. Xiao, C. Chang, D. Wu, H. Chi, L. Zheng, S. Gong, C. Uher, J. He, M.G. Kanatzidis, J. Am. Chem. Soc. **138**, 2366–2373 (2016)
79. R. Al Rahal Al Orabi, J. Hwang, C.-C. Lin, R. Gautier, B. Fontaine, W. Kim, J.-S. Rhyee, D. Wee, M. Fornari, Chem. Mater. **29**, 612–620 (2016)
80. Z. Zhou, J. Yang, Q. Jiang, Y. Luo, D. Zhang, Y. Ren, X. He, J. Xin, J. Mater. Chem. A **4**, 13171 (2016)
81. X. Zhang, Y. Zhou, Y. Pei, Y. Chen, B. Yuan, S. Zhang, Y. Deng, S. Gong, J. He, L.-D. Zhao, J. Alloys Compd. **709**, 575–580 (2017)
82. Z. Li, Y. Chen, J.-F. Li, H. Chen, L. Wang, S. Zheng, G. Lu, Nano Energy **28**, 78–86 (2016)
83. J. He, X. Jingtao, X. Tan, G.-Q. Liu, H. Shao, Z.H. Liu, H. Jiang, J. Jiang, J. Mater. **2**, 165–171 (2016)
84. J. Tang, B. Gao, S. Lin, J. Li, Z. Chen, F. Xiong, W. Li, Y. Chen, Y. Pei, Adv. Funct. Mater. **28**, 1803586 (2018). https://doi.org/10.1002/adfm.201803586
85. Y. Pei, A.D. LaLonde, H. Wang, G.J. Snyder, Energy Environ. Sci. **5**, 7963–7969 (2012)
86. H. Wang, Y. Pei, A.D. LaLonde, G.J. Snyder, Proc. Natl. Acad. Sci. U S A. **109**, 9705–9709 (2012)
87. M.A. Galushchak, D.M. Freik, I. Ivanyshyn, A. Lisak, M.J. Pyts, Thermoelectric **1**, 42 (2000)
88. A.F. Ioffe, *Semiconductor thermoelements and Thermoelectric cooling* (Infosearch, London, 1957)

89. X. Dong, H. Yu, W. Li, Y. Pei, Y. Chen, J. Mater. **2**, 158–164 (2016)
90. J.H. Tan, H.Z. Shao, J. He, G.Q. Liu, J.T. Xu, J. Jiang, H.C. Jiang, Phys. Chem. Chem. Phys. **18**, 7141 (2016)
91. D.K. Bhat, U.S. Shenoy, Mater. Today Phys. **4**, 12–18 (2018)
92. A. Banik, B. Vishal, S. Perumal, R. Datta, K. Biswas, Energy Environ. Sci. **9**, 2011 (2016)
93. R. Brebrick, A. Strauss, Phys. Rev. **131**, 104 (1963)
94. D.K. Bhat, S.U. Sandhya, J. Phys. Chem. C **121**, 7123–7130 (2017)
95. V.I. Kaidanov, S.A. Nemov, Y.I. Ravich, Sov. Phys. Semicond. **26**, 113–125 (1992)
96. P.G. Rustamov, C.I. Abilov, M.A. Alidzhan, Phys. Status Solid A-Appl. Res. **12**, K103–K10& (1972)
97. A.J. Rosenberg, F. Wald, J. Phys. Chem. Solids **26**, 1079–1086 (1965)
98. V.I. Kaidanov, S.A. Rykov, M.A. Rykova, O.V. Syuris, Sov. Phys. Semicond. **24**, 87–91 (1990)
99. A.A. Averkin, V.I. Kaidanov, R.B. Melnik, Sov. Phys. Semicond. **5**, 75–7& (1971)
100. S. Badrinarayanan, A. Mandale, J. Electron Spectrosc. Relat. Phenom. **53**, 87–95 (1990)
101. T.S. Yeoh, S.G. Teoh, H.K. Fun, J. Phys. Soc. Jpn. **57**, 3820–3823 (1988)
102. N.Z. Zhigareva, A.B. Ivanova, A.N. Melikhova, A.G. Obedkov, E.I. Rogacheva, Inorg. Mater. **17**, 1314–1316 (1981)
103. W.D. Johnston, D.E. Sestrich, J. Inorg. Nucl. Chem. **19**, 229–236 (1961)
104. L.E. Shelimova, O.G. Karpinskii, Inorg. Mater. **27**, 2170–2174 (1991)
105. D. Bashkirov et al., Sov. Phys. Crystallogr. **34**, 794–795 (1989)
106. J.Q. Li, S. Huang, Z.P. Chen, Y. Li, S.H. Song, F.S. Liu, W.Q. Ao, Phys. Chem. Chem. Phys. **19**, 28749–28755 (2017)
107. H. Wang, Y. Pei, A.D. LaLonde, G.J. Snyder, Adv. Mater. **23**, 1366–1370 (2011)
108. H. Wang, E. Schechtel, Y. Pei, G.J. Snyder, Adv. Energy Mater. **3**, 488–495 (2013)
109. K. Peng, X. Lu, H. Zhan, S. Hui, X. Tang, G. Wang, J. Dai, C. Uher, G. Wang, X. Zhou, Energy Environ. Sci. **9**, 454–460 (2016)
110. A. Banik, K. Biswas, J. Mater. Chem. A **2**, 9620–9625 (2014)
111. L. Zhang, J. Wang, Z. Cheng, Q. Sun, Z. Li, S. Dou, J. Mater. Chem. A **4**, 7936–7942 (2016)
112. Z.W. Zhou, J.Y. Yang, Q.H. Jiang, D. Zhang, J.W. Xin, X. Li, Y.Y. Ren, X. He, J. Am. Ceram. Soc. **100**, 5723–5730 (2017)
113. D.G. Cahill, S.K. Watson, R.O. Pohl, Phys. Rev. B **46**, 6131–6140 (1992)
114. Y. Pei, D. Morelli, Appl. Phys. Lett. **94**, 122112 (2009)
115. H. Okamoto, J. Phase Equilib. **21**, 496 (2000)
116. S. Katayama, Anomalous resistivity in structural phase transition of IV–VI. Solid State Commun. **19**(4), 381–383 (1976). https://doi.org/10.1016/0038-1098(76)91357-0
117. J. Li, Z. Chen, X. Zhang, H. Yu, Z. Wu, H. Xie, Y. Chen, Y. Pei, Adv. Sci. **4**, 1700341 (2017)

Chapter 5
Lead Chalcogenide Thermoelectric Materials

Shan Li, Xinyue Zhang, Yucheng Lan, Jun Mao, Yanzhong Pei, and Qian Zhang

Abstract Lead chalcogenides have long been studied as thermoelectric (TE) materials since the 1950s due to unique electrical and thermal transport properties. The unremitting efforts during the past two decades give rise to the high TE figure of merit ZT values in both n- and p-type lead chalcogenides. Many encouraging breakthroughs trigger us to systematically understand the underlying physical mechanism of the optimization strategies. In this chapter, we first discuss band engineering strategies for the recent advances in high-performance lead chalcogenides. Then the strategies for increasing average ZT from optimizing concentration of charge carriers will be elucidated. Finally, the microstructural manipulation for high thermoelectric performances of lead chalcogenides will be presented.

5.1 Band Engineering in Lead Chalcogenide

Lead chalcogenides $PbX(X = S, Se, Te)$ have a long history as thermoelectrics. Though PbS was found to have large Seebeck coefficient early in the nineteenth century, intensive studies in lead chalcogenides started in the 1960s. By optimizing concentration of charge carriers, both n- and p-type lead chalcogenides show

S. Li · Q. Zhang (✉)
Department of Materials Science and Engineering, Harbin Institute of Technology (Shenzhen), Shenzhen, Guangdong, P.R. China
e-mail: zhangqf@hit.edu.cn

X. Zhang · Y. Pei
Interdisciplinary Materials Research Center, School of Materials Science and Engineering, Tongji University, Shanghai, P.R. China

Y. Lan
Department of Physics, Morgan State University, Baltimore, MD, USA

J. Mao
Department of Physics and TcSUH, University of Houston, Houston, TX, USA

© Springer Nature Switzerland AG 2019
S. Skipidarov, M. Nikitin (eds.), *Novel Thermoelectric Materials and Device Design Concepts*, https://doi.org/10.1007/978-3-030-12057-3_5

intrinsically good thermoelectric performance (peak figure of merit ZT ~ 1.4 for PbTe [1, 2], ~1.2 for PbSe [3, 4], and ~0.7 for PbS [5]). More recently, the in-depth understanding of the band structure of lead chalcogenides has led to strategies of band engineering for further improving thermoelectric performance.

Due to the similar chemical composition in lead chalcogenides, band structures are very similar as well. For lead chalcogenides, principal valence band maximum locates at L point in Brillouin zone (L band, with band degeneracy $N_v = 4$), where secondary valence band maximum is along Σ direction (Σ band, $N_v = 12$) [6, 7]. Since conduction band minimum locates at L point, lead chalcogenides show direct bandgap at L point at room temperature ($E_g \sim 0.31$ eV for PbTe, ~0.29 eV for PbSe, and ~0.42 eV for PbS) [6], and the energy of L band is believed to decrease with increasing in temperature [8].

5.1.1 Band Convergence

The complexity in valence band structure of lead chalcogenides offers tremendous potential for improving thermoelectric performance through band engineering. Given that concentration of charge carriers can be fully optimized, maximum ZT of material can be determined by dimensionless thermoelectric quality factor B. $B \propto N_v / \left(m_I^* E_{def}^2 \kappa_L \right)$, where N_v is the band degeneracy, m_I^* is inertial effective mass, E_{def} is deformation potential coefficient, and κ_L is lattice thermal conductivity, respectively [9]. Large N_v offers multiple conducting channels for high electrical conductivity while inducing barely no detrimental effects on Seebeck coefficient, which is clearly beneficial for high thermoelectric performance. Converging different bands, which locate at different positions in reciprocal space, is a successful approach for increasing in N_v, therefore, ZT, in compounds with multiple bands, in which PbTe is a typical case.

With increasing in temperature, energy of L band decreases, while energy of Σ band remains roughly unchanged as the bandgap increases [8]. When L and Σ bands have comparable energy (within a few $k_B T$), these two bands are considered to be converged, meaning both L and Σ bands contribute to charge transport, as schematically shown in Fig. 5.1a. Among lead chalcogenides, the band offset between L and Σ bands for PbTe is the smallest [6, 7], leading to a relatively low convergence temperature as compared to that of PbSe and PbS, leaving availabilities for tuning convergence temperature to maximizing thermoelectric performance.

Due to similar band structure with different band offset among lead chalcogenides, formation of solid solution among PbX(X = S, Se, Te) can be used as effective approach to tune band offset for electronic performance enhancement. Figure 5.1b shows the improvement in ZT through this approach. A typical case is to partial substitute Te by Se in PbTe to increase in convergence temperature of L and Σ bands, resulting in significant improvement in ZT at high temperatures, with additional help of reduced κ_L by alloying defects for phonon scattering [10]. Further alloying with PbS, thermoelectric performance of PbTe can be further improved [11]. In addition,

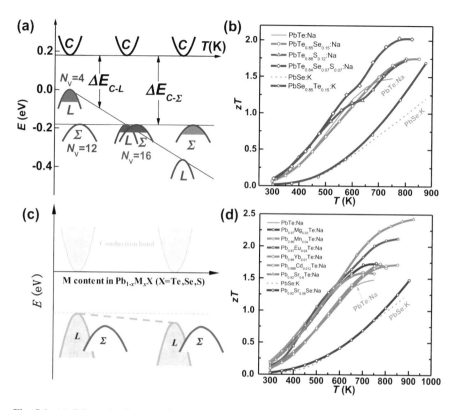

Fig. 5.1 (**a**) Schematic diagram of temperature-induced band convergence of L and Σ bands in PbTe; (**b**) temperature-dependent thermoelectric figure of merit ZT of solid solutions within lead chalcogenides [1, 10, 11, 22, 25]; (**c**) schematic diagram of tuning band offset between L and Σ bands through alloying with MX (M can be but not limited to Mg, Mn, Cd, Sr, Eu, Yb, X = Te, Se, S); (**d**) temperature-dependent ZT of lead chalcogenides through alloying with MX [12–17, 22, 23]

alloying with MTe (M = Mg [12], Cd [13], Mn [14], Eu [15], Yb [16], Sr [17]) can tune effectively the band offset between L and Σ bands as well, usually accompanied by increase in bandgap, as schematically shown in Fig. 5.1c. Improvements in ZT have also been demonstrated in these alloys, as shown in Fig. 5.1d.

It is worth noting that, for PbTe, band convergence temperature used to be believed to be around 450 K though both optical [18] and Hall measurements [10, 19, 20]. However, recent investigation indicated that convergence temperature should be higher [8, 21]. Nevertheless, existing studies show consistently that valence bands in PbTe tend to converge at high temperatures, which helps understand high ZT observed in this material.

In PbSe, Σ band is further away from L band maximum, suggesting band convergence at higher temperature as compared to that in PbTe [8]. Bringing this convergence temperature down to desired range through tuning of valence band structure is considered to be effective for improving thermoelectric p-type PbSe [22, 23], as shown in Fig. 5.1b. For PbS, however, $L - \Sigma$ energy separation is much larger than those in other two lead chalcogenides [24]; thus Σ band can be hardly involved even at high temperatures. This could be an electronic origin for the inferior thermoelectric performance of p-type PbS.

5.1.2 Resonant Levels

Converging electronic bands for improving thermoelectric performance relies on the enhanced band degeneracy, which leads to larger Seebeck coefficient, without lowering mobility of charge carriers. Similarly, some cases show resonant electronic states can enhance Seebeck coefficient at a given concentration of charge carriers in lead chalcogenides as well, which is due to the increase in density of states (DOS) effective mass [26–28]. To put it simply, resonant levels can be considered as additional energy states introduced by impurity lie inside conduction or valence band of host compound [27]. Existence of resonant levels would cause an increase in DOS effective mass in narrow energy range, which is favorable for large Seebeck coefficient. However, the benefits from resonant levels rely heavily on those energy position.

The position of resonant levels not only refers to the relative position to the band edge but also to the relative position to Fermi level. It is known that charge carriers with energy above or below Fermi level contribute to Seebeck coefficient oppositely [29], which indicates that resonant levels lying above or below Fermi level lead to possible increase or decrease in Seebeck coefficient. Taking p-type semiconductors as an example, an increase in Seebeck coefficient is more likely when the resonant levels are below Fermi level. Therefore, a careful doping for tuning Fermi level is necessary for taking the benefit from resonant levels.

Different from band convergence, realizing a beneficial resonant level in lead chalcogenides is much more element-selective. Majority of thermoelectric performance improvements was reported in Tl-doped PbTe (and its alloys) [28, 30] and Al-doped PbSe [26]. Some transition metals such as Ti [31], Cr [32], Sc [33], as well as In [34] are reported to be able to create resonant levels close to conduction band edge in lead chalcogenides. Readers are encouraged to consult the more detailed review by Heremans et al. [27] for more details on the effects of resonant levels in thermoelectric materials.

5.1.3 Low Effective Mass

Although the multiband conduction in p-type lead chalcogenides enabled many possibilities for improving ZT through band convergence, this strategy is limited in n-type lead chalcogenides due to existence of only one conduction band at L [24]. However, other band parameters involved in quality factor B, including m_I^* and E_{def}, provide additional possibilities for n-type.

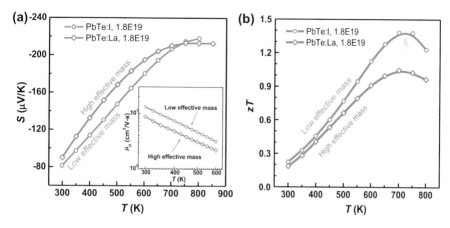

Fig. 5.2 Temperature-dependent Seebeck coefficient (**a**) and ZT (**b**) for I- and La-doped n-type PbTe, and the inset shows the temperature-dependent Hall mobility [39]

Straightforwardly, the quality factor B is inversely proportional to inertial mass m_I^*, which means that low m_I^* is actually beneficial for high ZT. In order to better understand, one needs to consider the underlying effect of m_I^* on electronic transport properties. Under approximation of single parabolic band conduction with acoustic scattering, maximal power factor $S^2\sigma$ is characterized by weighted mobility $(m_d^{*3/2}\mu_0)$, where m_d^* is DOS effective mass and μ_0 is nondegenerate mobility [35, 36]. Large $m_d^* = N_v^{2/3}m_b^*$, which can be arisen from either large N_v or large band effective mass m_b^* (equals to m_I^* for isotropic band), is beneficial for obtaining large Seebeck coefficient. However, these two cases have opposite effects on mobility of charge carriers, therefore, on electrical conductivity. Large N_v benefits usually electrical conductivity, while large m_b^* would cause a significant reduction in mobility of charge carriers via $\mu_0 \propto m_I^{*-1}m_b^{*-3/2}E_{def}^{-2}$ [37]. As a result, maximal power factor is actually proportional to N_v but inversely proportional to m_I^* and E_{def} via $S^2\sigma \propto N_v m_I^{*-1}E_{def}^{-2}$ [38].

A successful demonstration of low effective mass benefiting ZT is typified by comparison between I- and La-doped n-type PbTe [39]. DOS effective mass m_d^* is found to be ~0.25m_e and ~0.30m_e at room temperature for I- and La-doped PbTe, respectively. At the same concentration of charge carriers, higher Seebeck coefficient is obtained in La-doped sample due to its larger m_d^*, arising from the larger m_I^*. However, reduced mobility of charge carriers compensates largely enhancement in Seebeck coefficient and eventually leads to lower power factor and lower ZT in La-doped PbTe [39] (Fig. 5.2).

5.1.4 Low Deformation Potential Coefficient

In addition to low effective mass, low deformation potential coefficient E_{def} is also expected for high mobility. Low E_{def} corresponds to weak scattering of charge

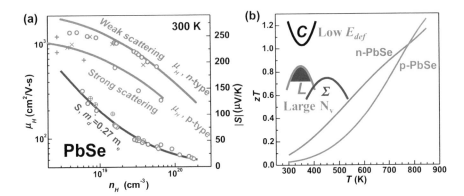

Fig. 5.3 Dependence of Seebeck coefficient on same concentration of charge carriers indicates the same effective mass at room temperature, while (**a**) different Hall mobility for *n*- and *p*-type PbSe reveals a lower deformation potential coefficient in PbSe which leads to (**b**) comparable high ZT in *n*-type PbSe with that of *p*-type PbSe having significantly larger number of conducting bands (N_v) at high temperatures

carriers by acoustic phonons, as has been observed in lead chalcogenides [6]. Successful demonstration on the importance of low E_{def} is achieved in *n*-type PbSe [4] (Fig. 5.3).

As mentioned before, high performance in *p*-type PbSe can be attributed to multiple bands conducting at high temperatures (with total $N_v > 12$ for valence bands). However, ZT of *n*-type PbSe, even having a relatively low $N_v = 4$ in conduction band, is actually comparable to that of *p*-type [4] at high temperatures. This is due to the weaker scattering of electrons as compared with that of holes under the same degree of lattice deformation waves, characterized by lower E_{def} in *n*-type PbSe. Quantitatively, $E_{def} \sim 25$ eV in *n*-type PbSe is much lower than ~ 35 eV in *p*-type PbSe [4], which leads to much higher mobility of charge carriers in *n*-type PbSe at room temperature. Though the difference in E_{def} between conduction and valence bands in PbTe is not as large as that in PbSe, PbTe also shows slightly lower $E_{def} \sim 22$ eV [2] for conduction band as compared to that of ~ 25 eV [10, 12] for valence band (*L* band).

5.2 Increase in Average ZT in Lead Chalcogenides

The strategies discussed in Sect. 5.1 mainly focus on band engineering to increase thermoelectric quality factor *B* which controls peak ZT. Continuously breakthroughs have been witnessed during the past two decades, pushing ZT values over 1.7 in PbSe [22], ~1.2 in PbS [40], and even breaking previously unheard-of levels of 2.0 in PbTe [15]. The encouraging progress in TE materials has taken an important step forward in broad practical applications of TE devices. However, the maximum conversion efficiency (η) of TE device depends on Carnot efficiency and materials' average ZT according to formula:

$$\eta = \frac{T_H - T_C}{T_H} \frac{\sqrt{1 + ZT_{avg}} - 1}{\sqrt{1 + ZT_{avg}} + T_C/T_H} \tag{5.1}$$

where $(T_H - T_C)/T_H$ is Carnot efficiency and T_H and T_C are temperature at hot junction and cold junction, respectively. It should be pointed out that average ZT of most lead chalcogenides is still low due to very low ZT below 500 K. Therefore, it is imperative to obtain better overall thermoelectric performance of lead chalcogenides.

We all know that Seebeck coefficient, electrical conductivity, and electronic thermal conductivity are strongly interdependent with each other via concentration of charge carriers. The optimized ZT of TE materials can be achieved at a given concentration of charge carriers. According to classical statistics approximation [41], optimized concentration of charge carriers (n_{opt}) is proportional to $\left(m_d^* T\right)^{3/2}$ for system with single-type of charge carriers, where m_d^* is the total DOS effective mass that can be expressed as follows: $m_d^* = N_v^{2/3} m_b^*$, where N_v is the band degeneracy and m_b^* is DOS effective mass of single valley. Apparently, n_{opt} increases rapidly with increase in temperature. To demonstrate clearly this temperature-dependent n_{opt}, here we take n-type PbTe as an example. By applying Kane single-band model, thermoelectric transports of n-type PbTe can be calculated as shown in Fig. 5.4a [42]. Value n_{opt} for PbTe is equal to ~0.4 × 10^{19} cm^{-3} at 300 K and ~3.0 × 10^{19} cm^{-3} at 800 K, nearly one order of magnitude difference, which demonstrates strong temperature dependence (blue solid line). Figure 5.4b shows clearly the comparison of temperature-dependent power factors with different Hall concentration of charge carriers. When Hall concentration of charge carriers equals to ~0.3 × 10^{19} cm^{-3}, high power factor can be obtained at lower temperature, but it

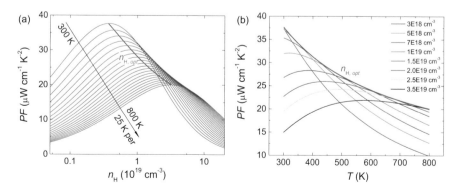

Fig. 5.4 (a) Power factors of n-type PbTe at different temperatures as function of Hall concentration of charge carriers, (b) temperature-dependent power factors at different Hall concentrations of charge carriers [42]

is quite low at higher temperature. Conversely, when Hall concentration of charge carriers is approaching $\sim 3.0 \times 10^{19}$ cm^{-3}, high power factor can be obtained at elevated temperature, whereas power factor at lower temperature is greatly reduced. An optimized temperature-dependent power factor is achieved at optimized concentration of charge carriers. However, concentration of charge carriers of most homogeneously doped semiconductors is produced by extrinsic doping and is constant with increasing temperature below the intrinsic excitation. Normally, concentration of charge carriers is optimized for high peak ZT, which is obviously much higher than optimized concentration at room temperature. In this section, we summarize the strategies to obtain temperature-dependent concentration of charge carriers in lead chalcogenides for enhanced average ZT.

5.2.1 Deep Defect Level Engineering

First-principle calculations found two types of localized defect states (deep defect states) caused by Indium (In) doping in PbTe as shown in Fig. 5.5b, which is different from shallow level doping by iodine (I) in PbTe (see Fig. 5.5a) [42] and the other deep defect state with energy much lower than the valence band top and shallow level with energy within the bandgap, both of which are capable of being occupied by two electrons with consideration of the spin (four electrons in total). However, In atom has only three valence electrons $s^2 p^1$, making defect state in the bandgap half-filled (see Fig. 5.5b). Fermi level is then pinned at the center of half-filled states (dashed line, Fig. 5.5b). This feature enables a unique design for deep level and shallow level co-doping system, where In builds up localized half-filled deep defect state in the bandgap and I acts as shallow donor level that supplies sufficient electrons. Deep defect state acting as electron repository is expected to trap electrons at lower temperature, and the trapped electrons will be thermally activated back to the conduction band with rising temperature as illustrated in Fig. 5.5d, e. As a result, temperature-dependent concentration of charge carriers will be realized. Based on this assumption, In/I co-doped In$_x$Pb$_{1-x}$Te$_{0.996}$I$_{0.004}$ ($x = 0$, 0.0025, 0.0035, and 0.005) were prepared, together with a corresponding first-principle calculation on In/I co-doped PbTe (shown in Fig. 5.5c). Fermi level lies within the conduction band of Pb$_{32}$Te$_{31}$I, while it locates deep in the bandgap for InPb$_{31}$Te$_{31}$I, indicating that incorporation of In will noticeably reduce the room temperature concentration of charge carriers. Temperature-dependent Hall concentration of charge carriers was subsequently achieved in In/I co-doped PbTe as presented in Fig. 5.5f. When doping by shallow level impurities (I doping, black squares), Hall concentration of charge carriers keeps constant $\sim 4.8 \times 10^{19}$ cm^{-3} below bipolar temperature. In contrast, Hall concentration of charge carriers of In/I co-doped In$_x$Pb$_{1-x}$Te$_{0.996}$I$_{0.004}$ exhibits an obvious temperature dependence, which is close to theoretically predicted optimized Hall concentration of charge carriers over the whole temperature range. On the other hand, optimized concentration of charge carriers could reduce the electronic thermal conductivity (in consistent with reduced

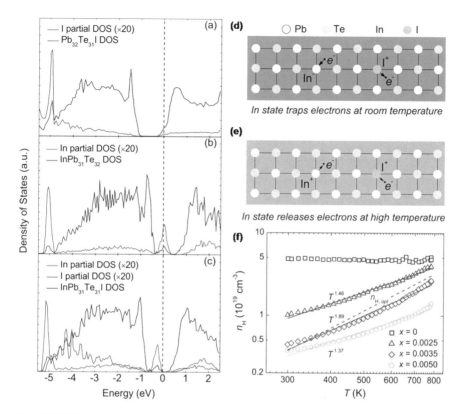

Fig. 5.5 DOS comparison in (**a**) I-doped PbTe, (**b**) In-doped PbTe, and (**c**) In/Ico-doped PbTe. Fermi level (dashed line) is set to be 0, and partial DOS associated with In atom (blue) and I atom (red line) is multiplied by factor of 20 for clearer demonstration. Schematic diagram of (**d**) In atom deep defect level that traps electrons at lower temperature and (**e**) trapped electrons being thermally activated at higher temperature. (**f**) Temperature-dependent Hall concentration of charge carriers of $In_xPb_{1-x}Te_{0.996}I_{0.004}$ ($x = 0$, 0.0025, 0.0035, and 0.005) [42]

electrical conductivity) and, thus, decrease the total thermal conductivity at lower temperature. Owing to simultaneously enhanced power factor and reduced thermal conductivity, room temperature ZT ~ 0.4, together with peak ZT ~ 1.4 at 773 K, were achieved in *n*-type In/I co-doped PbTe (Fig. 5.6a), leading to a record high average ZT ~ 1.04 in temperature range from 300 K to 773 K (Fig. 5.6b). Similar optimization was also realized in In-doped PbTe – PbS pseudo-binary system. Combined with decreased κ_L due to enhanced alloying scattering and spinodal decomposition, high room temperature ZT ~ 0.5 and peak ZT ~ 1.1 at 673 K were achieved in $Pb_{0.98}In_{0.02}Te_{0.8}S_{0.2}$ [44] (Fig. 5.6a).

Gallium (Ga) has also been considered as a deep level dopant in PbTe [43]. Both the first-principle band structure calculations and experimental results show that doping Ga on Pb site forms two types of impurity states: shallow impurity levels associated with Ga^{3+} state and deep impurity levels due to Ga^{1+} state. Compared

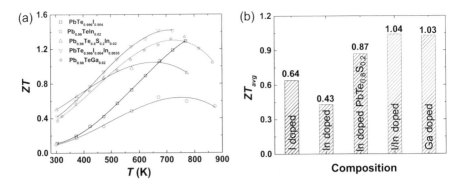

Fig. 5.6 (**a**) Temperature dependences of ZT and (**b**) average ZT values among the reports on *n*-type PbTe [42–44]

with Ga^{3+} shallow impurity levels that are completely ionized at lower temperatures, the presence of Ga^{1+} deep impurity levels delays the ionization temperature over 473 K, contributing to tunable concentration of charge carriers closer to optimized concentration required for maximizing the power factor over a wide temperature range. Moreover, Ga doping decreases in κ_L by enhancing the point defect phonon scattering. As a result the highest ZT value of 1.34 at 766 K and high average ZT value of 1.03 between 300 K and 865 K were obtained in *n*-type Ga-doped PbTe compounds as shown in Fig. 5.6 [43].

5.2.2 Dynamic Doping

Temperature-dependent concentration of charge carriers has also been realized by dynamic doping. With increasing temperature, dopant solubility increases, leading to increased concentration of charge carriers. This strategy needs a matrix with a reservoir of excess dopant (a second phase) in equilibrium and directly controls TE properties by temperature. To be noted that interstitial doping is easily to be formed by extra doping with increasing temperature. In Ag-doped PbTe/Ag_2Te composites [45], concentration of charge carriers was successfully controlled by excess Ag. Appropriate temperature-dependent solubility Ag in PbTe matrix enables increase in concentration of charge carriers with increasing in temperature as shown in Fig. 5.7a. This contributes to an average ZT enhancement of ~50% compared with the composites optimized for high-temperature performance by La doping, even though both have maximum ZT ~ 1.3 (Fig. 5.7b). Such effect due to increased dopant solubility with increasing temperature has also been observed in PbSe − Cu system [46]. As shown in Fig. 5.7c, optimized concentration of charge carriers in PbCu$_x$Se over a wide temperature range is achieved with higher Cu content when $x = 0.005$ and 0.0075, because more Cu ions continuously diffuse from Cu-rich second phase into interstitial sites of PbSe and provide extra electrons

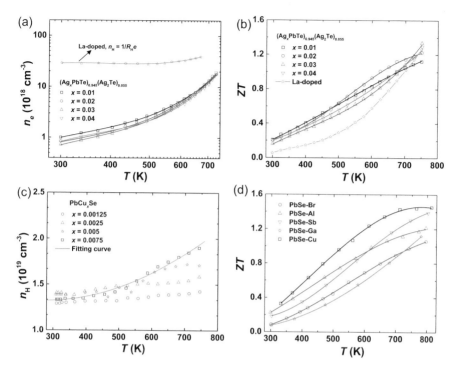

Fig. 5.7 (**a**) Temperature dependence of extrinsic concentration of charge carriers. (**b**) Comparison of ZT between Ag-doped and La-doped $(PbTe)_{0.945}(Ag_2Te)_{0.055}$ [45]. (**c**) Temperature dependence of Hall concentration of charge carriers for $PbCu_xSe$. Solid black line is the fitting curve of the sample with $x = 0.0075$. (**d**) Temperature dependence of ZT for reported n-type PbSe [46]

as temperature rises. In addition, the lattice thermal conductivity is significantly reduced due to the presence of Cu_2Se nanoprecipitates, and dislocations at low temperatures as well as vibrations of interstitial Cu ions at high temperatures, resulting in hierarchical phonon scattering. As a result, peak ZT ~ 1.45 and enhancement of average ZT were achieved for Cu-intercalated PbSe sample, which is superior to those for reported n-type PbSe-based materials (see Fig. 5.7d).

5.2.3 Stabilizing Optimized Concentration of Charge Carriers

Both deep defect doping and temperature-dependent doping aim at realizing temperature-dependent concentration of charge carriers. Alternatively, band engineering provides another strategy by altering temperature dependence of m_d^* and widening the bandgap to stabilize the optimized concentration of charge carriers. For example, alloying PbTe with MgTe decreases in energy gap between L and Σ bands (higher N_v) resulting in an increase in m_d^* and optimized concentration of charge

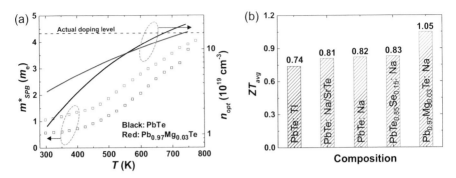

Fig. 5.8 (a) Calculated DOS effective mass and optimized concentration of charge carriers as a function of temperature. (b) Comparison of average ZT from 300 to 750 K among the reports on *p*-type PbTe [12]

carriers at lower temperature [12]. In addition, lower optimized concentration of charge carriers was obtained at higher temperature due to increased bandgap (much wider bandgap of MgTe (~3.5 eV) than PbTe (~0.3 eV at room temperature)). As a result, m_d^* of $Pb_{0.97}Mg_{0.03}Te$ shows weaker temperature dependence compared with that of PbTe, which results in an efficient compromise for $n_{opt} \sim T$ in both low and high temperature ranges to be closer to n_{opt} which is required in medium temperature range (Fig. 5.8a) without changing actual doping level. Significantly enlarged ZT over broad temperature range and enhancement of average ZT (from 300 K to 750 K) were achieved in PbTe alloyed with MgTe, even though without further improving maximum ZT (see Figs. 5.1d and 5.8b).

5.3 Microstructural Manipulation for High Thermoelectric Performance of Lead Chalcogenides

The lead chalcogenides have excellent electronic band structures and electrical properties. However, κ_L is relatively high, e.g., 2.85 $W \times m^{-1} \times K^{-1}$ for PbTe at room temperature, because of the scattering of longitudinal acoustic phonons and intrinsic short phonon mean free path as well as three-phonon scattering [47, 48]. First-principle calculations [49–51] and experimental investigation indicated that alloying and nanostructuring are effective ways to reduce κ_L of chalcogenides. Microstructures of chalcogenides with excellent thermoelectric performances are discussed below, covering zero-dimensional defects caused by doping/alloying, nanoprecipitations and grain boundaries, as well as second phases.

5.3.1 Alloying and Doping

The lead chalcogenides have a rock-salt structure with space group of $Fm\bar{3}m$. Thermal conductivity of ideal rock-salt structure is high. However, X-ray diffraction data indicated that Pb atoms locate at $(x, 1/2, 1/2)$ position, not $(1/1, 1/2, 1/2)$ position of ideal rock-salt structures [52]. Such kind of cation disorder induces a large degree of anharmonicity in chalcogenides, leading to low thermal conductivity. The disorder also modifies effective mass of charge carriers and affects short- and long-range acoustic phonon [53]. As a result, thermal conductivity of chalcogenides is usually low, and figure of merit is high, being a very important mid-temperature range thermoelectric material.

Besides pure chalcogenides, chalcogenides can be appropriately modified by chemical doping and substitution to reduce thermal conductivity. Figure 5.9a shows scanning tunnel microscopy (STM) image of Bi-doped SnSe ingots. It was reported that Bi doping in SnSe single crystals can enhance electrical conductivity while maintaining low thermal conductivity, achieving ZT ~ 2.2 in *n*-type SnSe single crystals [54].

Compared with doping, alloying is a better way to reduce thermal conductivity. Alloying can cause mass disorder and anharmonicity in chalcogenide lattice, increasing in effective phonon scattering rate. Therefore, PbTe has been alloyed with various compounds, such as PbS and PbSe, to enhance thermoelectric performances. Figure 5.9b shows transmission electron microscopy (TEM) images of PbTe-PbSe-PbS ingots. High-resolution TEM (HRTEM) images and selected area electron diffraction (SAED) patterns indicated these alloys are single crystals. Values ZT ~ 1.5 were achieved in such kind of alloys. Additionally, alloying also generates point defects in ingots, such as Te/S, Te/Se, and Se/S in crystalline solid solutions. The alloying and point defects produce disorder in crystalline ingots. As a result, thermal conductivity of the alloys, such as PbTe − PbSe and PbTe − PbS, is lower than that of PbTe [11, 49]. Figure 5.9c plots thermal conductivity of PbTe − PbSe alloys. Other kinds of chalcogenides, such as 2 mol. % Na-doped PbTe − PbSe − PbS solid solutions, have similar tendencies with alloying ratio. Those thermal conductivity is very low because of alloy scattering and point defect

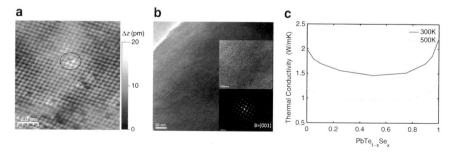

Fig. 5.9 (**a**) STM topographic image of Bi-doped SnSe. Dotted ellipses indicate Bi dopants [54]. (**b**) TEM image of $(PbTe)_{0.86}(PbSe)_{0.07}(PbS)_{0.07}$ ingots [11]. Inset: HRTEM (top) and SAED (bottom). (**c**) Thermal conductivity of PbTe − PbSe alloys [49]

scattering. First-principle investigations indicated [49] that there is large intrinsic anharmonicity in pristine PbTe and its anionic alloys (such as PbSe$_x$Te$_{1-x}$) to scatter optical and acoustic phonons, reducing thermal conductivity.

5.3.2 Precipitates and Strains

Nanoscale precipitates are always generated in chalcogenides because of low solubility. Figure 5.10a shows TEM image of two MgTe nanoprecipitates embedding in PbTe crystalline ingots. It was reported that 2–15 nm MgTe precipitates distributed throughout PbTe ingots, reducing κ_L from 4.1 W$\times$m$^{-1}\times$K^{-1} to 3.5 W$\times$m$^{-1}\times$K^{-1} at 300 K in 2 mol. % Na$_2$Te-doped PbTe ingots and resulting in ZT = 1.6 at 780 K [55].

Besides MgTe precipitates, other kinds of nanoscale precipitates were also generated in PbTe ingots [56–59]. Figure 5.10b, d, e, and g shows BaTe precipitates, PbS precipitates, SrTe precipitates, and CaTe nanocomposites embedding in PbTe ingots, respectively. The size distribution of these precipitates is from 1 nm to 10 nm.

Although no dislocations were observed at some small precipitate boundaries, dislocations and associated plastic strain are always observed at interfaces of large precipitates. Figure 5.10c shows dislocation at BaTe nanoprecipitate and PbTe matrix. These precipitates usually create strains in the immediate vicinity. Generally, it is believed that elastic strain is pervasive in and around all precipitates and there is additional plastic strain around dislocation cores. Figure 5.10f, h shows strain map profiles of SrTe precipitates and CaTe precipitates embedding in PbTe matrix.

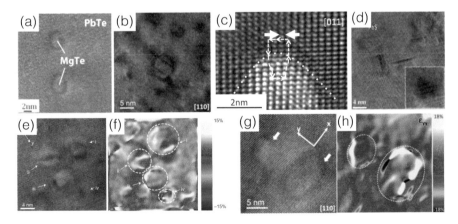

Fig. 5.10 TEM images of nanoprecipitates in ingots. (**a**) MgTe precipitates in (Pb, Mg)Te crystalline melt-grown ingots [55]. (**b**) BaTe precipitates in 1 mol. % Na$_2$Te-doped PbTe ingots [56]. (**c**) Misfit dislocation on the boundary between PbTe matrix and BaTe precipitate [57]. (**d**) PbS in PbTe matrix [58]. (**e**) SrTe precipitates in 1 mol. % Na$_2$Te-doped PbTe-2%SrTe ingots and (**f**) related elastic strains at and around these precipitates [59]. The color bar indicates −15% to 15% strain. (**g**) CaTe precipitates in 1 mol. % Na$_2$Te-doped PbTe-5%CaTe ingots and (**h**) related ε_{yy} strain map of these precipitates [56]. The color bar indicates −18% to +18% strain

Nanometer-scale precipitates working with related phase boundaries, interfacial dislocations, and strains together can reduce thermal conductivity of PbTe materials significantly. For example, thermal conductivity of $0.8 \, \text{W} \times \text{m}^{-1} \times \text{K}^{-1}$ was achieved at room temperate in PbTe ingots with PbS precipitate, approximately 35% and 30% of lattice thermal conductivity of either PbTe and PbS [58]. Calculations indicated that 60% reduction of κ_L comes from nanoprecipitates and 20% comes from dislocation and strains (point defects are responsible for about 20% reduction).

5.3.3   Composites

It is experimentally proved that thermal conductivity of PbTe is size dependent [60, 61]. The thermal conductivity of PbTe nanomaterials is significantly reduced by size effects because of enhanced phonon boundary scattering. Therefore, PbTe and related nanomaterials can be employed to reduce thermal conductivity further. Usually, PbTe-based nanoparticles are fabricated into nanocomposites. The grain boundaries of these PbTe-based nanoparticle can scatter more phonons. Up to now, various doped PbTe have been compacted into nanocomposites, for example, PbTe nanocomposites doped by Al [62], Na [63–65], Bi [65], and $(TlSbTe_2)_x(Tl_{0.02}Pb_{0.98}Te)_{1-x}$ [66]. Those grain sizes were in nm or in μm ranges. Figure 5.11a shows Al-doped PbTe composites with microscale grains. Figure 5.11b shows Na-doped PbTe composites consisting of nanograins. κ_L of these composites was reduced compared with doped PbTe ingots.

Besides grain boundaries, the first-principle and molecular dynamics simulations recently showed that κ_L of PbTe materials can be reduced to $0.4 \, \text{W} \times \text{m}^{-1} \times \text{K}^{-1}$ by nanotwin structures [67], about sevenfold lower than bulk value of 2.85 $\text{W} \times \text{m}^{-1} \times \text{K}^{-1}$. The nanotwin boundaries will induce anharmonicity to reduce thermal conductivity significantly. Unfortunately, experimental data of PbTe nanotwin structures have not been reported yet.

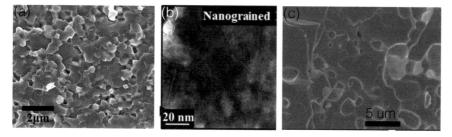

Fig. 5.11 SEM of PbTe-based nanocomposites. (**a**) Al-doped PbTe [62]. (**b**) Na-doped PbTe [63]. (**c**) Porous PbTe [68]

Porous PbTe composites were also reported. Figure 5.11c shows porous PbTe composites. Pores area is 200 nanometers - 2 microns and the porosity 7.3% in pure PbTe composites. An extremely low κ_L (0.56 W×m^{-1}×K^{-1} at 773 K) was achieved in porous PbTe composites [68], half of thermal conductivity of nonporous PbTe composites. It was believed that monodispersed nanopores with less than 50 nm in diameter should strongly scatter phonon while maintaining electronic transport properties to achieve PbTe-based phonon glass electron crystals.

5.3.4 Multiphase PbTe Materials

Because of limited solubility of PbTe, many two-phase PbTe composites were produced from PbTe-based alloys. Figure 5.12a–b shows SEM images of PbTe – PbS composites. There are two phases in the composites because of phase

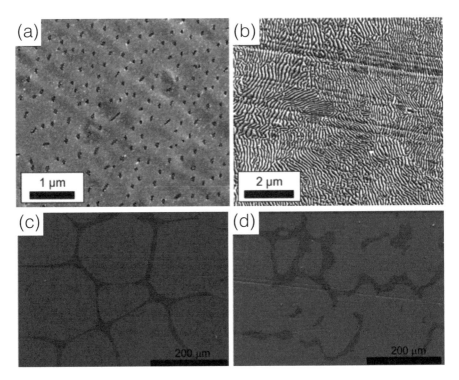

Fig. 5.12 SEM images of (**a**) PbTe-8 mol. % PbS compositions and (**b**) PbTe-30 mol. % PbS composites [72]. The second phases are embedding in PbTe matrix. SEM images of (**c**) $Pb_{0.9}Ag_{0.1}Te$ composites and (**d**) $Pb_{0.9}Sb_{0.1}Te$ alloys [69]. The second phases are located around PbTe grains. The brighter regions are PbTe

separation of Pb(Te, S) material. If other elements are existing in PbTe alloys, responding second phases were generated too. Figure 5.12c-d shows SEM images of (Pb, X)Te composites, where X is other elements. The different contrast comes from different phases. Ag_5Te_3 [69] or Ag_2Te [70] were generated at PbTe grain boundaries as the second phase in $Pb_{0.9}Ag_{0.1}Te$ composites (Figure 5.12c). Second phase of Sb_2Te_3 was observed in $Pb_{0.9}Sb_{0.1}Te$ alloys (Figure 5.12d). Second phase of PbS was observed in $PbTe_{0.38}S_{0.62}$ composites [71]. Second phases were also observed in metastable PbTe-based solid solutions, such as from phase separation of PbTe $-$ PbS materials [72, 73].

The second phase can reduce thermal conductivity of PbTe composites sometime [74]. It was reported that 20 mol. % PbS phase fraction in PbTe $-$ PbS composites reduced κ_L to the lowest.

However, depending on types of second phases, thermoelectric performances of multiphase PbTe-based composites are sometime improved or worsened.

5.3.5 All-Scale Nanocomposites

In order to reduce thermal conductivity efficiently, all the phonon scattering factors should be included, such as point defects, nanoscale precipitates, grain boundaries, and strains. Based on the idea, all-scale thermoelectric PbTe-based materials were fabricated recently, where alloying and doping, nanoscale precipitates, and mesoscale microstructures were contained in the composites, to scatter all-scale wavelength phonons [75, 76]. Figure 5.13 shows microstructures of such kind of all-scale thermoelectric PbTe-4 mol. % SrTe nanocomposites doped with 2 mol. % Na. Nanocomposites consisted of PbTe mesoscale grains (right panels in Fig. 5.13) with size of several microns (size distribution is shown as the inset in the panel). PbTe phase was also heavily alloyed with SrTe beyond thermodynamic solubility limit (1 mol. % [17]) and endotaxial SrTe nanostructures precipitated from PbTe grains when the content of SrTe exceeded 5 mol. % solubility limit. So, there were numerous SrTe nanoprecipitates existing in PbTe alloying grains (size distribution is shown as the inset in the middle panel of Fig. 5.13). These precipitates in PbTe matrix produced coherent strains to scatter phonons too. Additionally, PbTe was alloyed with SrTe and doped with Na (left panel in Fig. 5.13) to create defects and mass disorder. Various strains were also generated near these defects. As a result, PbTe $-$ SrTe alloying, SrTe nanoprecipitates, and grain boundaries worked together to suppress thermal conductivity of PbTe-based all-scale composites significantly, increasing ZT. It was reported that an extremely low κ_L of 0.5 $W \times m^{-1} \times K^{-1}$ was observed and ZT of 2.5 was achieved at 923 K in $Pb_{0.98}Na_{0.02}Te$-8 mol. % SrTe composites [75].

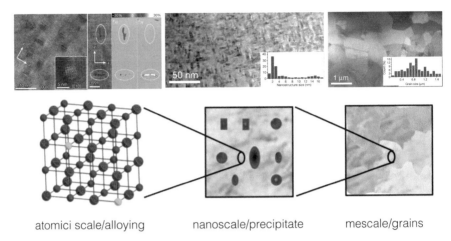

atomici scale/alloying   nanoscale/precipitate   mescale/grains

Fig. 5.13 TEM images of PbTe − SrTe (4 mol. %) doped with 2 mol. % Na [75]

Similar all-scale PbTe-based composites were fabricated in Na-doped PbTe − MgTe composites [76]. It was reported that ZT was enhanced to 2.0 at 823 K in $Pb_{0.98}Na_{0.02}Te$ + 6 mol. % MgTe composites from ZT ~ 1.1 of $Pb_{0.98}Na_{0.02}Te$ composites, resulting from reduction of κ_L of 2.83 $W \times m^{-1} \times K^{-1}$ ($Pb_{0.98}Na_{0.02}Te$) to 1.50 $W \times m^{-1} \times K^{-1}$ ($Pb_{0.98}Na_{0.02}Te$ + 6 mol. % MgTe) at room temperature.

κ_L of thermoelectric materials is dominated by various phonon scattering. The overall phonon scattering rate should be the sum of all scattering rates. Based on the above discussions of microstructures of PbTe-based materials, the overall scattering rate is:

$$\frac{1}{\tau_{total}} = \frac{1}{\tau_U} + \frac{1}{\tau_{e-p}} + \frac{1}{\tau_{md}} + \frac{1}{\tau_{inh}} + \frac{1}{\tau_{PD}} + \frac{1}{\tau_{precip}} + \frac{1}{\tau_{GB}} + \cdots \qquad (5.2)$$

where τ_U is the phonon-phonon Umklapp process, τ_{e-p} is electron-phonon interaction, τ_{md} is mass-disorder scattering, τ_{inh} is inharmonic scattering, τ_{PD} is point defect scattering, τ_{precip} is precipitate scattering, and τ_{GB} is grain boundary scattering. Among these scatterings, the point defect scattering can affect short- and mid-wavelength phonons and scattering rate $\tau_{PD}^{-1} \propto \omega$ [50]. The grain boundaries can scatter long-wavelength phonons with $\tau_{GB}^{-1} \propto L^{-1}$ (L is the grain size). The precipitate scattering is proportional to volume. Figure 5.14a demonstrates the phonon scattering rates of various scattering. Figure 5.14b illustrates accumulated κ_L of PbTe composites. It is obvious that solid solution defects and nanoprecipitates contribute most to reduction of κ_L. All those scattering mechanisms work together to reduce thermal conductivities of chalcogenides.

Fig. 5.14 (a) Frequency dependence of phonon scattering rates of various mechanisms [77]. (b) Accumulated lattice thermal conductivity of PbTe composites as a function of phonon mean free path [75]

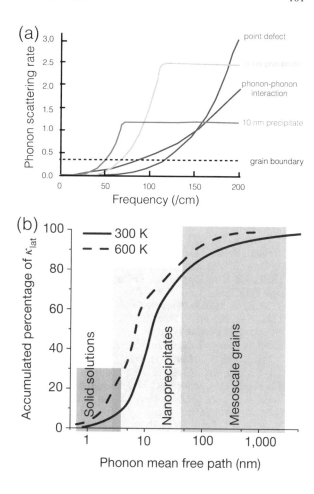

5.4   Summary and Outlook

Over the past two decades, a continuous enhancement in the figure of merit ZT values of various TE materials has been realized under the drive of application of modern theory and sophisticated equipment. Among these TE materials, lead chalcogenides have brought many intriguing features and kept producing surprises. To date, high peak ZT (~1.7 at 873 K for PbSe, ~1.2 at 923 K for PbS, and ~2.2 at 915 K for PbTe) and high average ZT over 1.0 in temperature range 300–800 K have been achieved in lead chalcogenide bulk alloys. The strategies for optimizing TE properties of lead chalcogenides have been demonstrated, including band engineering (band convergence, resonant states, low effective mass, and deformation potential coefficient) for high electrical properties, microstructural manipulation (alloying, precipitates and strains, composites, multiphase, and all-scale nanocomposites) for decreasing lattice thermal conductivity, and temperature-dependent doping (deep level doping, dynamic doping, and stabilizing optimized concentration of charge carriers) for high average ZT. The above strategies give an insight and comprehension of the interrelated parameters, which can in principle be applied to other materials systems for significantly improving TE performance. To realize future large-scale commercial applications of TE devices, much more attention should be put into the repeatability, mechanical properties, and longtime thermal stability of TE materials.

References

1. Y. Pei, A. LaLonde, S. Iwanaga, G.J. Snyder, Energy Environ. Sci. **4**, 2085 (2011)
2. A.D. LaLonde, Y. Pei, G.J. Snyder, Energy Environ. Sci. **4**, 2090 (2011)
3. H. Wang, Y. Pei, A.D. LaLonde, G.J. Snyder, Adv. Mater. **23**, 1366 (2011)
4. H. Wang, Y. Pei, A.D. LaLonde, G.J. Snyder, Proc. Natl. Acad. Sci. U. S. A. **109**, 9705 (2012)
5. H. Wang, E. Schechtel, Y. Pei, G.J. Snyder, Adv. Energy Mater. **3**, 488 (2013)
6. Y.I. Ravich, B.A. Efimova, I.A. Smirnov, *Semiconducting Lead Chalcogenides* (Plenum Press, New York, 1970)
7. G. Nimtz, B. Schlicht, *Springer Tracts in Modern Physics* **98**, 1 (1983)
8. Z.M. Gibbs, H. Kim, H. Wang, R.L. White, F. Drymiotis, M. Kaviany, G. Jeffrey Snyder, Appl. Phys. Lett. **103**, 262109 (2013)
9. Y. Pei, H. Wang, G.J. Snyder, Adv. Mater. **24**, 6125 (2012)
10. Y. Pei, X. Shi, A. LaLonde, H. Wang, L. Chen, G.J. Snyder, Nature **473**, 66 (2011)
11. R.J. Korkosz, T.C. Chasapis, S.H. Lo, J.W. Doak, Y.J. Kim, C.I. Wu, E. Hatzikraniotis, T.P. Hogan, D.N. Seidman, C. Wolverton, J. Am. Chem. Soc. **136**, 3225 (2014)
12. Y. Pei, A.D. LaLonde, N.A. Heinz, X. Shi, S. Iwanaga, H. Wang, L. Chen, G.J. Snyder, Adv. Mater. **23**, 5674 (2011)
13. Y. Pei, A.D. LaLonde, N.A. Heinz, G.J. Snyder, Adv. Energy Mater. **2**, 670 (2012)
14. Y. Pei, H. Wang, Z.M. Gibbs, A.D. LaLonde, G.J. Snyder, NPG Asia Mater. **4**, e28 (2012)
15. Z. Chen, Z. Jian, W. Li, Y. Chang, B. Ge, R. Hanus, J. Yang, Y. Chen, M. Huang, G.J. Snyder, Y. Pei, Adv. Mater. **29**, 1606768 (2017)
16. Z. Jian, Z. Chen, W. Li, J. Yang, W. Zhang, Y. Pei, J. Mater. Chem. C **3**, 12410 (2015)
17. G. Tan, F. Shi, S. Hao, L.D. Zhao, C. Hang, X. Zhang, C. Uher, C. Wolverton, V.P. Dravid, M.G. Kanatzidis, Nat. Commun. **7**, 12167 (2016)
18. A.F. Gibson, Proc. Phys. Soc. B **65**, 378 (1952)
19. I.A. Chernik, V.I. Kaidanov, M.I. Vinogradova, N.V. Kolomoets, Sov. Phys. Semicond. **2**, 645 (1968)
20. A.A. Andreev, V.N. Radionov, Sov. Phys. Semicond. **1**, 145 (1967)
21. C.M. Jaworski, M.D. Nielsen, H. Wang, S.N. Girard, W. Cai, W.D. Porter, M.G. Kanatzidis, J.P. Heremans, Phys. Rev. B **87**, 045203 (2013)
22. Q. Zhang, F. Cao, W. Liu, K. Lukas, B. Yu, S. Chen, C. Opeil, D. Broido, G. Chen, Z. Ren, J. Am. Chem. Soc. **134**, 10031 (2012)
23. H. Wang, Z.M. Gibbs, Y. Takagiwa, G.J. Snyder, Energy Environ. Sci. **7**, 804 (2014)
24. A. Svane, N.E. Christensen, M. Cardona, A.N. Chantis, M. van Schilfgaarde, T. Kotani, Phys. Rev. B **81**, 245120 (2010)
25. S.N. Girard, J. He, X. Zhou, D. Shoemaker, C.M. Jaworski, C. Uher, V.P. Dravid, J.P. Heremans, M.G. Kanatzidis, J. Am. Chem. Soc. **133**, 16588 (2011)
26. Q. Zhang, H. Wang, W. Liu, H. Wang, B. Yu, Q. Zhang, Z. Tian, G. Ni, S. Lee, K. Esfarjani, G. Chen, Z. Ren, Energy Environ. Sci. **5**, 5246 (2012)
27. J.P. Heremans, B. Wiendlocha, A.M. Chamoire, Energy Environ. Sci. **5**, 5510 (2012)
28. C.M. Jaworski, B. Wiendlocha, V. Jovovic, J.P. Heremans, Energy Environ. Sci. **4**, 4155 (2011)

29. X. Zhang, Y. Pei, NPG Quantum Mater. **2**, 68 (2017)
30. J.P. Heremans, V. Jovovic, E.S. Toberer, A. Saramat, K. Kurosaki, A. Charoenphakdee, S. Yamanaka, G.J. Snyder, Science **321**, 554 (2008)
31. J. Koenig, M. Nielsen, Y.B. Gao, M. Winkler, A. Jacquot, Phys. Rev. B **84**, 205126 (2011)
32. M.D. Nielsen, E.M. Levin, C.M. Jaworski, K. Schmidt Rohr, J.P. Heremans, Phys. Rev. B **85**, 045210 (2012)
33. E.P. Skipetrov, L.A. Skipetrova, A.V. Knotko, E.I. Slynko, V.E. Slynko, J. Appl. Phys. **115**, 133702 (2014)
34. V.I. Kaidanov, S.A. Nemov, Y.I. Ravich, Sov. Phys. Semicond. **26**, 113 (1992)
35. H.J. Goldsmid, *Introduction to Thermoelectricity* (Springer, Heidelberg, 2009)
36. D.M. Rowe, *Thermoelectrics Handbook: Macro to Nano* (CRC/Taylor & Francis, Boca Raton, 2006)
37. J. Bardeen, W. Shockley, Phys. Rev. **80**, 72 (1950)
38. H.J. Goldsmid, *Thermoelectric Refrigeration* (Plenum Press, New York, 1964)
39. Y. Pei, A.D. LaLonde, H. Wang, G.J. Snyder, Energy Environ. Sci. **5**, 7963 (2012)
40. L.D. Zhao, J. He, C.I. Wu, T.P. Hogan, X. Zhou, C. Uher, V.P. Dravid, M.G. Kanatzidis, J. Am. Chem. Soc. **134**, 7902 (2012)
41. A.F. Ioffe, Phys. Today **12**, 42 (1959)
42. Q. Zhang, Q. Song, X. Wang, J. Sun, Q. Zhu, K. Dahal, X. Lin, F. Cao, J. Zhou, S. Chen, G. Chen, J. Mao, Z. Ren, Energy Environ. Sci. **11**, 933 (2018)
43. X. Su, S. Hao, T.P. Bailey, S. Wang, I. Hadar, G. Tan, T.B. Song, Q. Zhang, C. Uher, C. Wolverton, X. Tang, M.G. Kanatzidis, Adv. Energy Mater. **8**, 1800659 (2018)
44. Q. Zhang, E.K. Chere, Y. Wang, H.S. Kim, R. He, F. Cao, K. Dahal, D. Broido, G. Chen, Z. Ren, Nano Energy **22**, 572 (2016)
45. Y. Pei, A.F. May, G.J. Snyder, Adv. Energy Mater. **1**, 291 (2011)
46. L. You, Y. Liu, X. Li, P. Nan, B. Ge, Y. Jiang, P. Luo, S. Pan, Y. Pei, W. Zhang, Energy Environ. Sci. **11**, 1848 (2018)
47. T. Shiga, J. Shiomi, J. Ma, O. Delaire, T. Radzynski, A. Lusakowski, K. Esfarjani, G. Chen, Phys. Rev. B **85**, 155203 (2012)
48. S. Lee, K. Esfarjani, T. Luo, J. Zhou, Z. Tian, G. Chen, Nat. Commun. **5**, 3525 (2014)
49. Z. Tian, J. Garg, K. Esfarjani, T. Shiga, J. Shiomi, G. Chen, Phys. Rev. B **85**, 45 (2012)
50. J.M. Skelton, S.C. Parker, A. Togo, I. Tanaka, A. Walsh, Phys. Rev. B **89**, 205203 (2014)
51. A.H. Romero, E.K.U. Gross, M.J. Verstraete, O. Hellman, Phys. Rev. B **91**, 214310 (2015)
52. S. Kastbjerg, N. Bindzus, M. Søndergaard, S. Johnsen, N. Lock, M. Christensen, M. Takata, M.A. Spackman, B.I. Bo, Adv. Funct. Mater. **23**, 5477 (2013)
53. H. Kim, M. Kaviany, Phys. Rev. B **86**, 3089 (2012)
54. A.T. Duong, V.Q. Nguyen, G. Duvjir, V.T. Duong, S. Kwon, J.Y. Song, J.K. Lee, J.E. Lee, S. Park, T. Min, J. Lee, J. Kim, S. Cho, Nat. Commun. **7**, 13713 (2016)
55. M. Ohta, K. Biswas, S.H. Lo, J. He, D.Y. Chung, V.P. Dravid, M.G. Kanatzidis, Adv. Energy Mater. **2**, 1117 (2012)
56. K. Biswas, J. He, G. Wang, S.H. Lo, C. Uher, V.P. Dravid, M.G. Kanatzidis, Energy Environ. Sci. **4**, 4675 (2011)
57. S.H. Lo, J. He, K. Biswas, M.G. Kanatzidis, V.P. Dravid, Adv. Funct. Mater. **22**, 5175 (2015)
58. J. He, S.N. Girard, M.G. Kanatzidis, V.P. Dravid, Adv. Funct. Mater. **20**, 764 (2010)
59. K. Biswas, J. He, Q. Zhang, G. Wang, C. Uher, V.P. Dravid, M.G. Kanatzidis, Nat. Chem. **3**, 160 (2011)
60. M. Fardy, A.I. Hochbaum, J. Goldberger, M.M. Zhang, P. Yang, Adv. Mater. **19**, 3047 (2010)
61. J.W. Roh, S.Y. Jang, J. Kang, S. Lee, Appl. Phys. Lett. **96**, 103101 (2010)
62. Q. Zhang, S. Yang, Q. Zhang, S. Chen, W. Liu, H. Wang, Z. Tian, D. Broido, G. Chen, Z. Ren, Nanotechnology **24**, 345705 (2013)
63. H. Wang, J. Hwang, M.L. Snedaker, I. Kim, C. Kang, J. Kim, G.D. Stucky, J. Bowers, W. Kim, Chem. Mater. **27**, 944 (2015)

64. H. Wang, J.H. Bahk, C. Kang, J. Hwang, K. Kim, J. Kim, P. Burke, J.E. Bowers, A.C. Gossard, A. Shakouri, Proc. Natl. Acad. Sci. U. S. A. **111**, 10949 (2014)
65. F.R. Sie, H.J. Liu, C.H. Kuo, C.S. Hwang, Y.W. Chou, C.H. Yeh, Intermetallics **92**, 113 (2018)
66. H. Wang, A. Charoenphakdee, K. Kurosaki, S. Yamanaka, G.J. Snyder, Phys. Rev. B **83**, 211 (2011)
67. Y. Zhou, J.Y. Yang, L. Cheng, M. Hu, Phys. Rev. B **97**, 085304 (2018)
68. J.Y. Hwang, E.S. Kim, S.W. Hasan, S.M. Choi, K.H. Lee, S.W. Kim, Adv. Cond. Matter. Phys. **6**, 496739 (2015)
69. H.S. Dow, M.W. Oh, B.S. Kim, S.D. Park, B.K. Min, H.W. Lee, D.M. Wee, J. Appl. Phys. **108**, 105 (2010)
70. T. Grossfeld, A. Sheskin, Y. Gelbstein, Y. Amouyal, T. Grossfeld, A. Sheskin, Y. Gelbstein, Y. Amouyal, T. Grossfeld, A. Sheskin, Crystals **6**, 1453 (2017)
71. S. Aminorroaya, A.Z. Williams, D. Attard, G.S.X. Dou, G.J. Snyder, Sci. Adv. Mater. **6**, 1453 (2014)
72. S.N. Girard, K. Schmidt. Rohr, T.C. Chasapis, E. Hatzikraniotis, B. Njegic, E.M. Levin, A. Rawal, K.M. Paraskevopoulos, M.G. Kanatzidis, Adv. Funct. Mater. **23**, 747 (2013)
73. H. Lin, E.S. Bozin, S.J.L. Billinge, J. Androulakis, C.D. Malliakas, C.H. Lin, M.G. Kanatzidis, Phys. Rev. B **80**, 045204 (2015)
74. D. Wu, L.D. Zhao, X. Tong, W. Li, L. Wu, Q. Tan, Y. Pei, L. Huang, J.F. Li, Y. Zhu, Energy Environ. Sci. **8**, 2056 (2015)
75. K. Biswas, J. He, I.D. Blum, C. Wu, T.P. Hogan, D.N. Seidman, V.P. Dravid, M.G. Kanatzidis, Nature **489**, 414 (2012)
76. L.D. Zhao, H.J. Wu, S.Q. Hao, C.I. Wu, X.Y. Zhou, K. Biswas, J.Q. He, T.P. Hogan, C. Uher, C. Wolverton, Energy Environ. Sci. **6**, 3346 (2013)
77. J. Yang, L. Xi, W. Qiu, L. Wu, X. Shi, L. Chen, J. Yang, W. Zhang, C. Uher, D.J. Singh, NPJ Comput. Mater. **2**, 15015 (2016)

Chapter 6
High Thermoelectric Performance due to Nanoprecipitation, Band Convergence, and Interface Potential Barrier in PbTe-PbSe-PbS Quaternary Alloys and Composites

Dianta Ginting and Jong-Soo Rhyee

Abstract Thermoelectric power generation is a direct heat to electric energy conversion technology and can be applied to waste heat power conversion as well. Among thermoelectric (TE) materials, $PbTe-PbSe-PbS$ quaternary alloys and composites are promising candidates for thermoelectric power generation applications in mid-temperature operating range from 500 to ~850 K. On the other hand, the thermoelectric performance of quaternary alloys and composites is not fully optimized regarding composition and synthesis process. Here we present results of investigation of quaternary system $PbTe-PbSe-PbS$. We found that PbS will form nanoprecipitation in the matrix of quaternary alloy for small content of PbS (≤ 0.07) which induces the reduction of lattice thermal conductivity. The power factor of $PbTe-PbSe-PbS$ quaternary alloys is significantly enhanced due to band convergence in $PbTe_{1-x}Se_x$. As the result of simultaneous PbS nanoprecipitation with coherent interface with the matrix and band structure modification, we obtained extremely high ZT value of 2.3 at 800 K for $(PbTe)_{0.95-x}(PbSe)_x(PbS)_{0.05}$. The chemical potential tuning by effective K doping ($x = 0.02$) and PbS substitution causes high power factor and low thermal conductivity, resulting in comparatively high ZT value of 1.72 at 800 K. The combination of high Seebeck coefficient and low thermal conductivity results in very high ZT value of 1.52 at 700 K for low Cl-doped ($x = 0.0005$) n-type $(PbTe_{0.93-x}Se_{0.07}Cl_x)_{0.93}(PbS)_{0.07}$ composites. Therefore, effective chemical potential tuning, band convergence, and nanoprecipitation give rise to

---

D. Ginting
Department of Applied Physics and Institute of Natural Sciences, Kyung Hee University, Yongin, Gyeonggi, Republic of Korea

Department of Mechanical Engineering, Universitas Mercu Buana, Meruya Selatan, Kota Jakarta Barat, Indonesia

J.-S. Rhyee (✉)
Department of Applied Physics and Institute of Natural Sciences, Kyung Hee University, Yongin, Gyeonggi, Republic of Korea
e-mail: jsrhyee@khu.ac.kr

© Springer Nature Switzerland AG 2019
S. Skipidarov, M. Nikitin (eds.), *Novel Thermoelectric Materials and Device Design Concepts*, https://doi.org/10.1007/978-3-030-12057-3_6

significant enhancement of thermoelectric performance of both p- and n-type PbTe – PbSe – PbS quaternary alloy and composite TE materials.

6.1 Introduction

Recently, much efforts have been devoted to search materials which can convert waste heat into electricity owing to growing global demand for green energy. Thermoelectric power generation is a direct heat to electric energy conversion technology and can be applied to waste heat power conversion as well. Effective TE materials are expected to have low thermal conductivity κ, high electrical conductivity σ, and high Seebeck coefficient S which are defined by dimensionless TE figure of merit $ZT = S^2\sigma T/\kappa$ [1]. The trade-off relationship among the parameters of S, σ, and κ makes it difficult to have high ZT value [2]. Among TE materials, semiconductor binary compound PbTe is the most promising and efficient TE material for mid-temperature range applications. TE performance of PbTe-based materials has been improved significantly through electronic structure modifications [3–6]. p-type PbTe can be produced by doping with Tl, Na, and K on Pb sites [5, 7–10]. On the other hand, n-type PbTe can be produced by Cl doping on Te site [11].

Additionally, low content in Earth's crust and high cost of tellurium Te inhibit the usage of PbTe in large-scale production. A lot of efforts has been made to replace Te by Se and S to form PbSe [12, 13], PbS [14], or ternary alloys PbTe– PbSe [3, 15, 16] and PbSe – PbS [17, 18]. In PbTe – PbSe system, high TE performance is mainly attributed from band structure engineering. Band structure engineering in $PbTe_{1-x}Se_x$ is beneficial to get high S while maintaining high σ by doping [3]. It is believed that low lattice thermal conductivity κ_L in ternary $PbTe_{1-x}Se_x$ alloys is explained mainly by formation of point defects, created by Te/Se mixed occupation in the rock salt structure [3]. In addition, TE performance in $PbTe_{1-x}Se_x$ ternary alloys can be increased by nanostructuring, resulting in low lattice thermal conductivity κ_L. Nanostructuring in PbTe – PbSe system can be made using bulk phase separation either by nucleation or spinodal decomposition depending on relative phase fraction [19].

Recently, it has been reported that κ_L in quaternary system of $(PbTe)_{1-x-y}(PbSe)_x(PbS)_y$ can be reduced by point defect which produced by triple disorder in the rock salt structure [20–22]. High $ZT \approx 2.2$ at 800 K was obtained in p-type $(PbTe)_{1-2x}(PbSe)_x(PbS)_x$ quaternary alloys due to band engineering and phonon scattering from point defects [20]. Quaternary alloys PbTe – PbSe – PbS are considered as effective n-type materials as well. For example, n-type $(PbTe)_{0.75}(PbSe)_{0.1}(PbS)_{0.15}$ exhibited high ZT value 1.1 at 800 K [22]. Therefore, quaternary alloy system PbTe – PbSe – PbS is a promising candidate for the use in TE power generation devices because alloy materials exhibit both n- and p-type properties with high TE performance [23]. Here, we introduce the simultaneous emergence of band convergence, nanostructuring, and chemical potential tuning in

PbTe – PbSe – PbS quaternary alloys and composites which provided extremely high ZT in p- and n-type materials.

6.2 Nanostructure in PbTe-PbSe-PbS Quaternary Alloys

In spite of high ZT values for Pb-based quaternary alloys, TE performances of alloys are not fully optimized in terms of content of binary compounds and synthesis process. Therefore, it needs systematic investigation of $(PbTe)_{0.95-x}(PbSe)_x(PbS)_{0.05}$ ($x = 0.0, 0.05, 0.10, 0.15, 0.20, 0.35,$ and 0.95) quaternary alloys with 1 at. % Na doping. Content of PbS enough for precipitation is fixed at 5 at. % in order to maintain nanoprecipitation and to eliminate modification of PbTe electronic band structure. The electronic band structure modification was associated with content of PbSe in $(PbTe)_{0.95-x}(PbSe)_x(PbS)_{0.05}$.

Figure 6.1 shows X-ray diffraction (XRD) pattern of $(PbTe)_{0.95-x}(PbSe)_x(PbS)_{0.05}$. XRD showed single phase with cubic structure. The lattice parameters decrease with increasing in Se content and follow Vegard's law indicating solid solution of PbSe and PbTe as shown in Fig. 6.1b. The lattice parameter decreased with increasing in Se content because atomic radius of Se is smaller compared to atomic radius of Te.

Figure 6.2a, b shows high-resolution transmission electron microscope (HRTEM) images of $(PbTe)_{0.75}(PbSe)_{0.20}(PbS)_{0.05}$ samples. It reveals numerous nanoprecipitates with size 10–20 nm which is comparable to previously reported PbTe-based nanocomposites [9, 24–27]. The nanoprecipitates form regular square lattice in magnified view for twice of the same area of Fig. 6.2b, as shown in Fig. 6.2c. The electron diffraction pattern (Fig. 6.2d) shows that nanoprecipitates have cubic structure.

The lattice parameters of PbTe, PbSe, and PbS are 3.22 Å, 3.07 Å, and 2.965 Å, respectively, along (200) plane [28, 29]. Numerous nanoprecipitates are found

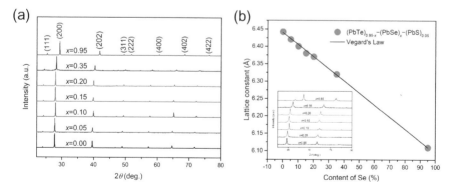

Fig. 6.1 (a) XRD data obtained on $(PbTe)_{0.95-x}(PbSe)_x(PbS)_{0.05}$ powder, (b) lattice parameter against content of Se and Vegard's law (line). Inset is the expanded XRD pattern from 60~80°. Reproduced with permission from reference [8], copyright 2017 Elsevier

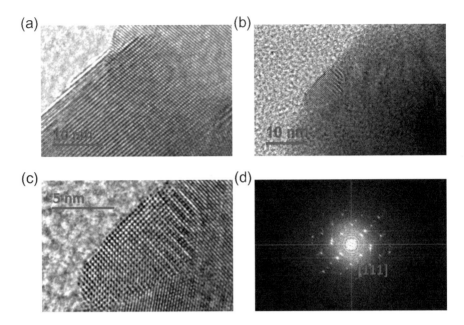

Fig. 6.2 (**a**) and (**b**) HRTEM images of $(PbTe)_{0.75}(PbSe)_{0.20}(PbS)_{0.05}$ at different location, (**b**) and (**c**) different magnification scales, (**d**) SAED (selected area electron diffraction) pattern with [111] zone axis direction of $(PbTe)_{0.75}(PbSe)_{0.20}(PbS)_{0.05}$ sample. Reproduced with permission from reference [8], copyright 2017 Elsevier

also in K-doped $(Pb_{0.98}K_{0.02}Te)_{0.70}(PbSe)_{0.25}(PbS)_{0.05}$ [7] and Cl-doped $(PbTe_{0.93-x}Se_{0.07}Cl_x)_{0.93}(PbS)_{0.07}$ ($x = 0.0005$) [11] which synthesized by the same process with $(PbTe)_{0.75}(PbSe)_{0.20}(PbS)_{0.05}$ as shown in Figs. 6.3a, b and 6.4a, b. Inverse fast Fourier transform (IFFT) images (Figs. 6.3c and 6.4d) along (200) plane show the disordering or dislocation of the precipitates as presented in Figs. 6.3d and 6.4f, respectively. IFFT images along plane show that there are many line dislocations at coherent interfaces between nanoprecipitates and PbTe matrix. In order to identify of lattice dislocation more clearly, we performed geometric phase analysis (GPA) of TEM images. The strain field map profile around the dislocation area shows clearly of lattice strain due to lattice parameter misfit between the matrix and the precipitates. It indicates that the lattice dislocation may generate a strain field between the matrix and nanoprecipitates as shown in Fig. 6.4f. Positive high strain filed is observed at the edge dislocation where there are interfaces between matrix and peripheral precipitates.

Several key questions arise when considering the nanoprecipitates in complex PbTe − PbSe − PbS quaternary alloys. The multiple element substitution in PbTe matrix with rapid quenching process results in lower free energy for creating nanoprecipitates in the matrix of PbTe [8]. The similar phenomena found also in ternary system PbTe − SrTe. Nanoprecipitates are formed in PbTe − SrTe due to the difference in lattice parameters between PbTe and SrTe. The limited SrTe doping

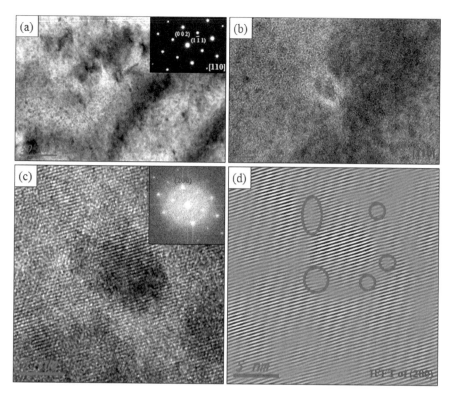

Fig. 6.3 TEM images of $(Pb_{0.98}K_{0.02}Te)_{0.70}(PbSe)_{0.25}(PbS)_{0.05}$ sample for low-magnified image showing numerous nanoprecipitations and spinodal decomposition region (**a**) and its enlarged image (**b**). Inset on (**a**) is electron diffraction pattern. HRTEM image of nanoprecipitation with semi-coherent interface (**c**) and electron diffraction (inset on (**c**)). IFFT image of nanoprecipitates along (200) plane (**d**). Reproduced with permission from reference [7], copyright 2016 the Royal Society of Chemistry

is segregated in PbTe matrix, resulting in nanoprecipitations [29]. In PbTe − PbSe − PbS quaternary alloys, the miscibility gap between PbTe and PbS phase is expected in thermodynamic phase diagram, implying the phase separation via metastable nucleation during grain growth or spinodal decomposition of nanoprecipitates [30] because, it is likely, of the thermodynamic nanoscale precipitation of PbS in PbTe − PbSe matrix [7, 28, 31].

In order to identify the phase of the nanoprecipitates, we performed elemental analysis by energy dispersive X-ray spectroscopy (EDS) in $(Pb_{0.98}K_{0.02}Te)_{0.70}(PbSe)_{0.25}(PbS)_{0.05}$ as shown in Fig. 6.5. Figure 6.5a and b shows nanoprecipitates and phase boundary due to spinodal decomposition displayed as dark region. The form of nanoprecipitates of $(Pb_{0.98}K_{0.02}Te)_{0.70}(PbSe)_{0.25}$ $(PbS)_{0.05}$ exhibits two different morphologies: (1) nanodot and (2) stripe phase due to spinodal decomposition as shown in Fig. 6.5b. When we identify elements from EDS spectrum of the matrix (region 1, white gray), nanodot (region 2, dark gray), and

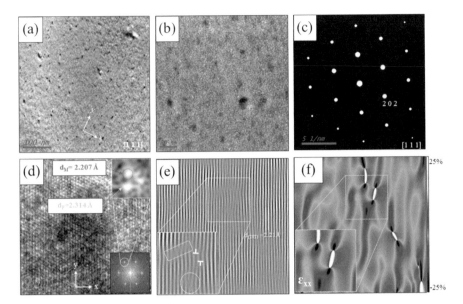

Fig. 6.4 TEM images of $(PbTe_{0.93-x}Se_{0.07}Cl_x)_{0.93}(PbS)_{0.07}$ $(x = 0.0005)$ sample along [111] zone axis for (**a**) low and (**b**) medium magnification with numerous and homogeneously distributed nanoprecipitates. (**c**) Electron diffraction pattern of selected area in the bulk matrix. (**d**) HRTEM image with lattice fringes in nanoprecipitate region. Lower right inset is electron diffraction pattern of the precipitate. Top right inset is IFFT analysis of enlarged view of a peak in electron diffraction, as indicated. (**e**) IFFT image of the lattice fringes in (**d**). Lower left inset is in enlarged view of line dislocation region, as indicated. (**f**) Strain field distribution along x-axis (ε_{xx}) around precipitate (color scale indicates -25% to 25% strain distribution). Reproduced with permission from reference [11], copyright 2017 the Royal Society of Chemistry

spinodal (dark phase boundary) regions, we found that the nanoprecipitates (nanodot and spinodal decomposition) correspond to PbS-rich phase, while PbTe corresponds to the matrix as shown Fig. 6.5d.

6.3 Thermoelectric Performances of PbTe-PbSe-PbS Quaternary Alloys

6.3.1 Simultaneous Occurrence of Nanostructuring and Band Convergence in p-Type $(PbTe)_{0.95-x}(PbSe)_x(PbS)_{0.05}$

We reported high TE performance in p-type $(PbTe)_{0.95-x}(PbSe)_x$ $(PbS)_{0.05}$ alloys by adopting simultaneous occurrence of band structure engineering and nanostructuring due to nanoprecipitation [8]. Band convergence is effective to increase in power factor due to both high σ and S by highly dispersive and degenerated energy bands [32]. According to theoretical band structure calculation of PbTe, PbSe, and PbS,

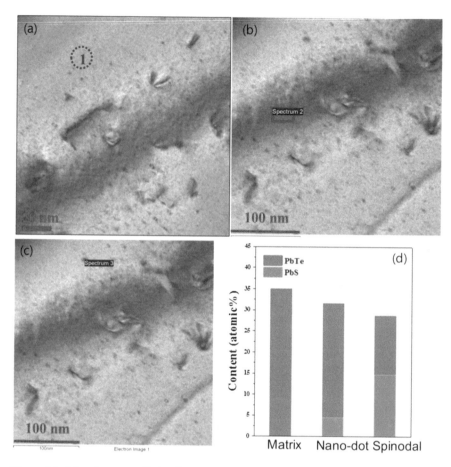

Fig.   6.5 (**a**)   Low-magnified   image   of   bright-field   HRTEM   micrograph   for
$(Pb_{0.98}K_{0.02}Te)_{0.70}(PbSe)_{0.25}(PbS)_{0.05}$ sample. (**b**) Enlarged view of HRTEM image of selected
area of $(Pb_{0.98}K_{0.02}Te)_{0.70}(PbSe)_{0.25}(PbS)_{0.05}$: indicating (1) matrix, (2) nanodot, and (3) spinodal
decomposition region. (**c**) X-ray EDS images at each different regions of matrix (region 1), nanodot
(region 2), and spinodal decomposition (region 3). (**d**) Expected compositional concentration of
PbTe and PbS for different regions. Reproduced with permission from reference [7], copyright 2016
the Royal Society of Chemistry

those compounds are known to have two valence bands which are light hole L and
heavy hole Σ bands. Experimental realization of band convergence in PbTe − PbSe
alloy enhances power factor by high valley degeneracy with dispersive energy bands
[3]. Energy difference between conduction band minimum and valence L band
maximum ΔE_{C-L} and Σ band maximum $\Delta E_{C-\Sigma}$ changes with temperature T and
content x as follows [33]:

$$\Delta E_{C-L} = 0.18 + \left(\frac{4T}{10000}\right) - 0.04x \qquad (6.1)$$

$$\Delta E_{C-\Sigma} = 0.36 + 0.10x \tag{6.2}$$

Total Seebeck coefficient is defined by coefficients and electric conductivity provided by two valence bands as follows [3, 8]:

$$S_{\text{tot}} = \frac{\sigma_L S_L + \sigma_\Sigma S_\Sigma}{\sigma_L + \sigma_\Sigma} \tag{6.3}$$

where S_L, S_Σ, σ_L, and σ_Σ are Seebeck coefficients and electric conductivity due to light hole L and heavy hole Σ valence bands, respectively. Light hole L band has a dispersive energy band, resulting in high mobility of charge carriers, while heavy hole Σ band is a flat band with heavy effective mass of charge carriers. When we align light L and heavy Σ hole bands, highly dispersive and degenerated energy bands give rise to high S with maintaining high mobility of charge carriers.

We demonstrated the effect of band alignment of L and Σ bands in 1 at. % Na-doped $(PbTe)_{0.95-x}(PbSe)_x(PbS)_{0.05}$, giving increase in value of Seebeck coefficient and power factor. Figure 6.6 shows dependences of Seebeck coefficient and electrical conductivity. Temperature-dependent Seebeck coefficient of $(PbTe)_{0.95-x}(PbSe)_x(PbS)_{0.05}$ alloys increased with increasing in temperature

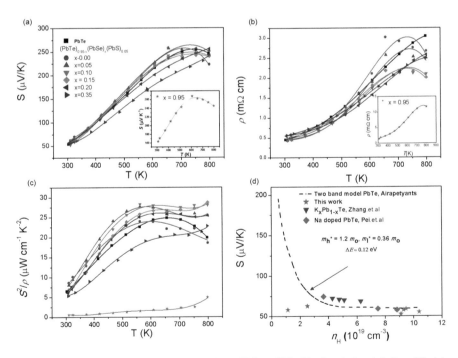

Fig. 6.6 (a) Temperature-dependent Seebeck coefficient $S(T)$, (b) electrical resistivity $\rho(T)$, (c) power factor S^2/ρ of PbTe and $(PbTe)_{0.95-x}(PbSe)_x(PbS)_{0.05}$ samples. (d) Room temperature Pisarenko plot based on two band models of PbTe (dashed line) with experimental data of alloys samples. Reproduced with permission from reference [8], copyright 2017 Elsevier

indicating metallic or degenerated semiconducting behavior of $S(T)$ as presented in Fig. 6.6a. Figure 6.6a shows broad shoulder in $S(T)$ near 700 K for alloys with $x < 0.2$. The plateau around 700 K is caused by band alignment of L and Σ bands in $(PbTe)_{0.95-x}(PbSe)_x(PbS)_{0.05}$ due to hole transfer from L to Σ band [8]. However, the plateau disappeared in alloys $(PbTe)_{0.95-x}(PbSe)_x(PbS)_{0.05}$ with $x \geq 0.2$ due to increasing band gap with growth of Se content, implying that hole excitation from L to Σ band may occur at temperature higher than 800 K [8]. Electrical resistivity $\rho(T)$ shows behavior similar to Seebeck coefficient, as shown in Fig. 6.6b, where broad shoulders of $\rho(T)$ are observed near 700 K for alloys with $x < 0.2$ [8].

Figure 6.6c shows temperature-dependent power factor for series of quaternary alloys $(PbTe)_{0.95-x}(PbSe)_x(PbS)_{0.05}$. The maximum power factor equals to $28.7\ \mu W \times m^{-1} \times K^{-2}$ at 800 K is achieved in alloy samples with $x = 0.15$ which is higher than those of pristine PbTe and PbTe-based alloys [3, 4, 28, 34–37]. So high power factor is due to contribution of band convergence of L and Σ bands. The lowest power factor is observed in alloy samples with $x = 0.35$ because increase in Se content results in decrease in mobility of charge carriers due to scattering of charge carriers by impurities that is not beneficial to the increase in power factor [3]. In order to confirm contribution of L and Σ band convergence to thermoelectric properties, we made Pisarenko plot based on two valence band model comparing with PbTe [20, 38, 39], $Pb_{1-x}K_xTe$ [10], and Na-doped PbTe [38] as shown in Fig. 6.6d. Instead of the single parabolic band model, Pisarenko plot corresponds to two valence band model with effective masses of light L and heavy Σ bands equal to $0.36m_e$ and $1.6m_e$, respectively, which means that Fermi levels lie deep within the valence band and the two valence bands contribute significantly to Seebeck coefficient [8].

Figure 6.7a shows temperature-dependent total thermal conductivity $\kappa_T(T)$ of $(PbTe)_{0.95-x}(PbSe)_x(PbS)_{0.05}$ alloys, which is reduced with increasing in Se content. The lowest $\kappa_T(T)$ is obtained at $x = 0.95$. The significant decrease in $\kappa_T(T)$ due to Se incorporation may come from the phonon scattering caused by alloying, which will be discussed in detail later. Figure 6.7b shows $\kappa_L(T)$ of alloys. The lowest $\kappa_L \sim 0.40\ W \times m^{-1} \times K^{-1}$ at 800 K is measured in alloy with $x = 0.20$. The value of κ_L in alloy with $x = 0.20$ is lower on 60% at 300 K and 55% at 800 K comparing to pristine PbTe compound, and at 800 K approaches the glass limit for bulk PbTe compound [40].

In order to clarify mechanism of κ_L in $(PbTe)_{0.95-x}(PbSe)_x(PbS)_{0.05}$ alloys, we calculate it based on Debye-Callaway model using expression [29, 41]:

$$\kappa_L = \frac{k_B}{2\pi^2 v}\left(\frac{k_B T}{\hbar}\right)^3 \left\{ \int_0^{\theta_D/T} \tau_C \frac{y^4 e^4}{(e^y-1)^2}\,dy + \frac{\left[\int_0^{\theta_D/T}\frac{\tau_C}{\tau_N}\times\frac{y^4 e^4}{(e^y-1)^2}\,dy\right]^2}{\int_0^{\theta_D/T}\frac{1}{\tau_N}\left(1-\frac{\tau_C}{\tau_N}\right)\frac{y^4 e^4}{(e^y-1)^2}\,dy} \right\}, \quad (6.4)$$

where k_B, $\hbar$, T, v, θ, and y are Boltzmann constant, reduced Plank constant, absolute temperature, sound velocity, Debye temperature, and $y = \frac{\hbar\omega}{k_B T}$, respectively. τ_N and

Fig. 6.7 (**a**) Temperature-
dependent total thermal
conductivity $\kappa_T(T)$ and
(**b**) lattice thermal
conductivity $\kappa_L(T)$ of
$(PbTe)_{0.95-x}$ $(PbSe)_x$
$(PbS)_{0.05}$ series of alloys.
Reproduced with
permission from reference
[8], copyright 2017 Elsevier

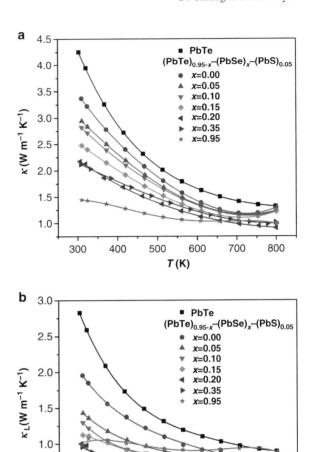

τ_C are relaxation times due to normal phonon-phonon scattering and combined relaxation time. From Matthiessen's rule, combined relaxation time is obtained by accounting of relaxation times from various scattering processes. In this case combined relaxation time is given as:

$$\frac{1}{\tau_C} = \frac{1}{\tau_U} + \frac{1}{\tau_N} + \frac{1}{\tau_B} + \frac{1}{\tau_S} + \frac{1}{\tau_D} + \frac{1}{\tau_P}, \tag{6.5}$$

where τ_U, τ_N, τ_B, τ_S, τ_D, and τ_P are relaxation time corresponding to scattering from Umklapp process, normal process, boundaries, strain, dislocations, and precipitates [14, 42–46]. In this work, we consider four different scattering mechanisms such as point defect scattering, phonon-phonon scattering, grain boundary scattering, and nanoprecipitates scattering with radius r and volume fraction n_V. Therefore, combined phonon relaxation time can be expressed as:

$$\frac{1}{\tau_C} = A\omega^4 + CT\omega^2 + \frac{v}{L_{gr}} + \frac{3}{2}\frac{n_V v}{r}, \qquad (6.6)$$

where L_{gr} is grain's average size and coefficients A and C are constant prefactors. The value of C depends only on crystal structure. Thus, we obtain the value of C by fitting of the above equation to lattice thermal conductivity of undoped $(PbTe)_{0.95-x}(PbSe)_x(PbS)_{0.05}$ alloy. Because we cannot distinguish grain boundary scattering, it is convenient to define an effective mean free path [14]:

$$\frac{1}{L_{eff}} = \frac{1}{L_{gr}} + \frac{3}{2}\frac{n_V}{r} \qquad (6.7)$$

Based on TEM observation, average size of nanoparticle is about 10 nm as shown in Fig. 6.2 and other appropriate parameters of Debye temperature and sound velocity are obtained from [14]. We have calculated theoretical $\kappa_L(T)$ and fitted it with the experimental data. Figure 6.8 shows two pairs of temperature dependences of $\kappa_L(T)$: experimental and theoretical of PbTe and $(PbTe)_{0.75}(PbSe)_{0.20}(PbS)_{0.05}$ alloy. In the case of PbTe, theoretical model of $\kappa_L(T)$ is solely determined by normal model process scattering, while for $(PbTe)_{0.75}(PbSe)_{0.20}(PbS)_{0.05}$, model is based on

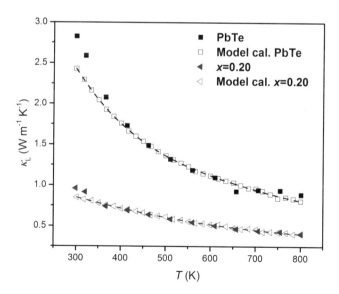

Fig. 6.8 Experimental (closed symbols) and theoretical (open symbols) lattice thermal conductivities $\kappa_L(T)$ of PbTe (black square) and $(PbTe)_{0.75}(PbSe)_{0.20}(PbS)_{0.05}$ (blue triangle). The data of PbTe and $(PbTe)_{0.75}(PbSe)_{0.20}(PbS)_{0.05}$ were calculated based on phenomenological effective medium theory (described in the text). Reproduced with permission from reference [8], copyright 2017 Elsevier

alloy scattering as well as on precipitates. Figure 6.8 clearly shows that the theoretical calculation fits very well with the experimental data of PbTe and $(PbTe)_{0.75}(PbSe)_{0.20}(PbS)_{0.05}$ alloys over a wide temperature range indicating that the strong phonon scattering in alloys comes from nanoprecipitates because of the shorter relaxation time as compared with other processes.

Low κ due to nanoprecipitates and high power factor caused by band engineering affect directly on TE performance of $(PbTe)_{0.95-x}(PbSe)_x(PbS)_{0.05}$ alloys which displayed clearly in dimensionless figure of merit ZT. Figure 6.9a show temperature

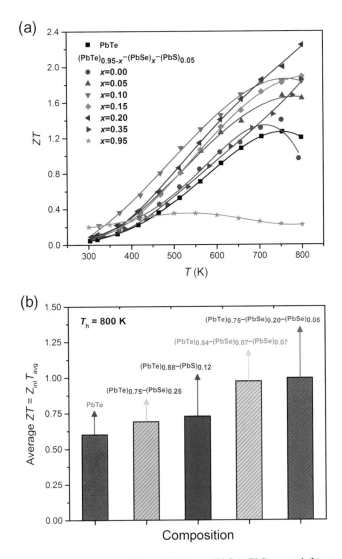

Fig. 6.9 (**a**) Temperature-dependent ZT for $(PbTe)_{0.95-x}(PbSe)_x(PbS)_{0.05}$ and (**b**) average ZT_{avg} for various Pb-based chalcogenides at $T_h = 800$ K and $T_c = 300$ K comparing with references. Reproduced with permission from reference [8], copyright 2017 Elsevier

dependences of ZT for $(PbTe)_{0.95-x}(PbSe)_x(PbS)_{0.05}$ alloys. Alloy with $x = 0.20$ shows the highest ZT value 2.3 at 800 K. Biswas et al. reported high ZT value of 2.0 at 800 K in hierarchical architecture of atomic and mesoscale in PbTe alloyed with 2 mol. % SrTe. In order to achieve a hierarchical architecture, additional processing is needed to produce fine powder and mesoscale microstructure. On the other hand, the direct synthesis of $(PbTe)_{0.95-x}(PbSe)_x(PbS)_{0.05}$ alloys realizes high ZT without multiple sintering process.

For practical application, we also estimated average ZT_{avg} value as defined by [47]:

$$ZT_{avg} = Z_{int}T_{avg} = T_{avg}\frac{1}{\Delta T}\int_{T_c}^{T_h} Z(T)dT, \qquad (6.8)$$

where Z is the figure of merit, T_c is cold side temperature, T_h is hot side temperature, T_{avg} is average temperature $(T_h + T_c)/2$, and ΔT is temperature difference between hot and cold sides $(T_h - T_c)$. We have calculated also ZT_{avg} values of PbTe, $(PbTe)_{0.75}(PbS)_{0.25}$ [31], $(PbTe)_{0.88}(PbS)_{0.12}$ [3], and $(PbTe)_{0.84}(PbSe)_{0.07}(PbS)_{0.07}$ doped with 2 at. % Na [20]. Figure 6.9b shows that average ZT_{avg} of $(PbTe)_{0.75}(PbSe)_{0.20}(PbS)_{0.05}$ alloy has the highest value ZT_{avg} comparing to the other compounds in the previous reports [3, 20, 31].

6.3.2 Thermoelectric Performance of p-Type $Pb_{1-x}K_xTe_{0.7}Se_{0.25}S_{0.05}$

Ahmad et al. have suggested that power factor of PbTe compound can be enhanced by a resonant-like density of states (DOS) distortions in p-type $Pb_{1-x}A_xTe$ (A = K, Rb, and Cs but not Na) [48]. However, Androulakis et al. showed that K doping does not form resonance states but has control over energy difference of the maxima of two primary valence sub-bands in PbTe [49]. The role of K in PbTe has been extensively analyzed, and ZT values reaching 1.3 at 673 K [10] were obtained which is comparable with that of Na-doped PbTe at the same temperature [38]. Furthermore, high thermoelectric performance (ZT $\approx$ 1.6 at 773 K) has been obtained in K-doped $PbTe_{1-x}Se_x$ ternary alloys due to increase in DOS around Fermi levels resulting in higher Seebeck coefficient for two valence bands of $PbTe_{1-x}Se_x$ [10]. Given the valence bands, convergence in PbTe $-$ PbSe [10] and the presence of nanostructure in PbTe $-$ PbSe [30, 31] are effective for increase in power factor and decrease in thermal conductivity, respectively. Therefore, we examined the role of K in enhancing ZT of quaternary alloys PbTe $-$ PbSe $-$ PbS with combination of band converge and nanostructure.

The room temperature Hall effect measurements show p-type conductivity of $Pb_{1-x}K_xTe_{0.70}Se_{0.25}S_{0.05}$ alloys. Figure 6.10 shows Hall concentration n_H and Hall mobility μ_H of charge carriers at room temperature in $Pb_{1-x}K_xTe_{0.70}Se_{0.25}S_{0.05}$

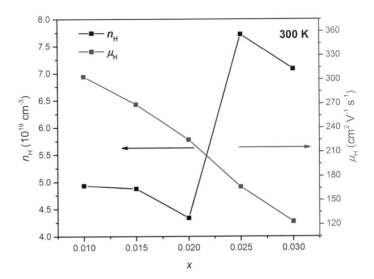

Fig. 6.10 Hall concentration n_H (black, left axis) and mobility μ_H (blue, right axis) of charge carriers in $Pb_{1-x}K_xTe_{0.70}Se_{0.25}S_{0.05}$ alloys depending on K doping level. Reproduced with permission from reference [7], copyright 2016 the Royal Society of Chemistry

alloys depending on K doping level. Hall concentration of charge carriers is reduced until $x = 0.02$ and then increased significantly at higher doping levels $x > 0.02$ from 4.33×10^{19} cm^{-3} ($x = 0.02$) to 7.08×10^{19} cm^{-3} ($x = 0.03$), which indicates that K doping effectively tunes charge carriers concentration in $Pb_{1-x}K_xTe_{0.70}Se_{0.25}S_{0.05}$ alloys. Hall mobility is monotonically decreased with increasing K doping level from 350.2 cm$^2 \times$ V$^{-1} \times$ s^{-1} ($x = 0.01$) to 120 cm$^2 \times$ V$^{-1} \times$ s^{-1} ($x = 0.03$). The reduction of mobility can be understood in terms of defect and alloy scattering due to K doping.

The thermoelectric properties are measured on $Pb_{1-x}K_xTe_{0.70}Se_{0.25}S_{0.05}$ alloys. Temperature-dependent Seebeck coefficient $S(T)$ and electrical resistivity $\rho(T)$ exhibited typical behavior of degenerate semiconductor. Figure 6.11a shows temperature-dependent Seebeck coefficient $S(T)$ of $Pb_{1-x}K_xTe_{0.70}Se_{0.25}S_{0.05}$ and $Pb_{0.98}K_{0.02}Te$ alloys. $S(T)$ increased with temperature and showed broad maximum near 700 K. The maximum Seebeck coefficient reached 313 μV/K at 750 K for K doping $x = 0.02$, which is much higher even than the previously reported for Na-doped PbTe ($\approx$260 μV/K at 775 K), which may be associated with the high effective mass m^*.

Here we estimated the effective masses in terms of single parabolic band model [50]:

$$m^* = \frac{h^2}{2k_B T}\left[\frac{n \times r_H}{4\pi F_{1/2}(\eta)}\right]^{2/3},\qquad(6.9)$$

where $F_{1/2}(\eta)$ is Fermi integral, η is reduced Fermi energy, $r_H \sim 1$ is Hall factor, and k_B is Boltzmann constant.

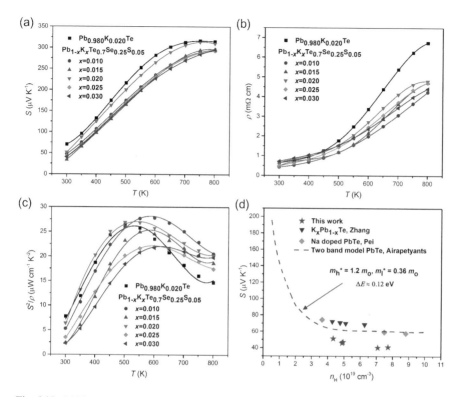

Fig. 6.11 (**a**) Temperature dependences of Seebeck coefficient, (**b**) resistivity, and (**c**) power factor of $Pb_{1-x}K_xTe_{0.70}Se_{0.25}S_{0.05}$ and $Pb_{0.98}K_{0.02}Te$ alloys. (**d**) Room temperature two band model Pisarenko plot of $Pb_{1-x}K_xTe_{0.70}Se_{0.25}S_{0.05}$ based on second band. Reproduced with permission from reference [7], copyright 2016 the Royal Society of Chemistry

The calculated effective masses of charge carriers in $Pb_{1-x}K_xTe_{0.70}Se_{0.25}S_{0.05}$ alloys are presented in Table 6.1.

Effective masses increased with increase in K doping level over $x \geq 0.015$. Effective masses of charge carriers in alloys are higher than in parent compound PbTe $m^* \sim (0.25-0.27)m_e$ [21]. The larger effective masses are related to two valence bands [3, 10]. Because L and Σ bands come closer to each other with increasing in temperature, holes are transferred from L to Σ band, which is the cause of the broad maximum near 700 K.

Electrical resistivity $\rho(T)$ of $Pb_{1-x}K_xTe_{0.70}Se_{0.25}S_{0.05}$ and $Pb_{0.98}K_{0.02}Te$ alloys increases with temperature and then reaches a broad maximum, as shown in Fig. 6.11b; variation of $\rho(T)$ with K doping level is attributed from changing concentration of charge carriers (Table 6.1). Another reason is impurity or alloy scattering due to K doping, which is confirmed by the decrease in Hall mobility (Table 6.1). Power factor of $Pb_{1-x}K_xTe_{0.70}Se_{0.25}S_{0.05}$ and $Pb_{0.98}K_{0.02}Te$ alloys displays the maximum value that equals to 27.78 $\mu W \times m^{-1} \times K^{-2}$ at 600 K when $x = 0.01$, which attributed from significant reduction in electrical resistivity.

Table 6.1 Lattice parameter a, Hall concentration n_H and Hall mobility μ_H of charge carriers, Seebeck coefficient S, and effective mass m^* for $Pb_{1-x}K_xTe_{0.70}Se_{0.25}S_{0.05}$ alloys

x	a	n_H cm^{-3}	μ_H cm$^2 \times$ V$^{-1} \times$ s^{-1}	S µV/K	m^*
0.01	6.355 Å	4.932×10^{19}	303.487	46.997	$0.309m_e$
0.015	6.353 Å	4.875×10^{19}	268.894	33.492	$0.220m_e$
0.02	6.353 Å	4.336×10^{19}	224.972	51.401	$0.310m_e$
0.025	6.356 Å	7.720×10^{19}	166.440	41.106	$0.365m_e$
0.3	6.355 Å	7.085×10^{19}	123.253	40.436	$0.339m_e$

Pisarenko plot made on two valence band model for $Pb_{1-x}K_xTe_{0.70}Se_{0.25}S_{0.05}$ alloys, Na-doped PbTe, and $Pb_{1-x}K_xTe$ is shown in Fig. 6.11d [10, 20, 38, 39] and obtained values of effective masses of heavy hole $m_h^* = 1.2m_e$ (Σ band) and light hole $m_l^* = 0.36m_e$ (L band) with energy band gap $\Delta E \approx 0.12$ eV. The experimental data points of $Pb_{1-x}K_xTe_{0.70}Se_{0.25}S_{0.05}$ alloys lie below theoretical curve plotted using two valence band model indicating lower effective mass of charge carriers in alloys. Based on Pisarenko plot made on two valence band model (Fig. 6.11d), it is clear that second valence band contributes to Seebeck coefficient and there is no any effect of resonance level.

The temperature-dependent $\kappa(T)$ of $Pb_{1-x}K_xTe_{0.70}Se_{0.25}S_{0.05}$ alloys decreased with increase in temperature as shown in Fig. 6.12a; $\kappa(T)$ decreased with increase in K doping level until $x \leq 0.02$ and then increased again for $x > 0.02$ as result of decreased electric resistivity. $\kappa(T)$ for alloy with $x \leq 0.02$ shows a weak increase at high temperature region (T $\geq$ 700 K) implying bipolar diffusion of charge carriers. This small upturn of $\kappa(T)$ is found also in 3 at. % Na-doped $(PbTe)_{1-x}(PbS)_x$ alloy [28]. The lowest κ is reached to 1.96 W $\times$ m$^{-1} \times$ K^{-1} at room temperature and 0.95 W $\times$ m$^{-1} \times$ K^{-1} at high temperature for $(PbTe)_{1-x}(PbS)_x$ ($x = 0.02$). We extracted κ_L by subtracting of electronic thermal conductivity κ_{el} determined by Wiedemann-Franz law as $\kappa_{el} = L_{num}T/\rho$; L_{num} is Lorenz number, T is absolute temperature, and ρ is electrical resistivity. Figure 6.12b shows very low κ_L in $Pb_{1-x}K_xTe_{0.70}Se_{0.25}S_{0.05}$ alloys of 0.91 W $\times$ m$^{-1} \times$ K^{-1} at room temperature and 0.69 W $\times$ m$^{-1} \times$ K^{-1} at high temperature.

We performed theoretical calculation of κ_L of PbTe $-$ PbSe alloy based on Klemens model [20, 51]:

$$\kappa_{L\ alloy} = \kappa_{L\ pure} \frac{\tan^{-1}(u)}{u}, \quad u^2 = \pi \frac{\Theta_D \Omega}{2\hbar v^2} \kappa_{L\ pure} \Gamma, \tag{6.10}$$

where Θ_D is Debye temperature, Ω is molar volume, v is velocity of sound, and Γ is disorder scaling parameter that depends on mass and strain field fluctuations ($\Delta m/m$ and $\Delta a/a$). Figure 6.13 shows theoretical κ_L dependence for $(PbTe)_{1-x}(PbSe)_x$ alloys including experimental data of $Pb_{1-x}K_xTe_{0.70}Se_{0.25}S_{0.05}$, $Pb_{0.98}K_{0.02}Te$, and $(Pb_{0.98}K_{0.02}Te)_{1-x}(PbSe)_x$ [10]. Experimental data of $(Pb_{0.98}K_{0.02}Te)_{1-x}(PbSe)_x$ is roughly followed well to theoretical κ_L values. It means that a strong phonon scattering in $(Pb_{0.98}K_{0.02}Te)_{1-x}(PbSe)_x$ contributes significantly to κ_L from point defect scattering created by Te/Se mixed occupation in rock salt structure. On the

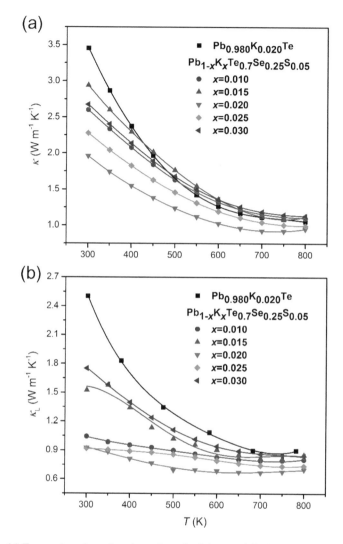

Fig. 6.12 (**a**) Temperature-dependent thermal conductivity κ and (**b**) lattice thermal conductivity κ_{L} of $Pb_{0.98}K_{0.02}Te$ and $Pb_{1-x}K_xTe_{0.25}Se_{0.25}S_{0.05}$ alloys. Reproduced with permission from reference [7], copyright 2016 the Royal Society of Chemistry

other hand, the lowest κ_{L} of $Pb_{1-x}K_xTe_{0.70}Se_{0.25}S_{0.05}$ alloys lies below theoretical κ_{L} of $(PbTe)_{1-x}(PbS)_x$, corresponding to 38% reduction. It shows that low κ_{L} of $Pb_{1-x}K_xTe_{0.70}Se_{0.25}S_{0.05}$ is due to another scattering mechanism beyond the alloy scattering. Possible reason for low κ_{L} of $Pb_{1-x}K_xTe_{0.70}Se_{0.25}S_{0.05}$ alloys is nanostructuring effect, which causes strong phonon scattering. Nanostructuring effect in $Pb_{1-x}K_xTe_{0.70}Se_{0.25}S_{0.05}$ can be understood by Debye-Callaway analysis for various scattering mechanism.

Figure 6.14 shows both theoretical calculation data of κ_{L} by Callaway's model [29, 41] and experimental data for alloys $Pb_{0.98}K_{0.02}Te$ and $Pb_{0.98}K_{0.02}Te_{0.70}$

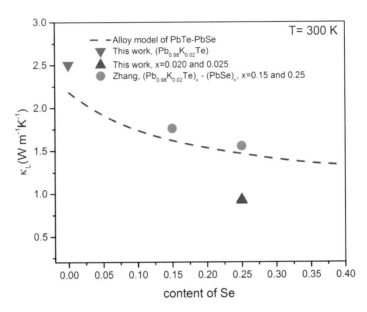

Fig. 6.13 Theoretical calculation of lattice thermal conductivity κ_L in terms of Callaway's alloy model (red dashed line) and experimental κ_L of $Pb_{0.98}K_{0.02}Te$, $Pb_{1-x}K_xTe_{0.70}Se_{0.25}S_{0.05}$ ($x = 0.02$ and 0.25), and $(Pb_{0.98}K_{0.02}Te)_{1-x}(PbSe)_x$ [10] depending on Se content. Reproduced with permission from reference [7], copyright 2016 the Royal Society of Chemistry

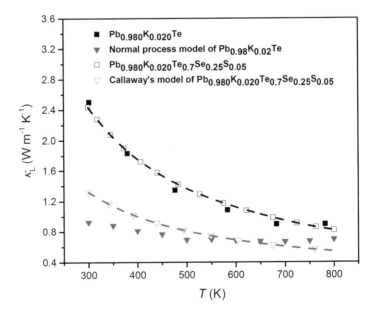

Fig. 6.14 Experimental and theoretical lattice thermal conductivity κ_L of $Pb_{0.98}K_{0.02}Te$ (closed black squares and pink triangles) and $Pb_{0.98}K_{0.02}Te_{0.70}Se_{0.25}S_{0.05}$ (open black squares and pink triangles). Reproduced with permission from reference [7], copyright 2016 the Royal Society of Chemistry

$Se_{0.25}S_{0.05}$. Experimental κ_L of alloys follows very well with theoretical calculation. However, experimental κ_L of $Pb_{0.98}K_{0.02}Te_{0.70}Se_{0.25}S_{0.05}$ is a little bit higher than those of theoretical calculation at high temperatures. The possible reasons of deviation from Callaway's model in $Pb_{0.98}K_{0.02}Te_{0.70}Se_{0.25}S_{0.05}$ alloy are as follows: (1) deviation of Lorenz number from conventional metallic one L_0 due to thermal excitation of charge carriers [52, 53], (2) the existence of inharmonic phonon excitation, and (3) partial local collapse of nanostructure at high temperature [54, 55]. Our calculation proves that the strong phonon scattering is attributed from the nanostructuring resulting in the reduction of κ_L.

Owing to very low κ and high power factor, $Pb_{1-x}K_xTe_{0.70}Se_{0.25}S_{0.05}$ alloys showed improved TE performance as shown in Fig. 6.15a. Alloy $Pb_{0.98}K_{0.02}Te_{0.70}Se_{0.25}S_{0.05}$ shows the highest ZT value of 1.72 at 750 K. We calculated and compared (Fig. 6.15b [20]) average value ZT_{avg} of $Pb_{0.98}K_{0.02}Te_{0.70}Se_{0.25}S_{0.05}$, $Pb_{0.98}Na_{0.02}Te$, $Pb_{0.98}K_{0.02}Te$, $Pb_{0.98}Na_{0.02}Te_{0.88}$ $S_{0.12}$ [31], $Pb_{0.98}Na_{0.02}Te_{0.75}Se_{0.25}$ [3], $Pb_{0.98}K_{0.02}Te_{0.75}Se_{0.25}$ [10], and $Pb_{0.98}Na_{0.02}Te_{0.84}Se_{0.07}S_{0.07}$. Figure 6.15b shows that $Pb_{0.98}K_{0.02}Te_{0.70}Se_{0.25}S_{0.05}$ alloy has the highest ZT_{avg} compared to other related alloys [3, 10, 20, 31].

6.3.3 High Thermoelectric Performance due to Nanoinclusions and Randomly Distributed Interface Potential in n-Type $(PbTe_{0.93-x}Se_{0.07}Cl_x)_{0.93}(PbS)_{0.07}$

While there are many reports on high TE performance of p-type PbTe-based thermoelectric materials [3, 8, 31, 56], it is still a big challenge to find n-type PbTe-based materials that can provide high TE performance over wide temperature range. The state-of-the-art ZT values for n-type Pb chalcogenide binary compounds and alloys are of 0.8 for n-type PbS [14]; 1.1 for n-type nanostructured PbS by introducing secondary phase [50]; 1.2 for n-type PbSe [50, 57], ternary alloys PbSe – PbS [17], and PbTe – PbSe [58]: and 1.1 for quaternary alloy $(PbTe)_{0.75}(PbS)_{0.15}(PbSe)_{0.10}$ [22]. It was reported that S in bulk PbTe can be enhanced significantly by participating a fine distribution of Pb nanoinclusions [59]. The increase in S is thought to be originating from an energy-filtering effect due to strongly energy-dependent electron scattering. In addition, nanoinclusions reduce κ_L resulting in the enhancement of ZT value of InGaAs [60]. These experimental evidences give a general understanding of the role of nanoinclusions in enhancing TE properties, particularly, to assess the relative importance of electronic and phonon scattering.

In a composite of two semiconductors with different energy band gaps or metal/semiconductor composites, we expect a band bending effect at the interfaces. The band bending (energy barrier) induced by different Fermi levels of two materials can scatter selectively charge carriers due to energy-dependent scattering time, resulting in enhancement of S [61]. In addition, nanoinclusions in a matrix will scatter

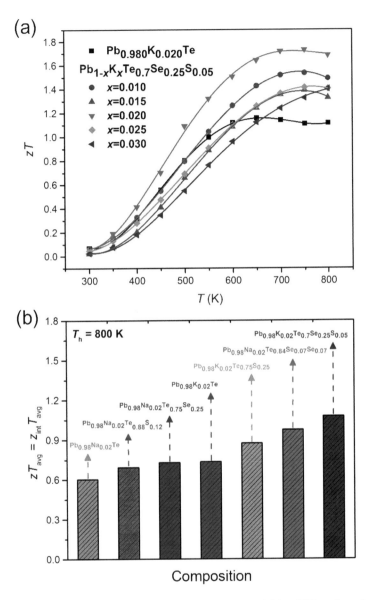

Fig. 6.15 (a) Temperature-dependent ZT of $Pb_{0.98}K_{0.02}Te$ and $Pb_{1-x}K_xTe_{0.70}Se_{0.25}S_{0.05}$, (b) average value ZT_{avg} of various Pb-based chalcogenides at $T_h = 800$ K and $T_c = 300$ K. Reproduced with permission from reference [7], copyright 2016 the Royal Society of Chemistry

effectively of phonons. Therefore, we demonstrated the effects of electron and phonon scattering by nanoinclusions in n-type $(PbTe_{0.93-x}Se_{0.07}Cl_x)_{0.93}$ $(PbS)_{0.07}$ ($x = 0.0005$, 0.01, 0.1 and 0.2) composites. Being in the matrix, nanoinclusions provide increase in S while reducing κ_L in $(PbTe_{0.93-x}Se_{0.07}Cl_x)_{0.93}(PbS)_{0.07}$.

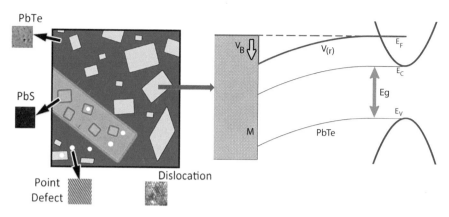

Fig. 6.16 Schematic image of PbS nanoinclusion in PbTe matrix and band bending (energy barrier) effect near interface. Reproduced with permission from reference [11], copyright 2017 the Royal Society of Chemistry

Doping by Cl lifts Fermi level toward the bottom of the conduction band resulting in increased electron concentration. The simultaneous emergence of high S and low κ resulted in high ZT value of 1.52 at 700 K for low Cl doping level ($x = 0.0005$), which is very high value in n-type thermoelectric materials. The randomly distributed interface potential induced by Fermi level tuning with nanoinclusions is a new technique for investigating TE properties [11]. Figure 6.4 shows nanoprecipitates and strong lattice strain near precipitates in $(PbTe_{0.93-x}Se_{0.07}Cl_x)_{0.93}(PbS)_{0.07}$ alloys from TEM images, electron diffraction patterns, and strain field map from IFFT mapping of TEM. In view of band gap difference between PbTe matrix and PbS nanoprecipitates, let us consider which chemical potential is tuned to the conduction band side by Cl doping. Figure 6.16 shows schematic image of inclusion in matrix (left panel) and induced band bending effect on interface between matrix and nanoinclusion (right panel). $PbTe_{1-x}Se_x$ matrix has intrinsic point defects and dislocations. Additional PbS nanoprecipitation further decreases κ_L by scattering of phonons. Cl doping is aiming to lift chemical potential to the conduction band bottom for n-type materials showing metallic behavior in electrical transport. Because PbS has sizable energy band gap, there should be a band bending effect near the interface between the matrix and PbS nanoprecipitates. In that case, filtering of charge carriers with energy lower than appropriate barrier height near the interface can be possible which will be discussed in detail later.

Figure 6.17a shows Hall concentration n_H and Hall mobility μ_H of charge carriers in $(PbTe_{0.93-x}Se_{0.07}Cl_x)_{0.93}(PbS)_{0.07}$ ($x = 0.0005, 0.01, 0.1, 0.2$). Negative Hall coefficient is indicating n-type conductivity. Electron concentration n_H increased systematically with increasing Cl doping level. Hall mobility μ_H decreased with increasing in Cl doping from 954 to lower 200 $cm^2 \times V^{-1} \times s^{-1}$ with increasing Cl doping which is due to enhanced charge carriers scattering. Using single parabolic band model, Hall mobility can be expressed as [62]:

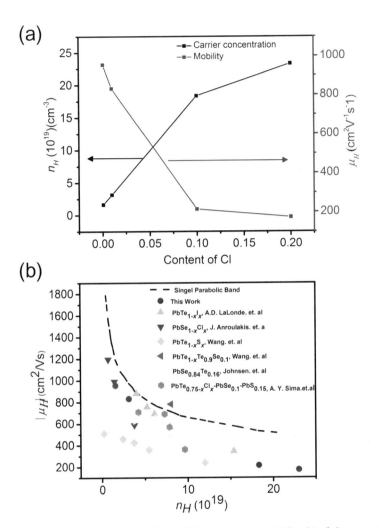

Fig. 6.17 (**a**) Hall mobility μ_H (right axis) and Hall concertation n_H (left axis) of charge carriers at 300 K versus Cl doping. (**b**) Hall mobility versus Hall concentration, single parabolic band model comparing with this work, $Pb_{1-x}I_xTe$ [63], $PbSe_{1-x}Cl_x$ [64], $PbS_{1-x}Cl_x$ [14], $PbTe_{0.9}Se_{0.1}$ [58], $PbSe_{0.84}Te_{0.16}$ [26], and $PbTe_{0.75-x}Cl_xSe_{0.10}S_{0.15}$ [22]. Reproduced with permission from reference [11], copyright 2017 the Royal Society of Chemistry

$$\mu_H = \frac{\sqrt{2}\pi\hbar^4 eC_lN_v^{5/3}}{3(m^*)^{5/2}(k_BT)^{3/2}(E_{def})^2} \times \frac{F_0(\eta)}{F_{1/2}(\eta)}, \tag{6.11}$$

where m^* is effective mass of charge carriers and $F_n(\eta)$ is Fermi integral defined as follows:

$$F_n(\eta) = \int_0^\infty \frac{x^n}{1+e^{x-\eta}} dx, \tag{6.12}$$

where η is reduced electrochemical potential, C_l is longitudinal elastic constant, N_v is valley degeneracy, and E_{def} is deformation potential.

Figure 6.17b shows Pisarenko plot of Hall mobility versus Hall concentration of charge carriers in terms of single parabolic band model for $Pb_{1-x}I_xTe$ [63], $PbSe_{1-x}Cl_x$ [64], $PbS_{1-x}Cl_x$ [14], $PbTe_{0.9}Se_{0.1}$ [58], $PbSe_{0.84}Te_{0.16}$ [26], and $PbTe_{0.75-x}Cl_xSe_{0.10}S_{0.15}$ [22]. Pisarenko plot of $(PbTe_{0.93-x}Se_{0.07}Cl_x)_{0.93}(PbS)_{0.07}$ lies below the value expected from single parabolic model. It can be understood by scattering of charge carriers by microscopic cracks, grain boundaries, and disorder and defect leading to high residual resistivity and low Hall mobility.

Figure 6.18 shows temperature-dependent TE properties of composites $(PbTe_{0.93-x}Se_{0.07}Cl_x)_{0.93}(PbS)_{0.07}$ ($x = 0.0005, 0.01, 0.1, 0.2$). Figure 6.18a displays $\sigma(T)$ in temperature range from 300 K to 800 K. The monotonic decrease in σ with temperature indicates on typical degenerated semiconducting or metallic behavior in all the samples. σ rises with increase in Cl doping content due to higher concentration of charge carriers. The temperature-dependent $S(T)$ (Fig. 6.18b) show negative values indicating n-type conductivity. Because S follows Mott relation [65] in which higher concentration of charge carriers leads to lower S, the behavior of σ showed trade-off relationship with S.

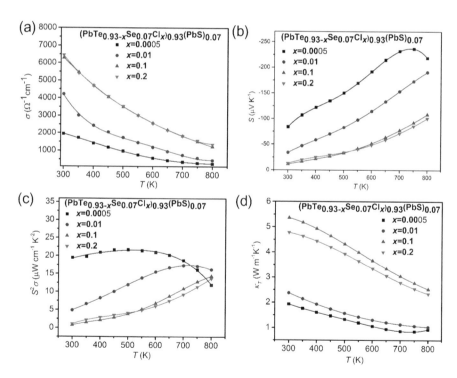

Fig. 6.18 (a) Temperature-dependent thermoelectric properties of $(PbTe_{0.93-x}Se_{0.07}Cl_x)_{0.93}$ $(PbS)_{0.07}$ ($x = 0.0005, 0.01, 0.1, 0.2$) composites: (a) electric conductivity $\sigma(T)$, (b) Seebeck coefficient $S(T)$, (c) power factor $S^2\sigma(T)$, and (d) total thermal conductivity $\kappa(T)$. Reproduced with permission from reference [11], copyright 2017 the Royal Society of Chemistry

The most negative value of S drops to -84.12 µV/K at 300 K for $x = 0.0005$, which corresponds to electron concentration of 1.62×10^{19} cm^{-3}, while the smallest negative value of S is observed on -11.14 µV/K at 300 K for $x = 0.1$ which corresponds to electron concentration of 1.83×10^{20} cm^{-3}. S of composite with $x = 0.0005$ is higher than those of n-type PbTe, PbSe, and PbS binary compounds over an entire temperature range. The peak S equals to -236.63 µV/K for $x = 0.0005$ which is higher than of PbTe, PbS, and PbSe as much as 7.57%, 15.77%, and 4.60%, respectively. There is a broad shoulder in S at $T \geq 750$ K for $x = 0.0005$ which could be mainly ascribed to the thermal excitation of bipolar diffusion effect. S for bipolar transport effect is given by [66]:

$$S = \frac{\sigma_e S_e + \sigma_h S_h}{\sigma_e + \sigma_h} = \frac{S_e nb + S_h p}{nb + p}, \tag{6.13}$$

where the subscripts e, h, n, and p denote the partial properties of electron, hole, electron density, and hole density, respectively. The b is the mobility ratio defined by $b = \mu_e/\mu_h$. S at high temperatures $T \geq 750$ K for $x = 0.0005$ come from the thermal activation of minority charge carriers. Figure 6.18c shows temperature-dependent power factor $S^2\sigma$. The maximum power factor is about 21.72 µW $\times$ cm^{-1} $\times$ K^{-2} at 500 K for $x = 0.0005$ owing to high absolute value of S. Even though the room temperature power factor decreased with increasing Cl doping content, attributed from the decrease in S, power factor at high temperature is not sensitive to Cl doping level.

Temperature-dependent κ decreased with increase in temperature which is a typical behavior of acoustic phonon scattering as presented in Fig. 6.18d. The lowest κ is found at $x = 0.0005$ ($\kappa \approx 0.84$ W $\times$ m^{-1} $\times$ K^{-1} at 750 K). The increase in κ over a wide temperature range with increasing Cl doping level is due to significant electron contribution to κ.

Low κ and high S of (PbTe$_{0.93-x}$Se$_{0.07}$ Cl$_x$)$_{0.93}$(PbS)$_{0.07}$ ($x = 0.0005$, 0.01, 0.1, 0.2) composites give rise to exceptionally high ZT value 1.52 at 700 K for $x = 0.0005$ (Fig. 6.19), which is higher than previously reported of n-type PbS [50], PbSe [57], and PbTe [63] binary compounds and (PbTe)$_{0.75}$(PbS)$_{0.15}$(PbSe)$_{0.10}$ alloy [22] as much as 27.36%, 21.52%, 7.89%, and 27.63%, respectively.

Here we argue that the interface energy barrier between the matrix and nanoprecipitates gives rise to charge carriers energy-filtering effect, resulting in the enhancement of S. We employed theoretical Boltzmann transport calculation with relaxation time approximation as following relations of σ, S, and κ_{el} [67]:

$$\sigma = \frac{e^2}{m_c^*} \times \frac{\left(2m_c^* k_B T\right)^{3/2}}{3a\pi^2 \hbar^3} \times \langle \tau(z) \rangle, \tag{6.14}$$

$$S = \frac{k_B}{e} \times \frac{\langle \tau(z)(z - z_f) \rangle}{\langle \tau(z) \rangle}, \tag{6.15}$$

$$\kappa_{el} = \sigma T \frac{k_B^2}{e^2} \left\{ \frac{\langle \tau(z)z^2 \rangle}{\langle \tau(z) \rangle} - \left[\frac{\langle \tau(z)z \rangle}{\langle \tau(z) \rangle} \right]^2 \right\}, \tag{6.16}$$

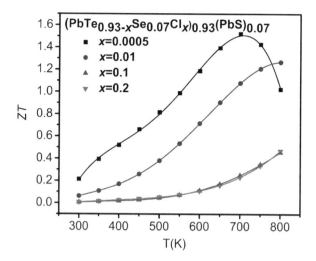

Fig. 6.19 Temperature-dependent ZT values of $(PbTe_{0.93-x}Se_{0.07}Cl_x)_{0.93}(PbS)_{0.07}$ ($x = 0.0005$, 0.01, 0.1, 0.2) composites. Reproduced with permission from reference [11], copyright 2017 the Royal Society of Chemistry

where $z = \frac{E}{k_B T}$ (dimensionless) and $z_f = \frac{E_F}{k_B T}$ are reduced energy and reduced Fermi energy, respectively, m_c^* is effective conductivity mass, and τ is relaxation time. The total relaxation time is given by:

$$\frac{1}{\tau_{bulk}(z)} = \frac{1}{\tau_{PO}(z)} + \frac{1}{\tau_a(z)} + \frac{1}{\tau_o(z)} + \frac{1}{\tau_v(z)}, \qquad (6.17)$$

where $\tau_{PO}(z)$, $\tau_a(z)$, $\tau_o(z)$, and $\tau_v(z)$ are relaxation times due to polar optical phonon scattering, acoustic phonon scattering, optical phonon scattering, and scattering on short-range potential of vacancies, respectively.

Figure 6.20 shows the experimental data (this work for $x = 0.0005$ and n-type PbTe [68]) and theoretical calculation (red line) of temperature-dependent $\sigma(T)$, $S(T)$, power factor $S^2\sigma(T)$, and $\kappa_T(T)$. Figure 6.20a shows clearly that the theoretical calculation of $\sigma(T)$ fits well with experimental data for $x = 0.0005$ composite.

Measured $\sigma(T)$ of composite with $x = 0.0005$ and theoretical calculation values are lower than those of PbTe over the entire temperature range. Such reduction in $\sigma(T)$ is caused by decrease in electron mobility. Electron mobility of PbTe and PbSe at room temperature equals to 1730 cm^2 $\times$ V^{-1} $\times$ s^{-1} ($n = 1.08 \times 10^{19}$ cm^{-3}) and 1045 cm^2 $\times$ V^{-1} $\times$ s^{-1} ($n = 2.38 \times 10^{19}$ cm^{-3}), respectively [23]. Compared with the pristine binary compound, the scattering produced by Se and Cl substitution in PbTe matrix scatters charge carriers additionally [22]. In addition, the interfaces and defects resulting from the distribution of PbS nanoinclusions increase in scattering of charge carriers near the interfaces [69], resulting in a decrease in charge carriers mobility to 954 cm^2 $\times$ V^{-1} $\times$ s^{-1} for $x = 0.0005$ ($n = 1.63 \times 10^{19}$ cm^{-3}).

As shown in Fig. 6.20b, theoretical calculation of $S(T)$ is in good agreement with experimental data ($x = 0.0005$) for $T \geq 750$ K, where the deviation for $T > 750$ K

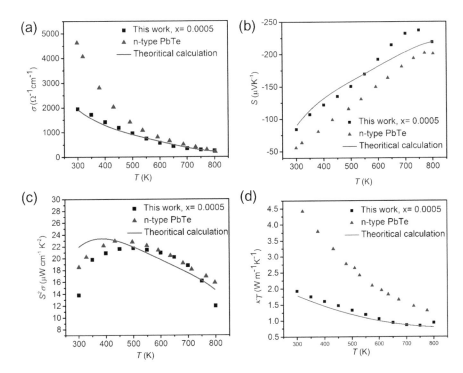

Fig. 6.20 Experimental data (this work for $x = 0.0005$ and n-type PbTe) and theoretical calculation (red line) of (**a**) electric conductivity $\sigma(T)$, (**b**) Seebeck coefficient $S(T)$, (**c**) power factor $S^2\sigma(T)$, and (**d**) total thermal conductivity $\kappa_T(T)$ versus temperature. Reproduced with permission from reference [11], copyright 2017 the Royal Society of Chemistry

originates from the bipolar diffusion effect. The power factor $S^2\sigma(T)$ of composite with nanoinclusions ($x = 0.0005$) is also comparable to that of n-type PbTe, as presented in Fig. 6.20c.

Figure 6.20d shows $\kappa_T(T)$ of composite with $x = 0.0005$ and n-type PbTe compound and theoretical calculation of Boltzmann transport. The theoretical calculation (red line) for $x = 0.0005$ composite fits relatively well with the experimental data. Small deviation for high temperature ($T \geq 750$ K) comes also from the bipolar effect. In $\kappa_T(T)$ calculation, we included the phonon thermal conductivity as well as electronic thermal conductivity as follows [67]:

$$\kappa_{ph} = \frac{k_B}{2\pi^2 v_s \hbar^3} \int_0^{k_B T_D} \tau_{ph}(\hbar\omega)^2 d(\hbar\omega), \tag{6.18}$$

$$\frac{1}{\tau_{ph}} = \frac{1}{\tau_U} + \frac{1}{\tau_D}, \tag{6.19}$$

$$\frac{1}{\tau_U} = cT\omega^2, \tag{6.20}$$

$$\frac{1}{\tau_D} = n_{incl} v_s \left(2\pi R^2\right) = \frac{3x}{2R} v_s \qquad (6.21)$$

where v_s is speed of sound in PbTe, $\hbar\omega$ is phonon energy, τ_U is scattering time by Umklapp process, τ_D is scattering time by nanoinclusion within geometric limit, n_{incl} is concentration of inclusions, R is nanoinclusion radius, x is mole fraction of nanoinclusions, and c is determined from κ_{ph} using experimental value $\kappa_{ph}^{bulk} = 2.0\ \mathrm{W} \times \mathrm{m}^{-1} \times \mathrm{K}^{-1}$ for PbTe at $T = 300\ \mathrm{K}$. Value of κ_T reduced significantly in composite with $x = 0.0005$ compared to nanoinclusion-free n-type PbTe.

In order to understand in detail role of nanoinclusions in enhancing TE properties, we calculated ZT values versus interface potential V_b, as shown in Fig. 6.21a.

Figure 6.16 shows schematic energy band diagram of band bending effect near the interfaces. The zero-point interface potential V_b (0 eV) indicates that there is no phonon scattering by nanoinclusions. Theoretical ZT values with accounting electron and electron + phonon scattering processes produced by nanoinclusions were calculated from the energy-dependent relaxation time and Born approximation. The electrostatic potential $V(r)$ for a single nanoinclusion can be solved by Poisson's equation, as follows [67]:

$$\frac{\varepsilon_0}{4\pi e^2} \times \frac{1}{r} \times \frac{d^2}{dr^2} r V(r) = n[E_F] - n[E_F - V(r)], \qquad (6.22)$$

where we assume the spherical nanoinclusion with radius $R = 2.4$ nm, $n = 2.5 \times 10^{19}\ \mathrm{cm}^{-3}$, and 7% volume fraction of nanoinclusions. The total relaxation time on nanoinclusions τ is expressed by:

$$\frac{1}{\tau} = \frac{1}{\tau_{bulk}} + \frac{1}{\tau_{incl}}, \qquad (6.23)$$

where the relaxation time for bulk τ_{bulk} is given by Eq. (6.17) and τ_{incl} is the relaxation time due to the scattering by $V(r)$ at randomly distributed nanoinclusions. The relaxation time due to the scattering can be expressed as:

$$\frac{1}{\tau_{incl}} = n_{incl} v \sigma_t, \qquad (6.24)$$

$$n_{incl} = \frac{3x}{4\pi R^3}, \qquad (6.25)$$

$$v = \delta_p E_p, \qquad (6.26)$$

where n_{incl} is concentration of inclusions, σ_t is electron transport scattering cross section, and v is electron velocity with p-quantum.

σ and κ_{el} and κ_L decreased with increasing mole fraction of nanoinclusions x in composite. S is more sensitive to interface potential height rather than nanoinclusions content. In order to get the maximum ZT, electrical transport should be retained, while κ should be minimized. In many cases, the decrease in σ affects more significantly on reduction of ZT than the decrease in κ. Therefore, content of nanoinclusions should not be a source of charge carriers scattering as much as possible. In other words, the mean distance between nanoinclusions should be longer than electron mean free path. In this sense, the optimal mole fraction of nanoinclusions x depends on electron mean free path of the material.

ZT value of n-type PbTe without nanoinclusions corresponds to zero-point of the interface potential. On the other hand, ZT value of composite with $x = 0.0005$ corresponds to the value which takes into account of electron + phonon scattering produced by nanoinclusions with negative interface potential of $V_b = -0.11$ eV. In addition, theoretical ZT values, which considered electron and phonon scatterings by nanoinclusions, are significantly higher than ZT values which considered electron scattering (15% or higher) only.

Therefore, the significant enhancement in ZT value (as much as 80.4%) for composite with $x = 0.0005$ is attributed to electron and phonon scattering by PbS nanoinclusions. Theoretical calculation of ZT value for composite with $x = 0.0005$ fits very well with the experimental data except of high temperature range $T \geq 700$ K, due to bipolar diffusion effect (Fig. 6.21b).

Fig. 6.21 (**a**) Calculated ZT values considering electron (green dashed line) and electron + phonon scatterings by nanoinclusions (purple dashed line) and experimental data (this work for $x = 0.0005$ and n-type PbTe) with respect to interface potential V_b and (**b**) temperature-dependent ZT values of experimental values and theoretical calculation (**b**). Reproduced with permission from reference [11], copyright 2017 the Royal Society of Chemistry

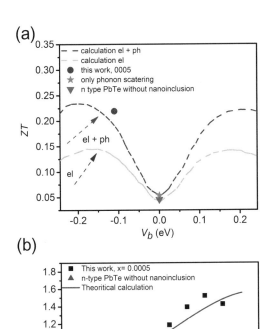

6.4 Conclusion

Combining emergence of nanoprecipitation and band engineering in bulk materials is effective way to enhance thermoelectric performance. In p-type $(PbTe)_{0.95-x}(PbSe)_x(PbS)_{0.5}$ quaternary alloys, PbS forms nanoprecipitates. The effective alloy scattering and phonon scattering by lattice dislocation and lattice distortion in nanostructured phase produce very low lattice thermal conductivity. We observed that band convergence of two valence heavy hole Σ and light hole L bands can increase the power factor (28.7 $\mu W \times cm^{-1} \times K^{-2}$ at 800 K for $x = 0.15$). High ZT value 2.3 at 800 K for $x = 0.20$ comes from high power factor and low thermal conductivity. The alloys have also the highest average ZT_{avg} value which is important for practical applications.

As n-type counterpart of p-type $(PbTe)_{0.95-x}(PbSe)_x(PbS)_{0.5}$ quaternary alloys, n-type $(PbTe_{0.93-x}Se_{0.07}Cl_x)_{0.93}(PbS)_{0.07}$ composites with nanoinclusions and chemical potential tuning by Cl doping showed high Seebeck coefficient due to randomly distributed interface potential and energy-dependent scattering by the interface potential. We have shown that nanoinclusions reduce thermal conductivity due to both electron and phonon scatterings by nanoinclusions. Due to high Seebeck coefficient and extremely low thermal conductivity, the figure of merit ZT reaches 1.52 at 700 K for very low Cl-doped ($x = 0.0005$) n-type $(PbTe_{0.93-x}Se_{0.07}Cl_x)_{0.93}(PbS)_{0.07}$ composite. The results showed that the randomly distributed interface potential, driven by the random distribution of nanoprecipitates, is an effective way to enhance thermoelectric performance.

References

1. K. Koumoto, T. Mori, *Thermoelctric Nano Materials: Material Design and Applications* (Springer, Heidelberg, 2013)
2. G.J. Snyder, E.S. Toberer, Complex thermoelectric materials. Nat. Mater. **7**, 105–114 (2008)
3. Y. Pei, X. Shi, A. LaLonde, H. Wang, L. Chen, G.J. Snyder, Convergence of electronic bands for high performance bulk thermoelectrics. Nature **473**, 66–69 (2011)
4. Y. Pei, A.D. LaLonde, N.A. Heinz, X. Shi, S. Iwanaga, H. Wang, L. Chen, G.J. Snyder, Stabilizing the optimal carrier concentration for high thermoelectric efficiency. Adv. Mater. **23**, 5674–5678 (2011)
5. J.P. Heremans, V. Jovovic, E.S. Toberer, A. Saramat, K. Kurosaki, A. Charoenphakdee, S. Yamanaka, G.J. Snyder, Enhancement of thermoelectric efficiency in PbTe by distortion of the electronic density of states. Science **321**, 554–557 (2008)
6. J. Androulakis, Y. Lee, I. Todorov, D.-Y. Chung, M. Kanatzidis, High-temperature thermoelectric properties of n-type PbSe doped with Ga, In, and Pb. Phys. Rev. B **83**, 195209 (2011)
7. D. Ginting, C.-C. Lin, L. Rathnam, B.-K. Yu, S.-J. Kim, R. Al Rahal Al Orabi, J.-S. Rhyee, Enhancement of thermoelectric properties by effective K-doping and nano precipitation in quaternary compounds of $(Pb_{1-x}K_xTe)_{0.70}(PbSe)_{0.25}(PbS)_{0.05}$. RSC Adv. **6**, 62958–62967 (2016)
8. D. Ginting, C.-C. Lin, R. Lydia, H.S. So, H. Lee, J. Hwang, W. Kim, R. Al Rahal Al Orabi, J.-S. Rhyee, High thermoelectric performance in pseudo quaternary compounds of

$(PbTe)_{0.95-x}(PbSe)_x(PbS)_{0.05}$ by simultaneous band convergence and nano precipitation. Acta Mater. **131**, 98–109 (2017)

9. J. He, L.-D. Zhao, J.-C. Zheng, J.W. Doak, H. Wu, H.-Q. Wang, Y. Lee, C. Wolverton, M.G. Kanatzidis, V.P. Dravid, Role of sodium doping in lead chalcogenide thermoelectrics. J. Am. Chem. Soc. **135**, 4624–4627 (2013)

10. Q. Zhang, F. Cao, W. Liu, K. Lukas, B. Yu, S. Chen, C. Opeil, D. Broido, G. Chen, Z. Ren, Heavy doping and band engineering by potassium to improve the thermoelectric figure of merit in p-Type PbTe, PbSe, and $PbTe_{1-y}Se_y$. J. Am. Chem. Soc. **134**, 10031–10038 (2012)

11. D. Ginting, C.-C. Lin, J.H. Yun, B.-K. Yu, S.-J. Kim, J.-S. Rhyee, High thermoelectric performance by nano-inclusion and randomly distributed interface potential in N-type $(PbTe_{0.93-x}Se_{0.07}Cl_x)_{0.93}(PbS)_{0.07}$ composites. J. Mater. Chem. A **5**, 13535–13543 (2017)

12. H. Wang, Y. Pei, A.D. LaLonde, G.J. Snyder, Heavily doped p-type PbSe with high thermoelectric performance: an alternative for PbTe. Adv. Mater. **23**, 1366–1370 (2011)

13. H. Wang, Z.M. Gibbs, Y. Takagiwa, G.J. Snyder, Tuning bands of PbSe for better thermoelectric efficiency. Energy Environ. Sci. **7**, 804–811 (2014)

14. H. Wang, E. Schechtel, Y. Pei, G.J. Snyder, High thermoelectric efficiency of n-type PbS. Adv. Energy Mater. **3**, 488–495 (2013)

15. Z. Tian, J. Garg, K. Esfarjani, T. Shiga, J. Shiomi, G. Chen, Phonon conduction in PbSe, PbTe, and $PbTe_{1-x}Se_x$ from first-principles calculations. Phys. Rev. B **85**, 184303 (2012)

16. T. Su, S. Li, Y. Zheng, H. Li, M. Hu, M. Ma, X. Jia, Thermoelectric properties of $PbTe_{1-x}Se_x$ alloys prepared by high pressure. J. Phys. Chem. Solids **74**, 913–916 (2013)

17. J. Androulakis, I. Todorov, J. He, D.-Y. Chung, V. Dravid, M. Kanatzidis, Thermoelectrics from abundant chemical elements: high-performance nanostructured PbSe–PbS. J. Am. Chem. Soc. **133**, 10920–10927 (2011)

18. H. Wang, J. Wang, X. Cao, G.J. Snyder, Thermoelectric alloys between PbSe and PbS with effective thermal conductivity reduction and high figure of merit. J. Mater. Chem. A **2**, 3169–3174 (2014)

19. Q. Zhang, E.K. Chere, Y. Wang, H.S. Kim, R. He, F. Cao, K. Dahal, D. Broido, G. Chen, Z. Ren, Nano Energy **22**, 572–582 (2012)

20. R.J. Korkosz, T.C. Chasapis, S.-H. Lo, J.W. Doak, Y.J. Kim, C.-I. Wu, E. Hatzikraniotis, T.P. Hogan, D.N. Seidman, C. Wolverton, V.P. Dravid, M.G. Kanatzidis, High ZT in p-type $(PbTe)_{1-2x}(PbSe)_x(PbS)_x$ thermoelectric materials. J. Am. Chem. Soc. **136**, 3225–3237 (2014)

21. S.A. Yamini, H. Wang, Z.M. Gibbs, Y. Pei, S.X. Dou, G.J. Snyder, Chemical composition tuning in quaternary p-type Pb-chalcogenides - a promising strategy for enhanced thermoelectric performance. Phys. Chem. Chem. Phys. **16**, 1835–1840 (2014)

22. S. Aminorroaya Yamini, H. Wang, D. Ginting, D.R.G. Mitchell, S.X. Dou, G.J. Snyder, Thermoelectric performance of n-type $(PbTe)_{0.75}(PbS)_{0.15}(PbSe)_{0.1}$ composites. ACS Appl. Mater. Interf. **6**, 11476–11483 (2014)

23. J. He, M.G. Kanatzidis, V.P. Dravid, High performance bulk thermoelectrics via a panoscopic approach. Mater. Today **16**, 166–176 (2013)

24. K. Biswas, J. He, I.D. Blum, C.I. Wu, T.P. Hogan, D.N. Seidman, V.P. Dravid, M.G. Kanatzidis, High-performance bulk thermoelectrics with all-scale hierarchical architectures. Nature **489**, 414–418 (2012)

25. I.D. Blum, D. Isheim, D.N. Seidman, J. He, J. Androulakis, K. Biswas, V.P. Dravid, M.G. Kanatzidis, Dopant distributions in PbTe-based thermoelectric materials. J. Electron. Mater. **41**, 1583–1588 (2012)

26. S. Johnsen, J. He, J. Androulakis, V.P. Dravid, I. Todorov, D.Y. Chung, M.G. Kanatzidis, Nanostructures boost the thermoelectric performance of PbS. J. Am. Chem. Soc. **133**, 3460–3670 (2011)

27. H. Wang, J. Hwang, M.L. Snedaker, I.-h. Kim, C. Kang, J. Kim, G.D. Stucky, J. Bowers, W. Kim, High thermoelectric performance of a heterogeneous PbTe nanocomposite. Chem. Mater. **27**, 944–949 (2015)

28. D. Wu, L.-D. Zhao, X. Tong, W. Li, L. Wu, Q. Tan, Y. Pei, L. Huang, J.-F. Li, Y. Zhu, M.G. Kanatzidis, J. He, Superior thermoelectric performance in PbTe-PbS pseudo-binary: extremely low thermal conductivity and modulated carrier concentration. Energy Environ. Sci. **8**, 2056–2068 (2015)

29. K. Biswas, J. He, Q. Zhang, G. Wang, C. Uher, V.P. Dravid, M.G. Kanatzidis, Strained endotaxial nanostructures with high thermoelectric figure of merit. Nat. Chem. **3**, 160–166 (2011)

30. S.N. Girard, K. Schmidt-Rohr, T.C. Chasapis, E. Hatzikraniotis, B. Njegic, E.M. Levin, A. Rawal, K.M. Paraskevopoulos, M.G. Kanatzidis, Analysis of phase separation in high performance PbTe–PbS thermoelectric materials. Adv. Funct. Mater. **23**, 747–757 (2013)

31. S.N. Girard, J. He, X. Zhou, D. Shoemaker, C.M. Jaworski, C. Uher, V.P. Dravid, J.P. Heremans, M.G. Kanatzidis, High performance Na-doped PbTe-PbS thermoelectric materials: electronic density of states modification and shape-controlled nanostructures. J. Am. Chem. Soc. **133**, 16588–16597 (2011)

32. L.D. Zhao, H.J. Wu, S.Q. Hao, C.I. Wu, X.Y. Zhou, K. Biswas, J.Q. He, T.P. Hogan, C. Uher, C. Wolverton, V.P. Dravid, M.G. Kanatzidis, All-scale hierarchical thermoelectrics: MgTe in PbTe facilitates valence band convergence and suppresses bipolar thermal transport for high performance. Energy Environ. Sci. **6**, 3346–3355 (2013)

33. T.C. Chasapis, Y. Lee, E. Hatzikraniotis, K.M. Paraskevopoulos, H. Chi, C. Uher, M.G. Kanatzidis, Understanding the role and interplay of heavy-hole and light-hole valence bands in the thermoelectric properties of PbSe. Phys. Rev. B **91**, 085207–0852211 (2015)

34. H. Wang, J.-H. Bahk, C. Kang, J. Hwang, K. Kim, J. Kim, P. Burke, J.E. Bowers, A.C. Gossard, A. Shakouri, W. Kim, Right sizes of nano- and microstructures for high-performance and rigid bulk thermoelectrics. Proc. Natl. Acad. Sci. U S A **111**, 10949–10954 (2014)

35. Y. Pei, H. Wang, Z.M. Gibbs, A.D. LaLonde, G.J. Snyder, Thermopower enhancement in $Pb_{1-x}Mn_xTe$ alloys and its effect on thermoelectric efficiency. NPG Asia Mater. **4**, e28 (2012)

36. J. He, S.N. Girard, M.G. Kanatzidis, V.P. Dravid, Microstructure-lattice thermal conductivity correlation in nanostructured $PbTe_{0.7}S_{0.3}$ thermoelectric materials. Adv. Funct. Mater. **20**, 764–772 (2010)

37. H.J. Wu, L.D. Zhao, F.S. Zheng, D. Wu, Y.L. Pei, X. Tong, M.G. Kanatzidis, J.Q. He, Broad temperature plateau for thermoelectric figure of merit ZT>2 in phase-separated $PbTe_{0.7}S_{0.3}$. Nat. Commun. **5**, 4515 (2014)

38. Y. Pei, A. LaLonde, S. Iwanaga, G.J. Snyder, High thermoelectric figure of merit in heavy hole dominated PbTe. Energy Environ. Sci. **4**, 2085–2089 (2011)

39. S.V. Airapetyants, N.M. Vinograd, I.N. Dubrovsk, N.V. Kolomet, Sov. Phys. Solid State USSR **8**, 1069–1072 (1966)

40. Y.K. Koh, C.J. Vineis, S.D. Calawa, M.P. Walsh, D.G. Cahill, Lattice thermal conductivity of nanostructured thermoelectric materials based on PbTe. Appl. Phys. Lett. **94**, 153101 (2009)

41. J. Callaway, H.C. von Baeyer, Effect of point imperfections on lattice thermal conductivity. Phys. Rev. **120**, 1149–1154 (1960)

42. W. Kim, A. Majumdar, Phonon scattering cross section of polydispersed spherical nanoparticles. J. Appl. Phys. **99**, 084306 (2006)

43. W. Kim, S.L. Singer, A. Majumdar, J.M.O. Zide, D. Klenov, A.C. Gossard, S. Stemmer, Reducing thermal conductivity of crystalline solids at high temperature using embedded nanostructures. Nano Lett. **8**, 2097–2099 (2008)

44. N. Mingo, D. Hauser, N.P. Kobayashi, M. Plissonnier, A. Shakouri, Nanoparticle-in-alloy approach to efficient thermoelectrics: silicides in SiGe. Nano Lett. **9**, 711–715 (2009)

45. A.J. Minnich, H. Lee, X.W. Wang, G. Joshi, M.S. Dresselhaus, Z.F. Ren, G. Chen, D. Vashaee, Modeling study of thermoelectric SiGe nanocomposites. Phys. Rev. B **80**, 155327 (2009)

46. R. Prasher, Thermal transport due to phonons in random nano-particulate media in the multiple and dependent (correlated) elastic scattering regime. J. Heat Trans. **128**, 627–637 (2006)

47. H.S. Kim, W. Liu, G. Chen, C.W. Chu, Z. Ren, Relationship between thermoelectric figure of merit and energy conversion efficiency. Proc. Natl. Acad. Sci. U. S. A. **112**, 8205–8210 (2015)

48. S. Ahmad, S.D. Mahanti, K. Hoang, M.G. Kanatzidis, Ab initiostudies of the electronic structure of defects in PbTe. Phys. Rev. B **74**, 155205 (2006)
49. J. Androulakis, I. Todorov, D.-Y. Chung, S. Ballikaya, G. Wang, C. Uher, M. Kanatzidis, Thermoelectric enhancement in PbTe with K or Na codoping from tuning the interaction of the light- and heavy-hole valence bands. Phys. Rev. B **82**, 115209 (2010)
50. L.-D. Zhao, S.-H. Lo, J. He, H. Li, K. Biswas, J. Androulakis, C.-I. Wu, T.P. Hogan, D.-Y. Chung, V.P. Dravid, M.G. Kanatzidis, High performance thermoelectrics from earth-abundant materials: enhanced figure of merit in PbS by second phase nanostructures. J. Am. Chem. Soc. **133**, 20476–20487 (2011)
51. P.G. Klemens, Thermal resistance due to point defects at high temperatures. Phys. Rev. **119**, 507–509 (1960)
52. W. Liu, X. Yan, G. Chen, Z. Ren, Recent advances in thermoelectric nanocomposites. Nano Energy **1**, 42–56 (2012)
53. H. Wang, J.-H. Bahk, C. Kang, J. Hwang, K. Kim, A. Shakouri, W. Kim, Large enhancement in the thermoelectric properties of $Pb_{0.98}Na_{0.02}Te$ by optimizing the synthesis conditions. J. Mater. Chem. A **1**, 11269–11278 (2013)
54. X.X. Li, J.Q. Li, F.S. Liu, W.Q. Ao, H.T. Li, L.C. Pan, Enhanced thermoelectric properties of $(PbTe)_{0.88}(PbS)_{0.12}$ composites by Bi doping. J. Alloys Compd. **547**, 86–90 (2013)
55. J. Androulakis, C.-H. Lin, H.-J. Kong, C. Uher, C.-I. Wu, T. Hogan, B.A. Cook, T. Caillat, K.M. Paraskevopoulos, M.G. Kanatzidis, Spinodal decomposition and nucleation and growth as a means to bulk nanostructured thermoelectrics: enhanced performance in $Pb_{1-x}Sn_xTe-PbS$. J. Am. Chem. Soc. **129**, 9780–9788 (2007)
56. C.M. Jaworski, M.D. Nielsen, H. Wang, S.N. Girard, W. Cai, W.D. Porter, M.G. Kanatzidis, J.P. Heremans, Valence-band structure of highly efficient P-type thermoelectric PbTe-PbS alloys. Phys. Rev. B **87**, 045203 (2013)
57. H. Wang, Y. Pei, A.D. LaLonde, G.J. Snyder, Weak electron-phonon coupling contributing to high thermoelectric performance in n-type PbSe. Proc. Natl. Acad. Sci. U. S. A. **109**, 9705–9709 (2012)
58. H. Wang, A.D. LaLonde, Y. Pei, G.J. Snyder, The criteria for beneficial disorder in thermoelectric solid solutions. Adv. Funct. Mater. **23**, 1586–1596 (2013)
59. J.P. Heremans, C.M. Thrush, D.T. Morelli, Thermopower enhancement in PbTe with Pb precipitates. J. Appl. Phys. **98**, 063703 (2005)
60. W. Kim, J. Zide, A. Gossard, D. Klenov, S. Stemmer, A. Shakouri, A. Majumdar, Thermal conductivity reduction and thermoelectric figure of merit increase by embedding nanoparticles in crystalline semiconductors. Phys. Rev. Lett. **96**, 045901 (2006)
61. D. Narducci, E. Selezneva, G. Cerofolini, S. Frabboni, G. Ottaviani, Impact of energy filtering and carrier localization on the thermoelectric properties of granular semiconductors. J. Solid State Chem. **193**, 19–25 (2012)
62. C.-L. Chen, H. Wang, Y.-Y. Chen, T. Day, G.J. Snyder, Thermoelectric properties of p-type polycrystalline SnSe doped with Ag. J. Mater. Chem. A **2**, 11171–11176 (2014)
63. A.D. LaLonde, Y. Pei, G.J. Snyder, Reevaluation of $PbTe_{1-x}I_x$ as high performance n-type thermoelectric material. Energy Environ. Sci. **4**, 2090–2096 (2011)
64. J. Androulakis, D.-Y. Chung, X. Su, L. Zhang, C. Uher, T.C. Hasapis, E. Hatzikraniotis, K.M. Paraskevopoulos, M.G. Kanatzidis, High-temperature charge and thermal transport properties of the n-type thermoelectric material PbSe. Phys. Rev. B **84**, 155207 (2011)
65. M. Cutler, N.F. Mott, Observation of Anderson localization in an electron gas. Phys. Rev. **181**, 1336–1340 (1969)
66. D.M. Rowe, *Thermoelectrics Handbook: Marco to Nano* (CRC Press, Boca Raton, 2005)
67. S.V. Faleev, F. Léonard, Theory of enhancement of thermoelectric properties of materials with nanoinclusions. Phys. Rev. B **77**, 214304 (2008)
68. K. Kishimoto, T. Koyanagi, Preparation of sintered degenerate n-type PbTe with a small grain size and its thermoelectric properties. J. Appl. Phys. **92**, 2544 (2002)
69. Y.I. Ravich, B.A. Efimova, I.A. Smirnov, *Semiconducting Lead Chalcogenides* (Pleanum, New York, 1970)

Chapter 7
Multicomponent Chalcogenides with Diamond-Like Structure as Thermoelectrics

Dan Zhang, Guangsheng Fu, and Shufang Wang

Abstract Due to the advantages of environment-friendly constituent elements, relatively large Seebeck coefficient, and low thermal conductivity, multicomponent diamond-like chalcogenides (MDLCs), such as $CuInTe_2$, Cu_2SnSe_3, Cu_3SbSe_4 and $Cu_2ZnSnSe_4$, have attracted intensive attention for energy conversion as promising thermoelectric (TE) materials in recent years. This chapter provides an overview of research on MDLCs in TE field. Commencing with the crystal structure and phase transition of MDLCs, we will introduce electronic structure and lattice dynamics of MDLCs through some typical TE compounds. We then discuss new methods (i.e., band engineering, entropy engineering, in situ displacement reaction, and mosaic nanostructure) developed in MDLCs for optimizing TE performance. Finally, in addition to the performance of TE device, investigations on stability and mechanical properties of MDLCs are also presented. For future practical applications of this potential material system, the problems needed to be solved and possible directions to further promote TE performance are also explored in the outlook part.

7.1 Introduction

As promising candidates to fight against energy crisis and environmental pollution, thermoelectric (TE) materials have drawn worldwide attention due to the ability to achieve direct and reversible energy conversion between heat and electricity [1–3]. The conversion efficiency of TE material is evaluated by the figure-of-merit $ZT = S^2\sigma T/\kappa$, where S is Seebeck coefficient, σ is electrical conductivity, T is absolute temperature, and κ is thermal conductivity (including electron component κ_e and lattice component κ_L). Therefore, to get high ZT, one needs large S, high σ, and low κ. However, the inherently coupled relationships among S, σ, and κ make the optimization of ZT value challenging, which means the improvement of one TE parameter is usually achieved at the cost of another [4, 5]. Recently, some effective

D. Zhang · G. Fu · S. Wang (✉)
Hebei Key Lab of Optic-Electronic Information and Materials, College of Physics Science and Technology, Hebei University, Baoding, China
e-mail: sfwang@hbu.edu.cn

© Springer Nature Switzerland AG 2019
S. Skipidarov, M. Nikitin (eds.), *Novel Thermoelectric Materials and Device Design Concepts*, https://doi.org/10.1007/978-3-030-12057-3_7

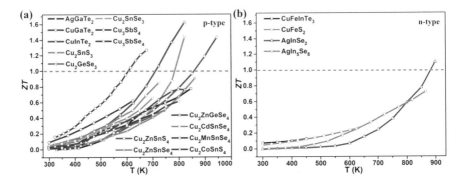

Fig. 7.1 Temperature-dependent ZT of (**a**) *p*- and (**b**) *n*-type MDLC TE materials: AgGaTe$_2$ [17], CuGaTe$_2$ [18], CuInTe$_2$ [15], Cu$_2$SnS$_3$ [19], Cu$_2$GeSe$_3$ [20], Cu$_2$SnSe$_3$ [21], Cu$_3$SbS$_4$ [22], Cu$_3$SbSe$_4$ [23], Cu$_2$ZnSnS$_4$ [24], Cu$_2$ZnSnSe$_4$ [25], Cu$_2$ZnGeSe$_4$ [26], Cu$_2$CdSnSe$_4$ [27], Cu$_2$MnSnSe$_4$ [28], Cu$_2$CoSnS$_4$ [29], CuFeInTe$_3$ [30], CuFeS$_2$ [31], AgIn$_5$Se$_8$ [32], and AgInSe$_2$ [16]

strategies have been adopted to obtain high ZT values greater than 2.0 in PbTe- and SnSe-based TE materials, such as electronic density of states (DOS) distortion [6], band convergence [7] and hierarchical architectures scattering [8] in polycrystal PbTe, shifting Fermi level (E_f) close to the edge of multibands, [9] and 3D charge and 2D phonon transports in single crystal SnSe [10]. However, high toxic Pb element or cleavage fracture issue in these high-efficiency TE materials will impede the commercialization process. Hence, it is of importance to alternatively develop mechanically robust TE materials with eco-friendly and earth-abundant elements [11–14].

Multicomponent diamond-like chalcogenides (MDLCs), characterized by the diamond-like structure with deformed tetrahedron building blocks, have ZT values as high as 1.6 for *p*-type CuInTe$_2$-based [15] and 1.1 for *n*-type AgInSe$_2$-based TE materials [16] (Fig. 7.1), implying great potential for TE power generation at mid-temperature range as eco-friendly TE material [17–31]. In this chapter, crystalline structure of MDLCs is analyzed concisely as well as the theoretical electron and phonon transport properties. Then, some new strategies to boost TE performance are discussed in MDLCs, which will provide useful reference for other TE materials. In the last part, TE device related advances in MDLCs are summarized.

7.2 Crystal Structure and Phase Transition

MDLCs and analogues (Fig. 7.2) can be viewed as derivatives from cubic zinc-blende binary II–VI compounds (e.g., ZnSe) through orderly substitution of II atoms with the same number of atoms of I + III, I$_2$ + IV, I$_3$ + V, and I$_2$ + II + IV or through other substitutions containing vacancy. Most MDLCs, such as CuInSe$_2$, Cu$_2$ZnSnSe$_4$, Cu$_3$SbSe$_4$, and AgIn$_5$Se$_8$ with cation deficiency, take in chalcopyrite

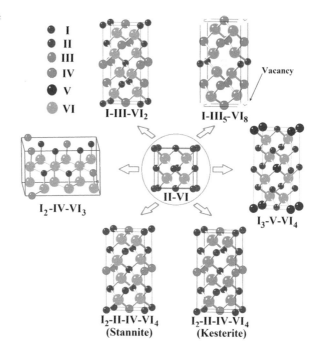

Fig. 7.2 The representative crystal structures of multicomponent diamond-like chalcogenides (MDLCs) derived from cubic zinc-blende structure

or stannite structure. But Cu_2SnSe_3-type semiconductor crystallizes in monoclinic structure with two types of anion-centered tetrahedron rather than one type of anion-centered tetrahedron in the whole structure [33]. As the substituting atoms own different ionic radius and electronegativity, the distorted lattice with diverse bonding length and angle will present in MDLCs [25, 27], which can disrupt the transport of phonons to suppress κ_L. Meanwhile, mobility of charge carriers of MDLCs will decrease also as compared with its binary counterparts. The special electron and phonon transport properties associated in compounds with diamond-like structure will be described below.

Phase transition phenomenon, which can be easily observed in some Cu- or Ag-based TE materials with non-diamond crystal structure like Cu_2Se [34], $AgBiSe_2$ [35], Cu_3SbSe_3 [36], etc., is often ignored in most MDLC TE materials. But information of temperature-dependent crystal structure is of critical significance to explain the changes of TE transport nature in different temperature ranges; therefore, it is necessary to have some general information on the phase transition of MDLCs. We focus only on some typical MDLC TE materials ($CuInTe_2$, $CuSnSe_3$, and $Cu_2ZnSnSe_4$) with phase transition in the following.

$CuInTe_2$ is I-III-VI_2 ternary compound and crystallizes normally in chalcopyrite structure which can be treated as a superstructure of zinc-blende type with two types of ordered cation sublattice positions. When temperature is higher than critical value T_c (~940 K), $CuInTe_2$ with ordered chalcopyrite structure will undergo a first-order phase transition and revert to zinc-blende phase with cation disorder [37]. As the disordered phase contains cross substitutions between non-isovalent cation atoms, it exhibits electronic compensation between donor and acceptor states as well as a decrease in band gap [38]. The reported $CuGaTe_2$-, $AgInSe_2$-, and $AgInTe_2$-based TE materials also have the similar phase transition phenomenon, but this is commonly excluded from consideration in TE research because measured temperatures are obviously below T_c [37]. In addition, abnormal orthorhombic $AgInSe_2$ can be experimentally obtained under some special synthesis condition [39], and this metastable phase may be retained to normal chalcopyrite state after heating as the case in $CuInS_2$ [40].

Cu_2SnSe_3 compound is reported to own diverse crystal structures, namely, monoclinic, tetragonal, orthorhombic, and cubic [41]. The actual type of phases for Cu_2SnSe_3 at low temperatures is still controversial as different preparation conditions and compositions may lead to a variety of results. But, among mentioned phases, monoclinic is stable structure, and cubic is metastable structure which occurs at high temperature. Recently Fan et al. reported that Cu_2SnSe_3 presented in monoclinic phase at low temperature and changed to cubic zinc-blende phase with cation disorder after a phase transition at ~930 K [41]. Siyar et al. reported also that stoichiometric Cu_2SnSe_3 samples annealed at low temperatures (720–820 K) mostly possessed of monoclinic phase, while those annealed at temperature above phase transition value (960 K) were mostly of cubic phase [42]. The analogous variety in crystal structure can also be observed in Cu_2SnS_3- and Cu_2GeSe_3-based TE materials, and dimorphic or polymorphic phases can be detected simultaneously in some cases [43, 44].

As cheap and nontoxic compound Cu_2ZnSnS_4 attracts intensive research in the field of solar cell and thermoelectricity, and also its temperature-dependent change in crystal structure is explored in both theoretical and experimental aspects. It usually tends to take in kesterite structure (tetragonal) with alternating Cu/Zn and Cu/Sn planes stacked along [001] direction. As the temperature increases (Fig. 7.3a), the order-disorder transition (second-order phase transition) will take place at ~530 K, and Cu_2ZnSnS_4 will crystallize in cation-disordered kesterite structure with no change in symmetry [45]. As the further increase in temperature (Fig. 7.3b), a first-order phase transition from tetragonal-disordered kesterite to cubic zinc-blende structure will appear at ~1149 K [46].

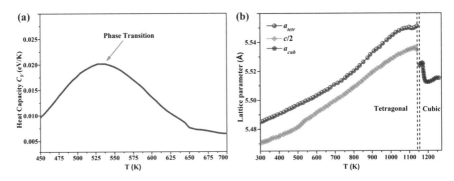

Fig. 7.3 (a) Temperature-dependent heat capacities and (b) lattice parameters of Cu_2ZnSnS_4 [45, 46]

Researchers observed also coexistence of stannite phase (which are alternatively stacked along [001] by Cu/Cu and Zn/Sn planes) in kesterite-phased Cu_2ZnSnS_4 due to small difference in enthalpies of formation between the phases [47], resulting in the variations in the band gap on the order of 100 meV [48]. However, it appears that $Cu_2GeSnSe_4$, $Cu_2ZnSnSe_4$, $Cu_2ZnSnTe_4$, etc. tend to occur in stannite structure as described in the International Centre for Diffraction Data (ICDD) of X-ray diffraction. The actual stable phase of $Cu_2ZnSnSe_4$, etc. presented at low temperature remains debatable because of the disagreements found in previous reports. On one hand, no significant difference was observed between the theoretically calculated energetic/dynamical properties of kesterite and stannite phases of either compound [49]; also, X-ray diffraction is not sensitive in distinguishing kesterite and stannite structures. On the other hand, although Cu/Zn disorder was observed by neutron diffraction or other ways [50], the deviation of stoichiometry by forming tiny second phase (like ZnSe) [51], the preserved high temperature-disordered tetragonal phase during preparation, or the existence of point defects (like V_{Cu}^- + Zn_{Cu}^+) in this quaternary compound [52] can also affect the reliability of results. So, most people in TE field simply treat stannite structure as ground states for most I_2-II-IV-VI_4 selenide compounds, and more detailed structural investigation and analysis are necessary as well as the mechanism for appeared phase transition in $Cu_2ZnGeSe_4$ and $Cu_2FeSnSe_4$ [26, 28].

Thus, before application in TE device, the phase transition behavior and probably related problems, for example, ionic migration, should receive utmost attention, especially in I_2-IV-VI_3 (e.g., Cu_2SnSe_3) and I_2-II-IV-VI_4 (e.g., $Cu_2ZnSnSe_4$) types of TE materials.

7.3 Electronic Structure and Lattice Dynamics of MDLCs

Analyzing electronic structure and lattice dynamic characteristics enables us to have an in-depth understanding of electron and thermal transport properties of MDLCs, thus providing insightful guidance for TE applications. In this section, electron and phonon structure of MDLCs will be simply explored by some typical chalcogenides.

Zhang et al. analyzed comparatively electronic structure of three representative diamond-like chalcogenides (ZnS, $CuGaS_2$, and Cu_2ZnSnS_4) as depicted in Fig. 7.4 and summarized main features as follows [53]. (1) Binary ZnS was a simple sp semiconductor with weak hybridization between S − $3p$ and Zn − $3d$ states. (2) Cu − $3d$ orbitals in CuS_4 tetrahedron will split into t_2 and e suborbitals due to crystal field effect. In ternary $CuGaS_2$, although Cu − $3d$-derived e states were localized, strong hybridization could occur between electrons in t_2 suborbitals and S − $3p$ states, leading to considerable decrease in band gap as well as an increase in the slop of density of states (DOS) near valence band maximum. Thus, intrinsically large Seebeck coefficient S in most MDLC TE materials can be attributed to this relatively steep slope of DOS. (3) The conduction band minimum of Cu

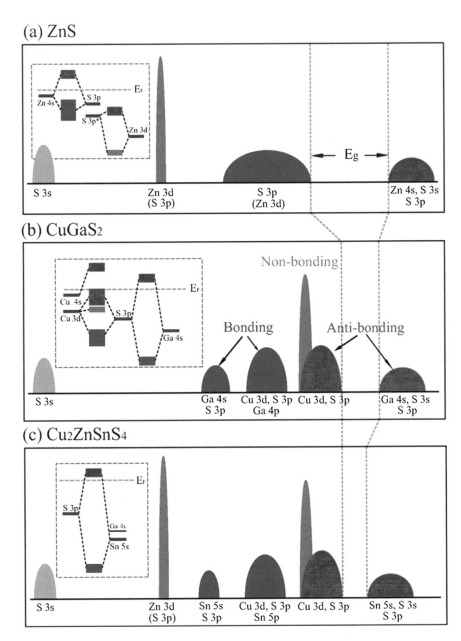

Fig. 7.4 Schematic diagram illustrating bonding for (**a**) ZnS, (**b**) CuGaS$_2$, and (**c**) Cu$_2$ZnSnS$_4$ [53]. Reproduced from Ref. [53] with permission from Elsevier

chalcogenides was primarily derived from s orbital of Sn (or Ga) and p orbitals of sulfur S, and the deeper Sn $-$ 5s states of Cu$_2$ZnSnS$_4$, as compared with Ga $-$ 4s orbitals of CuGaS$_2$, resulted in the further reduction of band gap. (4) The valence band maximum of Cu chalcogenide was dominated by $p - d$ anti-bonding states, in contrast to the bonding states in ZnS.

Ab initio lattice dynamics (Fig. 7.5) of AgInSe$_2$, CuInSe$_2$ and AgInTe$_2$ were summarized by Qiu et al. [16]. (1) The acoustic phonons (heat carrying phonons) owned low sound velocity and large Grüneisen parameter γ, which are beneficial for low κ_L in crystalline compounds [54]. (2) The low-frequency optical phonons could bring resonant scattering effect and then hindered the normal transport process of acoustic phonons with similar frequencies, leading to the further decrease of κ_L [54]. (3) The discovered Ag $-$ Se "cluster vibrations" with low frequencies by phonon animation analysis explained of extremely low κ_L of AgInSe$_2$. Shao et al. reported that low κ of Cu$_2$GeSe$_3$ was related to its low Debye temperature and high bonding anharmonicity ($\gamma \sim 1.2$), based on lattice dynamic calculations [55]. Cu$_3$SbSe$_4$ compound with relatively low κ_L was reported to possess relatively large Grüneisen parameter, $\gamma(\text{Cu}_3\text{SbSe}_4) = 1.2 > \gamma(\text{ZnSe}) = 0.75$, and strong coupling between low-lying optical branches and acoustic phonon modes [56].

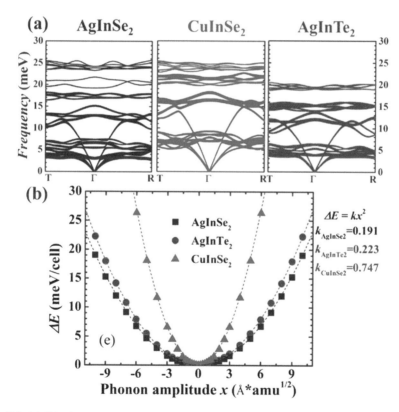

Fig. 7.5 (a) Calculated phonon spectra of AgInSe$_2$, CuInSe$_2$, and AgInTe$_2$. Line thickness denotes contributions from Ag or Cu atoms. (b) Energy difference ΔE as function of the phonon amplitude for the first optical mode caused by Ag $-$ Se (or Cu $-$ Se or Ag $-$ Te) clusters [16]. Reproduced from Ref. [16] with permission from John Wiley and Sons

7.4 New Routes for High-Performance MDLC Thermoelectric Materials

In this section, we revisit only the new routes for high-performance MDLC TE materials, and other approaches, such as optimizing concentration of charge carriers and reduction of κ_L, can be found in our previous work [57].

7.4.1 Band Engineering

Pseudocubic Approach

Band convergence with increased N_v number, which is generally reported in the crystal structures with high symmetry (e.g., cubic PbTe), is desired for gaining large S without substantially degrading σ [7]. In 2014, Zhang et al. expanded this strategy into chalcopyrite with low symmetry and designed a "pseudocubic approach" (Fig. 7.6) to achieve good electronic transport properties with effective band convergence by altering the lattice toward a more cubic crystal symmetry arrangement (i.e., $\eta = \frac{c}{2a}$ toward 1) [58]. It was also verified in experiment that CuInTe$_2$ in pseudocubic structure (Cu$_{0.75}$Ag$_{0.2}$InTe$_2$ or CuIn$_{0.64}$Ga$_{0.36}$Te$_2$) possessed of significantly enhanced power factor, as well as ZT values due to band convergence

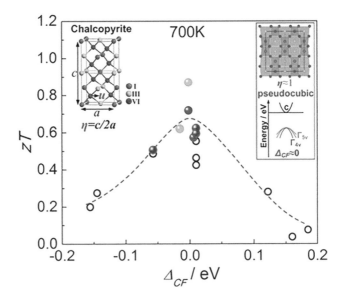

Fig. 7.6 Schematic diagram illustrating pseudocubic band convergence in ternary chalcopyrites. Symbols c and a are lattice constants. Γ_{4v} is nondegenerate band, and Γ_{5v} is double degenerate band. Δ_{CF} is the crystal field-induced energy split at top of bands Γ_{4v} and Γ_{5v}. The pseudocubic structure has cubic cation framework with Δ_{CF} embedded in noncubic distorted anion framework [58]. Reproduced from Ref. [58] with permission from John Wiley and Sons

($\Delta_{CF} \sim 0$). As reported by Zeier et al., increasing η close to unity in stannite $Cu_2ZnGeSe_4$ via Cu and Fe co-doping could increase in DOS effective mass and lead in the result to increase in power factor and ZT values, providing evidence for band convergence via crystal symmetry [59]. However, this promising approach is limited to low symmetry materials with ideal band gap and low κ_L [4].

Highly Efficient Doping (HED)

As for TE materials possessing intrinsically high band degeneracy and low concentration of charge carriers, heavy doping by guest atoms to reach the optimal concentration of charge carriers will damage more or less high band degeneracy, resulting in possible decrease of both S and power factor. Thus, selecting highly efficient dopants, which can provide optimal concentration of charge carriers at greatly lower doping content as compared with non-efficient dopants, should be a feasible attempt to improve σ while maintaining large S with little divergence of degenerate bands [60]. As exhibited in Fig. 7.7, serious separation of the degenerate bands in non-HED can be relieved by HED when the same level of high concentration of charge carriers is reached. One could clearly note the advantage of HED by Al/Ga/In compared with non-HED by Ge/Sn in high band degeneracy containing Cu_3SbSe_4: 0.5 mol. % Al/Ga/In substitution could provide comparable or even higher concentration of charge carriers as that of 1 mol. % Ge doping and also had little deviation from Pisarenko relation based on pristine Cu_3SbSe_4 due to nearly non-changed band structure. But high doping concentration (e.g., 5 mol. % Sn) made Seebeck coefficient S degrade obviously due to severe split of degenerate bands. Ultimately, ZT of samples with HED obtained is of higher value with respect to that of non-HED at 623 K.

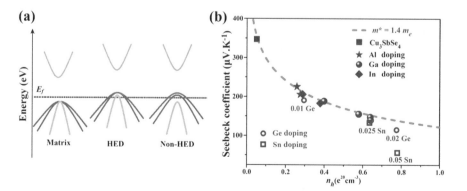

Fig. 7.7 (**a**) Schematic diagram illustrating the maintenance of highly degenerate electronic bands of matrix phase by highly efficient doping (HED). (**b**) Dependence (dashed curve) of Seebeck coefficient S on concentration of charge carriers at room temperature shows that nearly non-changed band structure is obtained via HED (i.e., Al, Ga, In doping) in Cu_3SbSe_4 with high band degeneracy [60]. Reprinted with permission from Ref. [60]. Copyright (2017) American Chemical Society

7.4.2 *Entropy Engineering*

Spin Entropy

It was suggested that the accompanying extra entropy in magnetic ions with degeneracy of electronic spin configurations in real space (e.g., transition metal ions with unpaired $3d$ electrons) could play an important role in getting large Seebeck coefficient S at high σ in cobalt oxides [24, 61]. This striking fact encourages many researchers to explore the possibility of substitution using magnetic ion to enhance TE performance of the other materials. Yao et al. first observed the simultaneous optimization of all three TE parameters in Mn-doped $CuInSe_2$ [62]. Mn substitution increased the concentration of charge carriers by shifting E_f downward into the region of valence band, and Mn $3d$ level hybridization with Se $4p$ level resulted in additional DOS near E_f, enabling simultaneous enhancement of σ and S for $CuInSe_2$. Introduced local disorder after Mn substitution, which was also verified by Raman spectroscopy, could contribute to the reduction of κ_L. However, the contribution of the spin entropy has been neglected by Yao et al. Up to 2013, Xiao et al. reported that magnetic ion could decouple the strongly interrelated three TE parameters in Cu_2ZnSnS_4 (Fig. 7.8) [24]. More importantly, electron paramagnetic resonance (EPR) results implied that Ni^{2+} doping truly brought extra spin into Cu_2ZnSnS_4 and, therefore, the remarkable enhancement of Seebeck coefficient S. After fully substituted with magnetic ions (e.g., Co) in Cu_2ZnSnS_4, electron and phonon structure would undergo significant change due to the strong hybridization between $3d$ states of magnetic ions and S $3p$ states, resulting in improved TE properties. Due to the distinct band structure features of Cu_2CoSnS_4 with magnetic component, high $ZT \sim 0.8$ could be achieved in this sulfide due to improved electron transport properties from enhanced effective mass and depressed κ_L by weakening of covalent bonding [29].

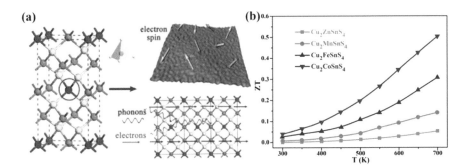

Fig. 7.8 (**a**) Schematic diagram illustrating various electron spin and phonon scatterings in quaternary magnetic sulfides. (**b**) Comparison of ZT for Cu_2XSnS_4 nanocrystals [24]. Reproduced from Ref. [24] with permission from The Royal Society of Chemistry

High-Entropy Alloys

Recently, high-entropy alloys (HEAs), which consist of multiple principal elements at equimolar or close to equimolar ratios, have been proposed as novel types of alloys with intriguing structural and functional properties [63]. Due to increased chemical complexity and configuration entropy from the occupation of multiple components at the identical atomic sites (Fig. 7.9a, b), HEA may bring significant extra phonon disorders and offer also potential way to modify band structure and result in tuning electrical and thermal transport properties [64]. HEAs of $Al_xCoCrFeNi$ ($0.0 \leq x \leq 3.0$) were firstly introduced in the context of TE by Shafeie et al. as potential and stable TE material [65]. In this scenario, Liu et al. devised a strategy of using entropy as the global gene-like performance indicator which showed how multicomponent TE materials with high configuration entropy could be designed (Fig. 7.9c, d) [64]. Firstly, parameter δ in a novel elastic model was defined by shear modulus, lattice parameters, and mismatch in the atomic radius as a criterion for judging the feasibility of forming a complete solid solution with high entropy. Guided by δ criterion, several multicomponent TE material systems were screened out. In the end, screened candidates were tested in experiment, and enhanced ZT values approached to 1.6 and 2.23 in respective $(Cu/Ag)(In/Ga)Te_2$- and $Cu_2(S/Se/Te)$-based multicomponent materials due to the optimization of entropy. Hence, entropy engineering by forming a solid solution with multiple components works as an effective guide to improve greatly TE performance through

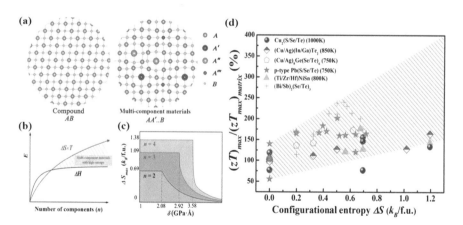

Fig. 7.9 (**a**) Schematic diagram of lattice framework in multicomponent materials compared to ordinary binary compound. (**b**) Schematic diagram of the entropy engineering with multicomponent TE materials. (**c**) The maximum configurational entropy as a function of material's solubility parameter δ for given multicomponent TE materials, where n is the number of components. (**d**) Maximum ZT as a function of configurational entropy in some multicomponent TE materials [64]. Reproduced from Ref. [64] with permission from John Wiley and Sons

the simultaneous optimization of electron and thermal transport properties, i.e., to lower κ_L by the presence of local mass and strain fluctuation and to improve S by increasing in crystal symmetry. Although this strategy holds promise for depressing κ_L, the simultaneous improvement of S is limited to some TE materials with increased crystal symmetry.

7.4.3   In Situ Displacement Reaction

In order to reduce κ_L of TE materials (e.g., skutterudite), the addition of oxide nanoinclusions is widely used to scatter phonons in nanoscale frequencies [66]. Luo et al. found firstly in situ displacement reaction between ZnO additive and CuInTe$_2$ matrix at 714 K by differential scanning calorimeter (DSC) and transmission electron microscopy (TEM) [15]. Consequently, this chemical reaction co-strengthened electron and thermal transport properties of CuInTe$_2$, i.e., to decrease in κ_L via extra phonon scattering from In$_2$O$_3$ nanoinclusions and to improve σ by formation of Zn$_{In}^-$ point defects. Then, 80% enhanced ZT value of 1.44 had been obtained in 1 wt. % ZnO + CuInTe$_2$ sample. Furthermore, by integrating anion substitution (i.e., Sb doping) and in situ displacement reaction, high power factor of 1445 μW $\times$ m^{-1} $\times$ K^{-2} and record-high ZT of 1.61 at 823 K were achieved in CuInTe$_2$-based TE material (Fig. 7.10). In addition, simultaneous optimization of electronic and thermal transport properties in TiO$_2$- or SnO$_2$-added CuInTe$_2$ illustrated also the general effectiveness of in situ displacement reaction by other oxides in enhancing TE performance [67, 68].

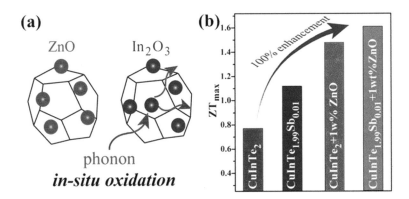

Fig. 7.10 (**a**) Schematic diagram of in situ displacement reaction. (**b**) Illustration of the stepping increment of ZT at 823 K in CuInTe$_2$ samples by Sb substitution, ZnO adding, and combined Sb substitution and ZnO adding [15]. Reproduced from Ref. [15] with permission from John Wiley and Sons

Fig. 7.11 Atomic-resolution HAADF-STEM (high-angle annular dark-field scanning transmission electron microscopy) images of mosaic nanostructures [70]. Reprinted with permission from Ref. [70]. Copyright (2018) American Chemical Society

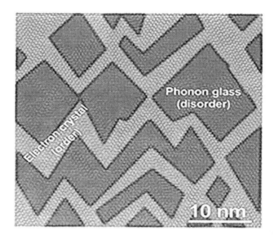

7.4.4 Mosaic Nanostructure

The strategy of nanoscale mosaic is reported to be a useful strategy to tune material's TE properties [69]. As the case in $Cu_2S_{0.5}Te_{0.5}$, high lattice coherence in mosaic crystal helps to gain excellent electron transport, whereas, the lattice strains or interfaces of mosaic nanograins are still very effective in scattering heat conducting phonons, resulting in exceptional ZT above 2 [69]. Recently, Shen et al. discovered that matrix phase of 20 mol. % Zn-doped Cu_2SnS_3 sample presented as a distinctive mosaic-type nanostructure consisting of well-defined cation-disordered domains (the "tesserae") coherently bonded to a surrounding network phase with semiordered cations (Fig. 7.11) [70]. The network phase $Cu_{4+x}Zn_{1-x}Sn_2S_7$ was found to have high mobility of charge carriers, and the tesserae had compositions closer to that of the nominal composition. More importantly, this distinctive mosaic nanostructure plays an important role in promoting TE properties due to simultaneous optimization of electron and phonon transport, i.e., charge carriers are smoothly transferred along crystalline network, while phonons are strongly scattered by atomic disorder and abrupt nanointerfaces of mosaic nanograins. Consequently, large power factor ($\sim$840 $\mu W \times m^{-1} \times K^{-2}$), ultralow κ ($\sim$0.4 $W \times m^{-1} \times K^{-1}$), and high ZT value ($\sim$0.6) at 723 K were obtained for Zn-doped Cu_2SnS_3 sample; therefore, this mosaic nanostructure represents a new type of phonon-glass electron-crystal topology, with high-mobility semiconducting phase interwoven with phonon scattering and low thermal conductivity phase.

7.5 MDLC-Based Thermoelectric Modules

Since the development of MDLC TE materials is still at the early stage, most attentions have been focused on boosting ZT; however, the feasibility to fabricate efficient TE device based on MDLCs is still an unanswered question. In this section, we will revisit the advances in TE device of MDLCs and discuss also some device-related properties which are of critical importance in practical application.

7.5.1 Fabrication and Performance Test

In year 2017, researchers started to face this inevitable question and finally opened the door for practical applications of MDLCs TE materials. Liu et al. used firstly (Sn, Bi) co-doped nanocrystalline Cu_3SbSe_4 samples to fabricate innovative ring-shaped TE generator, and thin insulator layers were placed between rings to alternatively contact the outer and inner rings [23]. The final assembled single p-type Cu_3SbSe_4 rings had inner and outer diameter of 28 mm and 39 mm. The test results showed that an open-circuit voltage near 20 mV was obtained for a single TE element when temperature gradient was maintained at 160 K (inner and outside ring temperatures were 250 and 90 °C, respectively) during test process. As electrical resistance of the ring <0.4 Ohm, the voltage translating into output power >1 mW for each single TE element could be obtained. Constricted by equipment setups, the actual output power density of abovementioned TE device is still unknown.

After that, the first truly TE module including both n- and p-type MDLCs had been fabricated by Qiu et al. as shown in Fig. 7.12 [16]. High-performance $Ag_{0.9}Cd_{0.1}InSe_2$ and $Cu_{0.99}In_{0.6}Ga_{0.4}Te_2$ TE materials were selected as respective n- and p-type legs to set up TE module. First, hot-pressed samples in cylinder shape were polished, diced, and electroplated in sequence. Then, obtained samples were welded to $Mo_{50}Cu_{50}$ alloy blocks with $Cu - P$ brazing filler metal at hot side and copper clad ceramic substrates with $Sn_{42}Bi_{58}$ at cold side. In order to lower the heat loss, the gap in assembled TE module with two couples was finally filled with thermal insulation material. After testing, fabricated TE module exhibited maximum output power of 0.06 W under temperature difference of 520 K. Although this output power value is still inferior to that of other well-known TE materials (e.g., PbTe) [71, 72], it opens clearly the door for applications of diamond-like TE materials.

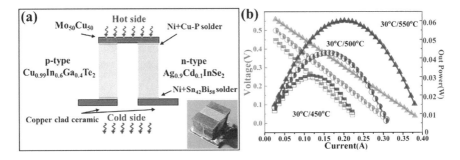

Fig. 7.12 (a) Schematic and photograph (inset) of fabricated diamond-like module. (b) Output voltage and power as function of current for TE module based on diamond-like materials [16]. Reproduced from Ref. [16] with permission from John Wiley and Sons

7.5.2 Operational Stability

Good stability with maintenance of desired TE properties upon cycling is essential to promote the widespread applications of TE materials. The issue of high-temperature instability in MDLC TE materials has attracted more and more attention in recent years. Skoug et al. reported that a dramatic decrease in both σ and ZT at high temperature was linked to physical deformation of $Cu_2Sn_{0.925}In_{0.075}(Se_{0.7}S_{0.3})_3$ sample during measurement [73]. Allowing for instability issue, the measurement of TE properties was just below 723 K in $Cu_2ZnSnSe_4$ samples despite the rising trend of ZT with temperature and high melting point (~1078 K) [74]. One could note also that variation of electronic transport properties after several high-temperature cycles was not suppressed by simply coating boron nitride layer on (Sn, Bi) co-doped Cu_3SbSe_4 sample [23]. After realizing the existence of instable properties in some MDLC TE materials, researchers start to face the problem and study it also.

In order to see temperature-dependent formation phases and crystal structure at 300–800 K of hot-pressed $CuGaTe_2$, Fujii et al. carried out synchrotron X-ray diffraction (SXRD) measurement as well as the crystal structure analysis [75]. It was revealed that Te and CuTe could precipitate from $CuGaTe_2$ matrix in temperature interval 500–650 K and this would also cause Cu and Te deficiencies in matrix phase. These impurity phases would exist in the form of liquid when temperature higher than 650 K and remained even when cooled to room temperature. Deficiencies of Cu and Te in $CuGaTe_2$ could have influence on the tetragonal distortion as well as x-coordinate of Te. Therefore, the above changes in composition as well as crystal structure will inevitably affect high-temperature stability of $CuGaTe_2$, which also can be reflected from the gradual increase in S and decrease in σ over ten measurement cycles from 300 to 800 K [76].

For the purpose of observing whether there is an occurrence of Ag-ion migration in $AgInSe_2$ compound or not, stability test was conducted by Qiu et al. [16]. The results (Fig. 7.13) showed that (1) electron transport properties were relatively reproducible during three independent cooling and heating rounds, and (2) high-temperature relative resistance (R/R_0, where R_0 is sample's initial resistance) under current density of 12 A/cm^2 was almost identical after a long time (~14 h) test, suggesting that $AgInSe_2$-based TE materials had a good stability, even under the condition of large current. To prevent oxidation, sublimation, or decomposition, a protective layer of glass sodium silicate (~1.5 mm in thickness) was coated on $Cu_{2.1}Zn_{0.9}SnSe_4$ sample, and coated sample exhibited relatively good reproducibility in electron transport properties during cooling and heating cycles [25]. In addition to protective layer, the stability can be strengthened by specific elemental doping (e.g., Ni doping in $Cu_{12}Sb_4Se_{13}$) [77] or incorporation of second phase (e.g., 3D graphene heterointerface in $Cu_{2-x}S$) [78] etc.

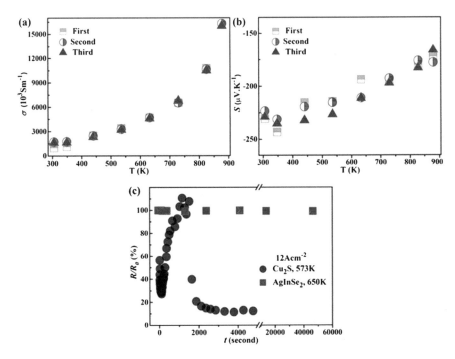

Fig. 7.13 (**a–b**) Reproducibility test for electron transport properties of $Ag_{0.9}Cd_{0.1}InSe_2$ sample. (**c**) Relative electrical resistance of $AgInSe_2$ as a function of current density of 12 A/cm^2 applied at 650 K for various time durations. The data for Cu_2S are included for comparison [16]. Reproduced from Ref. [16] with permission from John Wiley and Sons

7.5.3 *Mechanical Properties*

Crack or fatigue damage of TE materials, which can accelerate the deterioration of material performance as well as the failure of TE device, probably occurs because of thermomechanical stresses under temperature cycling and gradients condition [79]. So, enhancing such mechanical properties as mechanical strength and toughness is of significance for the development of MDLC TE devices.

In order to provide an insight into mechanical properties of Cu_3SbSe_4 compound, Tyagi et al. carried out hardness and fracture toughness measurements on Cu_3SbSe_4 by multimode atomic force microscope (AFM) and Vickers microhardness tester [80]. The hardness value (0.63 GPa) of Cu_3SbSe_4 compound measured by AFM agreed well with Vickers hardness value (0.65 GPa), and calculated fracture toughness was found to be 0.80 ± 0.03 MPa $\times$ m$^{1/2}$, which are comparable with those of some state-of-the-art TE materials (Fig. 7.14) mainly due to the nanoscale features in crystal size.

To understand the intrinsic mechanical properties and determine the deformation mechanism of $CuInTe_2$, Li et al. applied density functional theory (DFT) to investigate the shear deformation properties of single crystalline $CuInTe_2$ [81]. It was

Fig. 7.14 FESEM (field emission scanning electron microscopy) image of Vickers indentation cracks developed in Cu_3SbSe_4. Inset image is fracture toughness of Cu_3SbSe_4, compared with the reported values of state-of-the-art and competing TE materials [80]. Reprinted with permission from Ref. [80], with permission from AIP Publishing

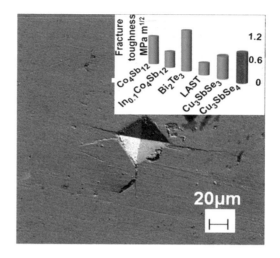

found that the lowest ideal shear strength was 2.43 GPa along (221) [11-1] slip system, which is obviously lower than those of other high-performance TE materials such as PbTe (3.46 GPa), $CoSb_3$ (7.19 GPa), and TiNiSn (10.52 GPa), suggesting that enhancing mechanical properties of $CuInTe_2$ is necessary for applications. The lowest ideal tensile strength was 4.88 GPa along [1–10] tension, and this value was higher than the lowest ideal shear strength. Moreover, the shear-induced failure arose mainly from the softening and breakage of covalent In − Te bond, whereas tensile failure was arisen from breakage of Cu − Te bond. The shear strength was 1.95 GPa under biaxial shear load along (221) [11-1], compression shrunk of In − Te bond and resulted in buckling of In − Te hexagonal framework. The fracture toughness of mode I (K_{Ic}), mode II (K_{IIc}), and mode III (K_{IIIc}) of $CuInTe_2$, which was estimated from ideal stress-strain relations, was equal to 0.19, 0.37, and 0.33 MPa $\times$ m$^{1/2}$, respectively. The fracture toughness of $CuInTe_2$ was much lower than those of ZnO, but greatly higher than that of such layered materials as $BiCuSeO$ and $NaCo_2O_4$. Additionally, relatively poor mechanical properties of $CuInTe_2$ can be strengthened through elemental substitution (e.g., S/Se doping in PbTe) [82], grain size reduction, crystal orientation regulation [83] or addition of nanosized secondary reinforcements [84], and so on.

7.6 Conclusion and Outlook

Recently, MDLCs have attracted an ever-increasing attention in TE field, and TE performance, measured in terms of ZT value, was significantly enhanced to 1.6 in 2017. The effective strategies including pseudocubic structure, highly efficient doping, spin entropy, high-entropy alloys, in situ displacement reaction, and mosaic nanostructure have been successfully utilized in improving TE properties of

MDLCs. Furthermore, diamond-like TE module with only two pairs can provide an output power of 0.06 W under a temperature difference of 520 K. Despite the major advances in ZT and device have been achieved in MDLCs with good prospects, more efforts are still needed to promote the widespread use of MDLC materials in power generation applications. First, ZT value, especially average one, cannot be as good as that of mainstream TE materials, so it is necessary to further increase in ZT by new strategies, such as dislocation engineering [85], resonant levels [86], and nanoheterostructure [87]. Second, as ZT reported in n-type MDLC is still inferior to that of p-type ones (Fig. 7.1), obtaining high-performance n-type MDLC is also urgently needed in future research. Third, in addition to excellent TE properties, mechanical properties, thermal stability, and choosing contact layer, which are of equal importance in commercial applications, should be also investigated intensively for MDLC materials [88]. Finally, MDLCs, not only TE material but also light absorber, probably have good application in photovoltaic-thermoelectric hybrid device if Seebeck effect and photovoltaic effect could be incorporated together in a proper way [89, 90].

Acknowledgments This work is financially supported by the National Natural Science Foundation of China (Grant Nos. 51372064, 61704044, 51802070), the Key Project of Natural Science Foundation of Hebei Province of China (Grant No. E2017201227), and the Start-up Fund for Young Researcher of Hebei University (Grant No. 801260201182).

References

1. L.E. Bell, Science **321**(5895), 1457–1461 (2008)
2. D. Champier, Energy Convers. Manag. **140**, 167–181 (2017)
3. S. Twaha, J. Zhu, Y. Yan, B. Li, Renew. Sust. Energ. Rev. **65**, 698–726 (2016)
4. J. He, T.M. Tritt, Science **357**(6358), eaak9997 (2017)
5. L. Yang, Z.G. Chen, M.S. Dargusch, J. Zou, Adv. Energy Mater. **8**(6), 1701797 (2018)
6. J.P. Heremans, V. Jovovic, E.S. Toberer, A. Saramat, K. Kurosaki, A. Charoenphakdee, S. Yamanaka, G.J. Snyder, Science **321**(5888), 554–557 (2008)
7. Y. Pei, X. Shi, A. LaLonde, H. Wang, L. Chen, G.J. Snyder, Nature **473**(7345), 66 (2011)
8. K. Biswas, J.Q. He, I.D. Blum, C.I. Wu, T.P. Hogan, D.N. Seidman, V.P. Dravid, M.G. Kanatzidis, Nature **489**(7416), 414 (2012)
9. L.D. Zhao, G.J. Tan, S.Q. Hao, J.Q. He, Y.L. Pei, H. Chi, H. Wang, S.K. Gong, H.B. Xu, V.P. Dravid, C. Uher, G.J. Snyder, C. Wolverton, M.G. Kanatzidis, Science **351**, 141–144 (2015)
10. C. Chang, M. Wu, D. He, Y. Pei, C.F. Wu, X. Wu, H. Yu, F. Zhu, K. Wang, Y. Chen, L. Huang, J.F. Li, J. He, L.D. Zhao, Science **360**(6390), 778–783 (2018)
11. P. Qiu, X. Shi, L. Chen, Energy Storage Materials **3**, 85–97 (2016)
12. D. Zhang, J. Yang, Q. Jiang, L. Fu, Y. Xiao, Y. Luo, Z. Zhou, J. Mater. Chem. A **4**, 4188–4193 (2016)
13. D. Zhang, J. Yang, Q. Jiang, L. Fu, Y. Xiao, Y. Luo, Z. Zhou, Mater. Design **98**, 150–154 (2016)
14. Z. Ge, L. Zhao, D. Wu, X. Liu, B. Zhang, J. Li, J. He, Mater. Today **19**(4), 227–239 (2016)
15. Y. Luo, J. Yang, Q. Jiang, W. Li, D. Zhang, Z. Zhou, Y. Cheng, Y. Ren, X. He, Adv. Energy Mater. **6**(12), 1600007 (2016)

16. P. Qiu, Y. Qin, Q. Zhang, R. Li, J. Yang, Q. Song, Y. Tang, S. Bai, X. Shi, L. Chen, Adv. Sci. **5** (3), 1700727 (2018)
17. A. Yusufu, K. Kurosaki, A. Kosuga, T. Sugahara, Y. Ohishi, H. Muta, S. Yamanaka, Appl. Phys. Lett. **99**(6), 061902 (2011)
18. T. Plirdpring, K. Kurosaki, A. Kosuga, T. Day, S. Firdosy, V. Ravi, G.J. Snyder, A. Harnwunggmoung, T. Sugahara, Y. Ohishi, H. Muta, S. Yamanaka, Adv. Mater. **24**(27), 3622–3626 (2012)
19. H. Zhao, X. Xu, C. Li, R. Tian, R. Zhang, Y. Lyu, D. Li, X. Hu, L. Pan, Y. Wang, J. Mater. Chem. A **5**(44), 23267–23275 (2017)
20. J.Y. Cho, X. Shi, J.R. Salvador, G.P. Meisner, J. Yang, H. Wang, A.A. Wereszczak, X. Zhou, C. Uher, Phys. Rev. B **84**(8), 085207 (2011)
21. Y. Li, G. Liu, T. Cao, L.M. Liu, J. Li, K. Chen, L. Li, Y. Han, M. Zhou, Adv. Funct. Mater. **26** (33), 6025–6032 (2016)
22. K. Chen, B. Du, N. Bonini, C. Weber, H. Yan, M. Reece, J. Phys. Chem. C **120**(48), 27135–27140 (2016)
23. Y. Liu, G. García, S. Ortega, D. Cadavid, P. Palacios, J. Lu, M. Ibáñez, L. Xi, J. De Roo, A.M. López, S. Martí-Sánchez, I. Cabezas, M. Mata, Z. Luo, C. Dun, O. Dobrozhan, D.L. Carroll, W. Zhang, J. Martins, M.V. Kovalenko, J. Arbiol, G. Noriega, J. Song, P. Wahnón, A. Cabot, J. Mater. Chem. A **5**(6), 2592–2602 (2017)
24. C. Xiao, K. Li, J. Zhang, W. Tong, Y. Liu, Z. Li, P. Huang, B. Pan, H. Su, Y. Xie, Mater. Horiz. **1**(1), 81–86 (2014)
25. M. Liu, F. Huang, L. Chen, I. Chen, Appl. Phys. Lett. **94**(20), 202103 (2009)
26. W.G. Zeier, A. Lalonde, Z.M. Gibbs, C.P. Heinrich, M. Panthofer, G.J. Snyder, W. Tremel, J. Am. Chem. Soc. **134**(16), 7147–7154 (2013)
27. M.L. Liu, I.W. Chen, F.Q. Huang, L.D. Chen, Adv. Mater. **21**(37), 3808–3812 (2009)
28. Q. Song, P. Qiu, F. Hao, K. Zhao, T. Zhang, D. Ren, X. Shi, L. Chen, Adv. Electron Mater. **2** (12), 1600312 (2016)
29. D. Zhang, J. Yang, Q. Jiang, Z. Zhou, X. Li, J. Xin, A. Basit, Y. Ren, X. He, Nano Energy **36**, 156–165 (2017)
30. H. Cabrera, I. Zumeta Dub, D. Korte, P. Grima Gallardo, F. Alvarado, J. Aitken, J.A. Brant, J.H. Zhang, A. Caldern, E. Marn, M. Aguilar Frutis, J.E. Erazo, E. Perez Cappe, M. Franko, J. Alloys Compd. **651**, 490–496 (2015)
31. Y. Pei, G. Tan, D. Feng, L. Zheng, Q. Tan, X. Xie, S. Gong, Y. Chen, J. Li, J. He, M. Kanatzidis, L. Zhao, Adv. Energy Mater. **7**(3), 1601299 (2017)
32. J.L. Cui, Y.Y. Li, Y. Deng, Q.S. Meng, Y.L. Gao, H. Zhou, Y.P. Li, Intermetallics **31**, 217–224 (2012)
33. L. Xi, Y.B. Zhang, X.Y. Shi, J. Yang, X. Shi, L.D. Chen, W. Zhang, J. Yang, D.J. Singh, Phys. Rev. B **86**(15), 155201 (2012)
34. H. Liu, X. Shi, F. Xu, L. Zhang, W. Zhang, L. Chen, Q. Li, C. Uher, T. Day, G.F. Snyder, Nat. Mater. **11**(5), 422 (2012)
35. L. Pan, D. Berardan, N. Dragoe, J. Am. Chem. Soc. **135**(13), 4914–4917 (2013)
36. K. Samanta, N. Gupta, H. Kaur, L. Sharma, S. Dogra Pandey, J. Singh, T.D. Senguttuvan, N. Dilawar Sharma, A.K. Bandyopadhyay, Mater. Chem. Phys. **151**, 99–104 (2015)
37. C. Rincon, Solid State Commun. **64**(5), 663–665 (1987)
38. S.H. Wei, L.G. Ferreira, A. Zunger, Phys. Rev. B **45**(5), 2533 (1992)
39. M.T. Ng, C.B. Boothroyd, J.J. Vittal, J. Am. Chem. Soc. **128**(22), 7118–7119 (2006)
40. Y.X. Qi, Q.C. Liu, K.B. Tang, Z.H. Liang, Z.B. Ren, X.M. Liu, J. Phys. Chem. C **113**(10), 3939–3944 (2009)
41. J. Fan, W. Carrillo-Cabrera, L. Akselrud, I. Antonyshyn, L. Chen, Y. Grin, Inorg. Chem. **52** (19), 11067–11074 (2013)
42. M. Siyar, J.Y. Cho, Y. Youn, S. Han, M. Kim, S.H. Bae, C. Park, J. Mater. Chem. C **6**(7), 1780–1788 (2018)

43. M. Ibáñez, R. Zamani, W. Li, D. Cadavid, S. Gorsse, N.A. Katcho, A. Shavel, A.M. López, J.R. Morante, J. Arbiol, A. Cabot, Chem. Mater. **24**(23), 4615–4622 (2012)
44. Y. Shen, C. Li, R. Huang, R. Tian, Y. Ye, L. Pan, K. Koumoto, R. Zhang, C. Wan, Y. Wang, Sci. Rep. **6**, 32501 (2016)
45. K. Yu, E.A. Carter, Chem. Mater. **28**(3), 864–869 (2016)
46. S. Schorr, G. Gonzalez Aviles, Phys. Stat. Solid A **206**(5), 1054–1058 (2009)
47. S. Nakamura, T. Maeda, T. Wada, Jpn. J. Appl. Phys. **49**(12R), 121203 (2010)
48. J.J.S. Scragg, L. Choubrac, A. Lafond, T. Ericson, C. Platzer-Björkman, Appl. Phys. Lett. **104** (4), 041911 (2014)
49. T. Gürel, C. Sevik, T. Çağın, Phys. Rev. B **84**(20), 205201 (2011)
50. S. Schorr, Sol. Energy Mater. Sol. Cells **95**(6), 1482–1488 (2011)
51. R. Djemour, M. Mousel, A. Redinger, L. Gütay, A. Crossay, D. Colombara, P. Dale, S. Siebentritt, Appl. Phys. Lett. **102**(22), 222108 (2013)
52. S. Chen, A. Walsh, X.G. Gong, S.H. Wei, Adv. Mater. **25**(11), 1522–1539 (2013)
53. Y.B. Zhang, L.L. Xi, Y.W. Wang, J.W. Zhang, P.H. Zhang, W.Q. Zhang, Comput. Mater. Sci. **108**, 239–249 (2015)
54. J. Yang, L. Xi, W. Qiu, L. Wu, X. Shi, L. Chen, J. Yang, W. Zhang, C. Uher, D. Singh, NPG Comput. Mater. **2**, 15015 (2016)
55. H. Shao, X. Tan, T. Hu, G.-Q. Liu, J. Jiang, H. Jiang, EPL Europhys. Lett. **109**(4), 47004 (2015)
56. Y. Zhang, E. Skoug, J. Cain, V. Ozoliņš, D. Morelli, C. Wolverton, Phys. Rev. B **85**(5), 054306 (2012)
57. D. Zhang, H.C. Bai, Z.L. Li, J.L. Wang, G.S. Fu, S.F. Wang, Chin. Phys. B **27**(4), 047206 (2018)
58. J. Zhang, R. Liu, N. Cheng, Y. Zhang, J. Yang, C. Uher, X. Shi, L. Chen, W. Zhang, Adv. Mater. **26**(23), 3848–3853 (2014)
59. W.G. Zeier, H. Zhu, Z.M. Gibbs, G. Ceder, W. Tremel, G.J. Snyder, J. Mater. Chem. C **2**(47), 10189–10194 (2014)
60. D. Zhang, J. Yang, Q. Jiang, Z. Zhou, X. Li, J. Xin, A. Basit, Y. Ren, X. He, W. Chu, J. Hou, ACS Appl. Mater. Interfaces **9**(34), 28558–28565 (2017)
61. P. Limelette, S. Hébert, V. Hardy, R. Frésard, C. Simon, A. Maignan, Phys. Rev. Lett. **97**(4), 046601 (2006)
62. J. Yao, N.J. Takas, M.L. Schliefert, D.S. Paprocki, P.E.R. Blanchard, H. Gou, A. Mar, C.L. Exstrom, S.A. Darveau, P.F.P. Poudeu, J.A. Aitken, Phys. Rev. B **84**(7), 075203 (2011)
63. Y.F. Ye, Q. Wang, J. Lu, C.T. Liu, Y. Yang, Mater. Today **19**(6), 349–362 (2016)
64. R.H. Liu, H.Y. Chen, K.P. Zhao, Y.T. Qin, B.B. Jiang, T.S. Zhang, G. Sha, X. Shi, C. Uher, W.Q. Zhang, L.D. Chen, Adv. Mater. **29**(38), 1702712 (2017)
65. S. Shafeie, S. Guo, Q. Hu, H. Fahlquist, P. Erhart, A. Palmqvist, J. Appl. Phys. **118**(18), 184905 (2015)
66. X.Y. Zhou, G.W. Wang, L.J. Guo, H. Chi, G.Y. Wang, Q.F. Zhang, C.Q. Chen, T. Thompson, J. Sakamoto, V.P. Dravid, G.Z. Cao, C. Uher, J. Mater. Chem. A **2**(48), 20629–20635 (2014)
67. Y.B. Luo, J.Y. Yang, Q.H. Jiang, Y. Xiao, L.W. Fu, W.X. Li, D. Zhang, Z.W. Zhou, Y.D. Cheng, Nano Energy **18**, 37–46 (2015)
68. W. Li, Y. Luo, Y. Zheng, C. Du, Q. Liang, B. Zhu, L. Zhao, J. Mater. Sci. Mater. El **29**(6), 4732–4737 (2017)
69. Y. He, P. Lu, X. Shi, F.F. Xu, T.S. Zhang, G.J. Snyder, C. Uher, L.D. Chen, Adv. Mater. **27** (24), 3639–3644 (2015)
70. C. Li, Y. Shen, R. Huang, A. Kumamoto, S. Chen, C. Dai, M. Yoshiya, S. Fujii, K. Funai, C.A.J. Fisher, Y. Wang, R. Qi, C.G. Duan, L. Pan, J. Chu, T. Hirayama, Y. Ikuhara, ACS Appl. Nano Mater. **1**(6), 2579–2588 (2018)
71. P.A. Zong, R. Hanus, M. Dylla, Y. Tang, J. Liao, Q. Zhang, G.J. Snyder, L. Chen, Energy Environ. Sci. **10**, 183–191 (2017)
72. X. Hu, P. Jood, M. Ohta, M. Kunii, K. Nagase, H. Nishiate, M.G. Kanatzidis, A. Yamamoto, Energy Environ. Sci. **9**(2), 517–529 (2016)

73. E.J. Skoug, J.D. Cain, D.T. Morelli, J. Electron. Mater. **41**, 1232–1236 (2012)
74. D. Chen, Y. Zhao, Y. Chen, B. Wang, Y. Wang, J. Zhou, Z. Liang, ACS Appl. Mater. Interfaces **7**, 24403–24408 (2015)
75. Y. Fujii, A. Kosuga, J. Electron. Mater. **47**(6), 3105–3112 (2018)
76. A. Kosuga, Y. Fujii, Rare Metals **37**(4), 360–368 (2018)
77. T. Barbier, P. Lemoine, S. Gascoin, O. Lebedev, A. Kaltzoglou, P. Vaqueiro, A. Powell, R. Smith, E. Guilmeau, J. Alloys Compd. **634**, 253–262 (2015)
78. H. Tang, F.H. Sun, J.F. Dong, H.L. Zhuang, Y. Pan, J.F. Li, Nano Energy **49**, 267–273 (2018)
79. M.T. Barako, W. Park, A.M. Marconnet, M. Asheghi, K.E. Goodson, J. Electron. Mater. **42**, 372–381 (2013)
80. K. Tyagi, B. Gahtori, S. Bathula, V. Toutam, S. Sharma, N.K. Singh, A. Dhar, Appl. Phys. Lett. **105**, 261902 (2014)
81. G. Li, Q. An, S.I. Morozov, B. Duan, P. Zhai, Q. Zhang, W.A. Goddard, G.J. Snyder, J. Mater. Chem. A **6**, 11743–11750 (2018)
82. G. Li, U. Aydenir, B. Duan, M.T. Agne, H. Wang, M. Wood, Q. Zhang, P. Zhai, W.A. Goddard, G.J. Snyder, ACS Appl. Mater. Interfaces **9**(46), 40488–40496 (2017)
83. Y. Zheng, Q. Zhang, X.L. Su, H.Y. Xie, S.C. Shu, T.L. Chen, G.J. Tan, Y.G. Yan, X.F. Tang, C. Uher, G.J. Snyder, Adv. Energy Mater. **5**, 1401391 (2015)
84. K. Yin, X. Su, Y. Yan, H. Tang, M.G. Kanatzidis, C. Uher, X. Tang, Scr. Mater. **126**, 1–5 (2017)
85. Z. Chen, Z. Jian, W. Li, Y. Chang, B. Ge, R. Hanus, J. Yang, Y. Chen, M. Huang, G.J. Snyder, Y. Pei, Adv. Mater. **29**, 1606768 (2017)
86. Q. Zhang, B. Liao, Y. Lan, K. Lukas, W. Liu, K. Esfarjani, C. Opeil, D. Broido, G. Chen, Z. Ren, Proc. Natl. Acad. Sci. **110**(33), 13261–13266 (2013)
87. H. Yang, J.H. Bahk, T. Day, A.M.S. Mohammed, G.J. Snyder, A. Shakouri, Y. Wu, Nano Lett. **15**(2), 1349–1355 (2015)
88. W. Liu, Q. Jie, H.S. Kim, Z. Ren, Acta Mater. **87**, 357–376 (2015)
89. N. Wang, L. Han, H. He, N.H. Park, K. Koumoto, Energy Environ. Sci. **4**(9), 3676–3679 (2011)
90. P. Huen, W.A. Daoud, Renew. Sust. Energ. Rev. **72**, 1295–1302 (2017)

Chapter 8
1-2-2 Layered Zintl-Phase Thermoelectric Materials

Jing Shuai, Shan Li, Chen Chen, Xiaofang Li, Jun Mao, and Qian Zhang

Abstract As promising thermoelectric (TE) materials, Zintl-phase compounds exhibit relatively large power factor due to high mobility of charge carriers and possess very low lattice thermal conductivity κ_L that benefits from its complex crystal structure. Among all reported Zintl-phase thermoelectric materials, 1-2-2 layer Zintl compounds (e.g., $CaZn_2Sb_2$, Mg_3Sb_2, $YbMg_2Bi_2$, etc.) have been extensively studied. In this chapter, the thermoelectric properties of p-type 1-2-2 Zintl compounds will be summarized, and strategies in improving thermoelectric performance of Zintl materials will be discussed. Subsequently, recently discovered n-type Mg_3Sb_2-based Zintl materials will be discussed. In addition, theoretical calculation on the thermoelectric performance of other n-type 1-2-2 Zintl compounds will be presented.

8.1 Introduction

Zintl phases have enjoyed a rich history of scientific study since initial investigation of these phases by Eduard Zintl [1–7] and have recently gained interest for potential thermoelectric applications because of intrinsic "electron-crystal, phonon-glass" nature [8]. Zintl compounds are made up of cations and anions with significant difference in electronegativity to allow electron transfer completely. Zintl anions provide "electron-crystal" electronic structure through the covalently bonded network of complex anion or metalloids, whereas cations in Zintl phases act as "phonon-glass" characteristic. Low κ_L can be expected in this complex structure,

J. Shuai
School of Materials, Sun Yat-sen University, Guangzhou, P.R. China

S. Li · C. Chen · X. Li · Q. Zhang (✉)
Department of Materials Science and Engineering, Harbin Institute of Technology (Shenzhen), Shenzhen, Guangdong, P.R. China
e-mail: zhangqf@hit.edu.cn

J. Mao
Department of Physics and TcSUH, University of Houston, Houston, TX, USA

© Springer Nature Switzerland AG 2019
S. Skipidarov, M. Nikitin (eds.), *Novel Thermoelectric Materials and Device Design Concepts*, https://doi.org/10.1007/978-3-030-12057-3_8

and the transport properties can be tuned by alloying in terms of a wide variety of compositions in the same type structure.

An outstanding example of Zintl phase, which has aroused significant interest in thermoelectric studies, is layered $CaAl_2Si_2$-type Zintl phase [9]. The highest figure-of-merit ZT ~ 1.3 achieved for p-type in this 1-2-2 Zintl family [10] is competitive with other Zintls (e.g., $Yb_{14}MnSb_{11}$ [11]) and even other good p-type skutterudites [12] and half-Heuslers [13] within 873 K. Meanwhile, the recent and surprising discovery was realized in n-type Mg_3Sb_2-based Zintls with ZT ~ 1.5 [14, 15]. Recent years have witnessed extensive studies on this Zintl family. We present overview of both p-type and n-type of layered Zintl phases.

8.2 Background of 1-2-2 Zintl Family as Thermoelectrics

More than 30 years ago, this structure has been fully understood by being formulated in AB_2X_2 stoichiometry [16, 17]. A sites contain alkaline-earth or divalent rare-earth element atoms, such as Ca, Ba, Sr, Yb, and Eu; B sites contain d^0, d^5, and d^{10} transition metal atoms (e.g., Mn, Zn, and Cd) or main group element atoms like Mg, while X comes from groups 14 and 15, such as Sb and Bi. Following Zintl-Klemm concept (Fig. 8.1), B and X form anionic sheets with covalent bonding due to the similar electronegativity, while A^{2+} cations donate electrons to $(B_2X_2)^{2-}$ framework. Figure 8.1 shows structure of AB_2X_2 Zintl phases with bondings within $B - X$ layers between planes of A^{2+}.

The ternary AB_2X_2 Zintl phases as TE material candidates are intrinsically p-type semiconductors with small bandgaps and complex structures [18, 19], rich crystal chemistry allows tuning of transport properties, which depends sensitively on the

Fig. 8.1 The structure of AB_2X_2 Zintl phases showing the bonding within $B - X$ layers between planes of A^{2+}. A, B, and X atoms are plotted as yellow, blue, and purple spheres, respectively [16, 17]

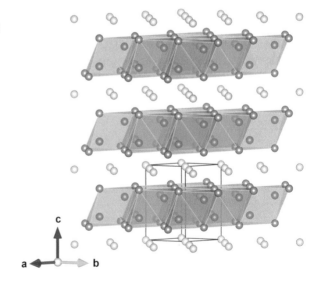

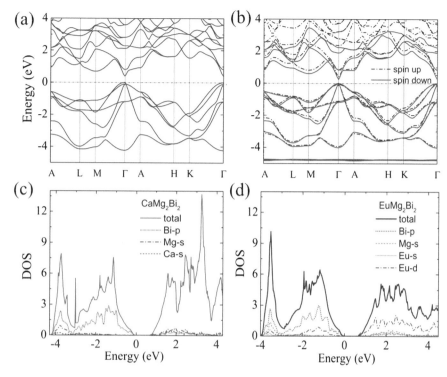

Fig. 8.2 Band structures calculated by DFT. (**a**) $CaMg_2Bi_2$ and (**b**) $EuMg_2Bi_2$ with total and projected density of states (DOS) shown in (**c**) $CaMg_2Bi_2$ and (**d**) $EuMg_2Bi_2$. Reproduced with permission from references [10], Copyright at 2016 National Academy of Science

degeneracy of the valence band edges around Brillouin zone centers. Employing the first-principles calculation based on density functional theory (DFT), two representative band structures (i.e., $CaMg_2Bi_2$, $EuMg_2Bi_2$) investigated are shown in Fig. 8.2 [10]. The band edge of valence bands consists of light and heavy hole bands with peak at zone center as shown in Fig. 8.2a, b. Overlaps of $B - s$ and $X - p$ states in the valence region are observed (Fig. 8.2c, d), indicating hybridization between these states. This hybridization supports polyanionic $(B_2X_2)^{2-}$ Zintl nature of these materials.

Different from direct bandgap in ternary AMg_2Bi_2 samples, bandgap of binary Mg_3Sb_2 is indirect gap around 0.6 eV between valence band maximum (VBM) located at Γ point and conduction band minimum (CBM) at K point (Fig. 8.3a). The better performance of n-type comes partially from multi-pocket character of the bands (K band and ML band) near CBM as shown in Fig. 8.3a, b [20]. This feature leads to larger DOS effective mass, larger Seebeck coefficient S, and higher power factor for n-type materials. In contrast, there is only one near-edge valence band at Γ point, which suppresses electron transport properties in p-type Mg_3Sb_2-based materials.

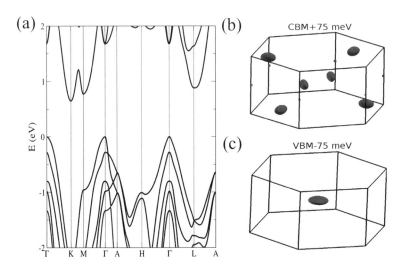

Fig. 8.3 (**a**) Band structures of Mg_3Sb_2. Calculated Fermi surfaces (**b**) of n-type and (**c**) p-type Mg_3Sb_2 at Fermi level. Six equivalent pockets of charge carriers are located near midpoint between M and L' for n-type, whereas p-type Mg_3Sb_2 possesses only one highly anisotropic pocket of charge carriers at Γ point. Reproduced with permission from references [20], Copyright at 2018 Elsevier

Thermoelectric material candidate $Ca_{1-x}Yb_xZn_2Sb_2$ with $CaAl_2Si_2$-type structure was first synthesized by Gascoin et al. in 2005 showing promising ZT ~ 0.56 [9]. Inspired by this work, other $CaAl_2Si_2$-type Zintl phases have been synthesized and studied within last decade. Table 8.1 summarizes most of reported 1-2-2 Zintl layered structures. Data indicate that better thermoelectric performance is expected in solid solutions, which yield low κ_L. Among those Zintl phases, antimonides (e.g., $YbCd_{1.6}Zn_{0.4}Sb_2$, $EuZn_{1.8}Cd_{0.2}Sb_2$, etc.) were reported to have ZT above unity [21, 22]. Recently, high thermoelectric performance of rarely studied bismuth-based Zintl phase $(Eu_{0.5}Yb_{0.5})_{1-x}Ca_xMg_2Bi_2$ has been reported with record ZT ~ 1.3 at 873 K [10]. Besides, ZT values of other Bi-based Zintl phases $Ca_{0.5}Yb_{0.5}Mg_2Bi_2$, $Eu_{0.5}Ca_{0.5}Mg_2Bi_2$, and $Eu_{0.5}Yb_{0.5}Mg_2Bi_2$ are also above unity [23]. The combination of relatively high power factor ~18 $\mu W \times m^{-1} \times K^{-2}$ and low κ_L ~ 0.5 $W \times m^{-1} \times K^{-1}$ makes this layered material much better in comparison with other Zintls.

For $A = B =$ Mg antimonide, binary Mg_3Sb_2 has transport properties different from those of ternary $CaAl_2Si_2$-type Zintls. Although it exhibits low κ_L ~ 1.4 and ~ 0.8 $W \times m^{-1} \times K^{-1}$ at 300 K and 723 K, respectively, and high S above 400 $\mu V/$K, ZT value of parent phase p-type Mg_3Sb_2 is too low (~0.3) due to its high electrical resistivity (ρ ~ 1 Ohm $\times$ m at 300 K) [24]. Mg_3Sb_2 has moderate ZT at high temperatures as is for most alloyed or doped materials. Relatively high ZT ~ 0.84 for p-type was achieved in $Mg_3Pb_{0.2}Sb_{1.8}$ by Bhardwaj et al. [25]. However, extremely high resistivity (ρ ~ 1 Ohm $\times$ m at 300 K) and low concentration of

Table 8.1 Experimental thermoelectric properties of AB_2X_2 at temperature T

Year	AB_2X_2 [Ref.]	ρ mOhm × cm	S μV/K	κ_L W × m^{-1} × K^{-1}	ZT	T K
2005	$Ca_{0.25}Yb_{0.75}Zn_2Sb_2$ [9]	3.7	170	1.4	0.56	773
2007	$BaZn_2Sb_2$ [28]	6.1	185	1.25	0.33	673
2008	$YbZn_{1.9}Mn_{0.1}Sb_2$ [29]	1.5	150	1.6	0.65	726
2008	$EuZn_2Sb_2$ [30]	1.8	180	1.45	0.9	713
2009	$YbCd_{1.6}Zn_{0.4}Sb_2$ [21]	1.66	180	1.1	1.2	650
2010	$Yb_{0.6}Ca_{0.4}Cd_2Sb_2$ [31]	4.4	240	0.9	0.96	700
2010	$Yb_{0.75}Eu_{0.25}Cd_2Sb_2$ [22]	4	240	1	0.97	650
2010	$EuZn_{1.8}Cd_{0.2}Sb_2$ [22]	2	200	1.4	1.06	650
2011	$YbCd_{1.85}Mn_{0.15}Sb_2$ [32]	5.7	245	0.6	1.14	650
2012	$YbMg_2Bi_2$ [33]	5	180	1.8	0.44	650
2013	$Mg_3Bi_{0.2}Sb_{1.8}$ [34]	40	400	0.58	0.6	750
2014	$Mg_3Pb_{0.2}Sb_{1.8}$ [25]	28.6	280	0.28	0.84	773
2014	$Yb_{0.99}Zn_2Sb_2$ [26]	1.3	160	1.7	0.85	800
2015	$Mg_{2.9875}Na_{0.0125}Sb_2$ [35]	5.4	200	0.95	0.6	773
2016	$YbCd_{1.9}Mg_{0.1}Sb_2$ [36]	3.3	230	1.02	1.08	650
2016	$Ca_{0.5}Yb_{0.5}Mg_2Bi_2$ [23]	2.8	187	1.08	1	873
2016	$Ca_{0.995}Na_{0.005}Mg_2Bi_{1.98}$ [37]	3	200	1.25	0.9	873
2016	$Eu_{0.2}Yb_{0.2}Ca_{0.6}Mg_2Bi_2$ [10]	3.5	215	0.92	1.3	875
2017	$Mg_{2.985}Ag_{0.015}Sb_2$ [38]	9	205	0.65	0.51	725
2017	$Mg_{3.2}Sb_{1.5}Bi_{0.49}Te_{0.01}$ [14]	6	−280	0.8	1.5	716
2017	$Mg_3Sb_{1.48}Bi_{0.48}Te_{0.04}$ [15]	10	−205	0.73	1.6	750
2017	$Mg_{3.05}Nb_{0.15}Sb_{1.5}Bi_{0.49}Te_{0.01}$ [39]	4.5	−277	0.84	1.57	700

charge carriers ($\sim 10^{15}$ cm^{-3}) of stoichiometric Mg_3Sb_2, significantly different from other ternary Zintls (e.g., 10^{19} cm^{-3} for $YbZn_2Sb_2$) [26, 27], make it possible for realizing n-type. However, it was overlooked for a long time; till recently, the surprising discovery of n-type thermoelectric performance with ZT of ~ 1.5 was made in $Mg_3Sb_{1.5}Bi_{0.5}$-based samples [14, 15].

8.3 p-Type 1-2-2 Zintl Phases

We would describe diverse approaches to enhancing thermoelectric performance of p-type 1-2-2 Zintl phases, including adjusting concentration of vacancies, optimizing concentration of charge carriers via doping, and improving performance through solid solution (alloying).

8.3.1 Enhanced Performance by Optimizing Concentration of Charge Carriers in p-Type 1-2-2 Zintl Phases

Zintl chemistry suggests that materials' properties may be adjusted by defects (e.g., vacancies). The reason for observed extremely high hole concentrations in most p-type ternary Zintl phases (e.g., AZn_2Sb_2, AMg_2Bi_2) could be resulted from vacancies on electron-donating A cation sites, as might be expected in most Zintl phases [26, 27, 40]. DFT calculation has also predicated that A sites are most energetically favorable for point defects and concentration of vacancies depends strongly on electronegativity of A [40].

In order to investigate the vacancy effect and optimize electrical properties, Zevalkink et al. selected $Yb_xZn_2Sb_2$ to study because Yb has the largest predicted vacancy concentration among AZn_2Sb_2 [26, 40]. Experimental Hall concentration of charge carriers can be effectively controlled by adjusting x when $x < 1$, while concentration of charge carriers remains almost constant when $x > 1$ (Fig. 8.4a). As a result, electron transport properties such as ρ and S were also adjusted accordingly. Peak ZT ~ 0.85 was obtained in the sample with $x = 0.99$. In other samples with higher Yb content ($x > 1$), average ZT improved even though peak ZT was not enhanced. The results indicate that varying concentration of vacancies

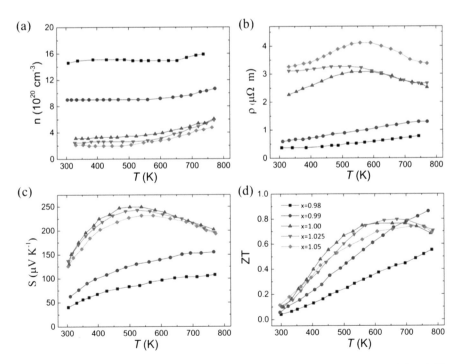

Fig. 8.4 Temperature dependences of (**a**) Hall concentration of charge carriers, (**b**) ρ, (**c**) S, and (**d**) ZT of $Yb_xZn_2Sb_2$ [26]

provides a potential route to controlling electron transport properties, hence the overall TE performance.

Another widely used strategy to control concentration of charge carriers is through introducing appropriate dopants. Generally, good thermoelectric performance is found in doped semiconductors, so it is important to be able to control concentration of charge carriers in Zintl compounds via doping. Pure Bi-based $CaMg_2Bi_{1.98}$ Zintl phase, which involves nontoxic, cheap, and abundant elements, has recently attracted some attention with competitive ZT ~ 0.9 at 873 K. The effect of Na doped into Ca sites on thermoelectric properties of $Ca_{1-x}Na_xMg_2Bi_{1.98}$ is studied due to relatively low concentration of charge carriers (in the order of 10^{18} cm^{-3}) in $CaMg_2Bi_{1.98}$ [37]. Figure 8.5a shows ρ as a function of temperature up to 873 K. Note, ρ of Na-doped samples is much smaller than that of undoped. Specifically, at room temperature, ρ of sample with $x = 0.0075$ is 15 times lower than that of undoped sample. Concentration of charge carriers has increased by more than one order of magnitude from 3.46×10^{18} cm^{-3} for $x = 0$ to 4.4×10^{19} cm^{-3} for $x = 0.0075$. For $x = 0.75$ at %, it is observed that resistivity exhibits a metallike behavior, indicating that doping level is high enough to suppress the intrinsic excitation. Figure 8.5b indicates doping of Na into Ca sites decreases in S, and

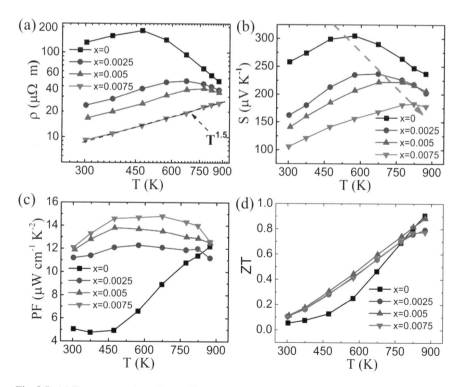

Fig. 8.5 (a) Temperature-dependent ρ, (b) S, (c) power factor, and (d) ZT of $Ca_{1-x}Na_xMg_2Bi_{1.98}$ ($x = 0, 0.0025, 0.005,$ and 0.0075). Reproduced with permission from references [37], Copyright at 2016 AIP Publishing

peak S shifts to higher temperature with more Na content, indicating that concentration of majority charge carriers suppresses the bipolar effect. With increasing Na content, power factor increases monotonically over entire temperature range, especially at lower temperatures (Fig. 8.5c). The figure-of-merit ZT vs. temperature and Na content x is plotted in Fig. 8.5d, showing better ZT values of Na-doped samples when temperature is below 750 K mainly due to improved power factor.

In contrast to ternary 1-2-2 Zintls discussed above, binary Mg_3Sb_2 sample exhibits poor electric transport properties with very low concentration of charge carriers (10^{15} cm^{-3}) and high ρ, which is far from optimization. Several dopants have been tried in order to optimize electron transport performance so far. Investigations include Zn-substituted $Mg_{3-x}Zn_xSb_2$ [41], Cd and Ag co-substituted $Mg_{3-x-y}Cd_xAg_ySb_2$ [42], Na-doped $Mg_{3-x}Na_xSb_2$ [35], Ag-substituted $Mg_{3-x}Ag_xSb_2$ [38], and substitution of Sb by Bi or Pb [25, 34]. Specifically, in Na-doped samples, concentration of charge carriers increased from 2.6×10^{15} cm^{-3} for $x = 0$ to 1.7×10^{20} cm^{-3} for $x = 0.025$. As a result, electrical transport properties of $Mg_{3-x}Na_xSb_2$ were enhanced by dramatically decreasing in ρ, optimizing S, and increasing in power factor [35].

8.3.2 Enhanced Performance Through Formation of Solid Solution

Although thermal conductivity in pristine Zintl family is intrinsically low, there is still room for further reduction. As shown in Table 8.1, better performance of $CaAl_2Si_2$-type Zintl phases with ZT above unity is directly related to low κ_L caused by strong phonon scattering in alloys or solid solutions.

For example, Fig. 8.6a shows experimental and calculated results of κ_L for $Ca_xYb_{1-x}Mg_2Bi_2$ alloys [23]. At room temperature, κ_L in alloys are nearly half of that of the end compounds, mainly due to mass and size differences. At higher temperature, contribution from alloying effect gets smaller compared to low temperature because Umklapp scattering becomes more and more efficient. At 873 K, ZT reaches maximum of ~1 for $Yb_{1-x}Ca_xMg_2Bi_2$ with $x = 0.5$ (Fig. 8.6b), owing to lower κ_L. The same is also observed in $Yb_{1-x}Ca_xZn_2Bi_2$ and other alloys [9].

Besides double element atoms in A site of AMg_2Bi_2, further enhanced performance by using triple element atoms in A site to form $(Eu_{0.5}Yb_{0.5})_{1-x}Ca_xMg_2Bi_2$ ($x = 0.4$, 0.5, 0.6, and 0.7) alloys has been reported [10]. It has been found that electron transport properties of $EuMg_2Bi_2$ and $Eu_{0.5}Yb_{0.5}Mg_2Bi_2$ are similar, while $Eu_{0.5}Yb_{0.5}Mg_2Bi_2$ has lower κ_L. Therefore, Ca is introduced to substitute some of Eu and Yb, while the ratio of Eu and Yb is kept the same. Figure 8.7a, b shows temperature-dependent electron transport properties. With increasing Ca content, ρ increases gradually, and S decreases because of reduction in concentration of charge carriers. Figure 8.7c shows temperature dependence of total thermal conductivity κ for all $(Eu_{0.5}Yb_{0.5})_{1-x}Ca_xMg_2Bi_2$ samples. The lowest value of κ is observed in

Fig. 8.6 (**a**) Dependence of κ_L on alloy composition x in $Yb_{1-x}Ca_xMg_2Bi_2$ at 300, 575, and 875 K. Lines show predicted reduction in κ_L based on mass contrast between Yb and Ca for respective temperatures with experimental results showing by dots. (**b**) ZT temperature dependences of $Yb_{1-x}Ca_xMg_2Bi_2$ ($x = 0$, 0.3, 0.5, 0.7, and 1) [23]

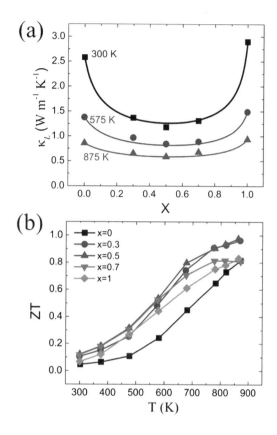

sample with $x = 0.6$. Specifically, at room temperature, κ of optimized $Eu_{0.2}Yb_{0.2}Ca_{0.6}Mg_2Bi_2$ sample reduces dramatically to ~1.5 W × m^{-1} × K^{-1}, half of that for three base compounds AMg_2Bi_2 (A = Eu, Yb, Ca). Combining electron and thermal transport properties, record peak ZT ~ 1.3 in p-type $Eu_{0.2}Yb_{0.2}Ca_{0.6}Mg_2Bi_2$ is achieved mainly due to the further reduction in thermal conductivity (Fig. 8.7d).

8.4 n-Type 1-2-2 Zintl Phase

8.4.1 Mg_3Sb_2 Zintl Compounds

Mg_3Sb_2-based material has long been regarded as persistent p-type compound [24, 34, 43–45]. Recently, n-type Mg_3Sb_2 is first reported in Mn-doped single-crystal specimen [46]. Later on, Te-doped n-type $Mg_3Sb_{1.5}Bi_{0.5}$-based materials with high thermoelectric performance are reported independently by Tamaki et al. [14] and Zhang et al. [47]. The crossover from p-type to n-type in Mg_3Sb_2 is a quite

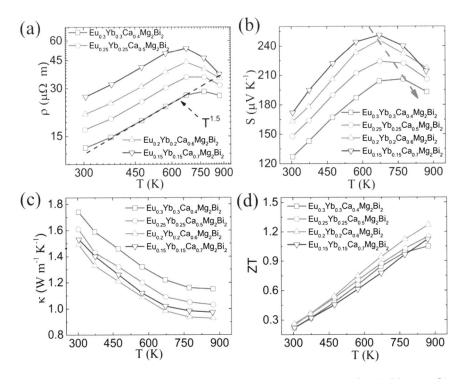

Fig. 8.7 Temperature-dependent thermoelectric properties of $(Eu_{0.5}Yb_{0.5})_{1-x}Ca_xMg_2Bi_2$: (**a**) ρ, (**b**) S, (**c**) κ, (**d**) ZT. Reproduced with permission from reference [10], Copyright at 2016 National Academy of Science

interesting topic that could shed some new lights on other Zintl phases. According to defect formation energy calculation (as shown in Fig. 8.8a, b) [14, 48], Mg vacancies have the lowest formation energies around Fermi level among all of defects (as shown in Fig. 8.8a). Therefore, Mg_3Sb_2 compound will have intrinsically high concentration of Mg vacancies. Acceptor level like Mg vacancy pins Fermi level to valence band, hence explaining persistent p-type conduction of pristine Mg_3Sb_2. After introducing extra Mg, defect formation energy of Mg vacancies will be markedly raised (as shown in Fig. 8.8b). In other words, extra Mg in Mg_3Sb_2-based materials is conducive and also critical for reducing concentration of Mg vacancies, thus enabling n-type conduction [14, 48]. The importance of extra Mg was also confirmed by studies from different groups [48–52]. As shown in Fig. 8.8c, by tuning extra Mg content in the range of $0 < x < 0.2$, S crossovers from p-type to n-type, where $x = 0.025$ is critical point. Correspondingly, Hall concentration of charge carriers demonstrates also similar transition from p-type to n-type (as shown in Fig. 8.8d) [53]. The results indicate that amount of extra Mg plays a vital role in tuning concentration of charge carriers, similar to the case of n-type Mg_2Sn-based materials [54, 55]. It is noteworthy that extra Mg was first assumed to occupy the interstitial site [14]; however, neutron diffraction characterization on

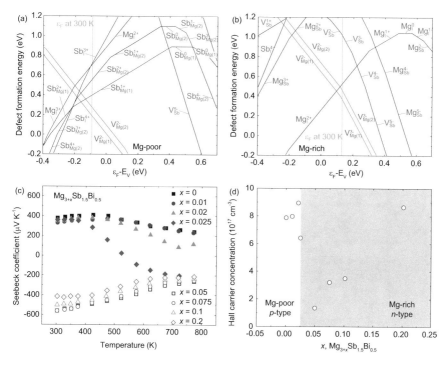

Fig. 8.8 Defect formation energy in Mg_3Sb_2-based materials at (**a**) Mg-poor condition and (**b**) Mg-rich condition, (**c**) temperature-dependent S of $Mg_{3+x}Sb_{1.5}Bi_{0.5}$ ($x = 0, 0.01, 0.02, 0.025, 0.05, 0.075, 0.1,$ and 0.2), and (**d**) composition-dependent Hall concentration of charge carriers in $Mg_{3+x}Sb_{1.5}Bi_{0.5}$. Reproduced with permission from reference [14, 53]. Copyright at 2016 John Wiley & Sons, Inc. and 2018 American Chemical Society

$Mg_{3.2}Sb_{1.5}Bi_{0.49}Te_{0.01}$ indicates that Mg occupancy at proposed interstitial site tends to zero [50]. Although powder X-ray diffraction did not reveal any impurity phases, however, spherical aberration-corrected (C_S-corrected) high-angle annular dark-field scanning transmission electron microscopy (HAADF-STEM) technique indicates that extra Mg precipitates as microscale secondary phases. Since Mg impurity phases will inevitably enhance thermal conductivity of the host material, slightly reducing in excess Mg will be beneficial for enhancing thermoelectric performance but still maintaining n-type conduction [53, 56].

It is worth noting that Mg vacancy not only determines n- or p-type conduction but plays also pivotal role in thermoelectric properties. The ionized impurity scattering dominated at lower temperature range in $Mg_{3.2}Sb_{1.5}Bi_{0.49}Te_{0.01}$ was noted from temperature-dependent electrical conductivity σ and Hall mobility [14, 49–52]. As shown in Fig. 8.8a, b, Mg vacancy will easily become negatively charged in n-type Mg_3Sb_2-based materials in both Mg-poor and Mg-rich conditions. The most stable charging Mg vacancy states are -2 and -3 [14]. Ionization of Mg vacancy should be mainly ascribed to unsaturated bonding capability of vacancy that facilitates the charge transfer between host and defect [57]. Therefore, ionized impurity scattering

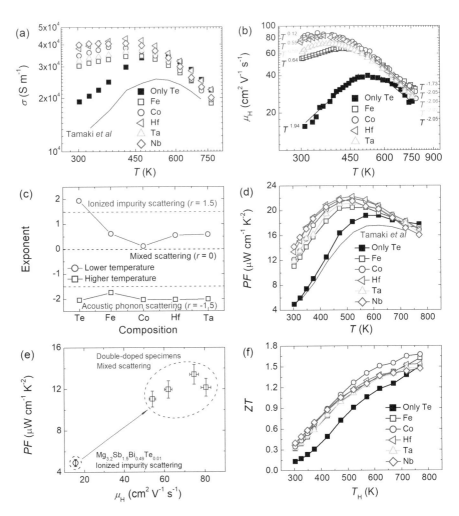

Fig. 8.9 Temperature dependence of (**a**) σ and (**b**) Hall mobility, (**c**) temperature exponent of Hall mobility, (**d**) temperature dependence of power factor, (**e**) Hall mobility-dependent power factor, and (**f**) temperature dependence of ZT of $Mg_{3.1}A_{0.1}Sb_{1.5}Bi_{0.49}Te_{0.01}$ (A = Fe, Co, Hf, Ta, and Nb). Reproduced with permission from references [49, 50]. Copyright at 2017 The Royal Society of Chemistry and National Academy of Sciences

is very likely due to the presence of charged Mg vacancy [50, 52]. Transition metal element atoms (e.g., Fe, Co, Hf, Ta, and Nb) doping at Mg site of $Mg_{3.2}Sb_{1.5}Bi_{0.49}Te_{0.01}$ can shift ionized impurity scattering to mixed scattering of ionized impurities and acoustic phonons [49, 50], as shown in Fig. 8.9a–c. Noticeable enhancements in room temperature Hall mobility (from ~20 to ~80 $cm^2 \times V^{-1} \times s^{-1}$) and σ (from ~1.9 $\times$ 10^4 to ~4 $\times$ 10^4 S/m) are observed, therefore, improving substantially the power factor. Room temperature power factor of $Mg_{3.2}Sb_{1.5}Bi_{0.49}Te_{0.01}$ is ~4 $\mu W \times cm^{-1} \times K^{-2}$ and ~14 $\mu W \times cm^{-1} \times K^{-2}$ for $Mg_{3.1}Nb_{0.1}Sb_{1.5}Bi_{0.49}Te_{0.01}$, an increase of ~250% (Fig. 8.9d). Such power factor enhancement can be mainly attributed to improved Hall mobility due to variation in scattering mechanism of charge carriers (Fig. 8.9e). Benefiting from additional reduction in thermal conductivities, substantial improvements in ZT can be achieved (Fig. 8.9f). Crossover from ionized impurity scattering to mixed scattering can be explained by the reason that transition metal element atoms will preferentially dope into vacant Mg site thus reducing concentration of Mg vacancies [50]. It indicates that thermoelectric properties of Mg_3Sb_2-based materials can be manipulated by controlling concentration of Mg vacancies. Since point defects are highly sensitive to preparation conditions, it is feasible to control concentration of Mg vacancies by directly tuning processing details. Indeed, similar transition from ionized impurity scattering to mixed scattering can also be realized by reducing concentration of Mg vacancies via simply increasing in hot-pressing temperature and elongating holding time [52].

It is noteworthy that similar enhancement of thermoelectric performance is also obtained by Kanno et al. [58] by the same approach. However, they propose that it is reduced grain boundary scattering due to enlarged grain size that is responsible for ZT enhancement instead of reduced ionized impurity scattering. Effect of grain boundaries on thermoelectric properties of Mg_3Sb_2-based materials is modelled by Kuo et al. [59], where grain boundary region is considered as a separate phase with band offset. However, grain boundary phase is regarded to be dominated by ionized impurity scattering in its model. The samples with and without doping by transition metal element atoms are prepared by the same procedure [49, 50]; therefore, grain size should be comparable, but thermoelectric properties are distinctly different. In addition, even when doping concentration of transition metal element atoms is rather low (which should not be able to modify the trap states on grain boundary), variation in thermoelectric properties is still noticeable [49]. Therefore, further study on this topic is necessary.

8.4.2   Theoretical Investigation on n-Type 1-2-2 Zintl Compounds

The layered $CaAl_2Si_2$-type Zintl compounds AB_2X_2, which have the same structure type as Mg_3Sb_2, have been extensively studied as p-type thermoelectric materials [18, 21, 28, 60–73]. A site is alkaline-earth or rare-earth element atom, B is usually element from main group or transition metal atom, and X is element from group 14 or 15 atom. To date, all synthesized specimens for AB_2X_2 compounds are p-type due to cation vacancies [67]. Theoretically, optimization of power factor of p-type AB_2X_2 Zintl compounds has been rationalized using crystal field scheme. The rational stems from splitting of degenerate p orbitals of X anions that dominate the valence band edge. Smaller splitting energies yield larger orbital convergence, therefore, higher band degeneracy, which is favorable for improving power factor [69]. Although n-type AB_2X_2 compounds have not been experimentally realized yet, it is meaningful to probe electronic structures of these compounds via the first-principles calculation. According to recent calculations [69, 70], valence band edge is dominated by

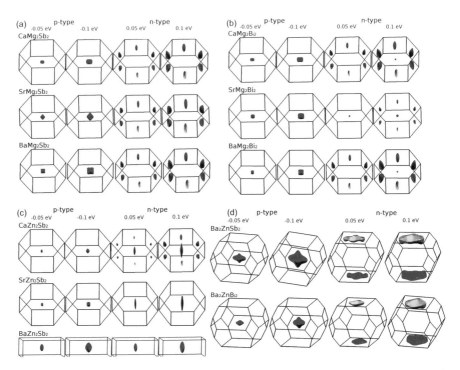

Fig. 8.10 Calculated constant energy surfaces at 0.05 eV and 0.1 eV below VBM (*p*-type) and above CBM (*n*-type) for (**a**) $[Mg_2Sb_2]^{2-}$, (**b**) $[Mg_2Bi_2]^{2-}$, (**c**) $[Zn_2Sb_2]^{2-}$, and (**d**) 212 phases. Reproduced with permission from reference [70]. Copyright at 2017 The Royal Society of Chemistry

p orbitals of *X* anions at Γ point, while CBM is mainly from hybridization between cation *d* states and *s* states from *B* site. Importantly, CBM is shown to have larger band degeneracy than VBM, as depicted in iso-energy surfaces of AMg_2P_2 (A = Ca, Sr, Ba; P = Sb, Bi) and $CaZn_2Sb_2$ in Fig. 8.10 [70]. One can find six-half pockets at zone boundary in *n*-type materials, while *p*-type counterparts have only one pocket of charge carriers at zone center.

Additionally, there is clear charge carriers pocket anisotropy in *n*-type compounds. This is important to realize high σ and S simultaneously [74]. Effects of charge carriers pocket anisotropy on thermoelectric properties can be better understood through electronic fitness function $\left(\text{EFF} = \frac{(\sigma/\tau)S^2}{N^{2/3}}\right)$ that describes electronic aspect of thermoelectric performance [70, 75]. This function finds materials that overcome inverse relationship between σ and S based on complexity of electronic structures regardless of specific origin. As shown in Fig. 8.11a, several *n*-type AB_2X_2 phases have larger EFF values than those of *p*-type counterparts, at least for basal plane transport. Combining anisotropic charge carriers pockets with relatively high band degeneracy, some of *n*-type AB_2X_2 phases show electronic structures that are favorable for thermoelectric performance, even comparable with *n*-type Mg_3Sb_2 (Fig. 8.11b).

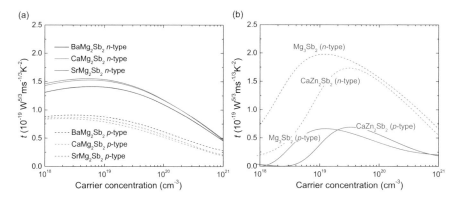

Fig. 8.11 (**a**) EFF for *p*- and *n*-type AMg$_3$Sb$_2$ (A = Ca, Sr, Ba). (**b**) Comparison of EFF between CaZn$_2$Sb$_2$ and Mg$_3$Sb$_2$. Reproduced with permission from reference [70]. Copyright at 2017 The Royal Society of Chemistry

8.5   Conclusions

Research efforts by several groups have unequivocally demonstrated that CaAl$_2$Si$_2$-type Zintl phases are promising thermoelectric materials. In this chapter, thermoelectric properties of 1-2-2 layered Zintl phases were discussed, and strategies such as tuning concentration of charge carriers and formation of solid solution for improving thermoelectric properties of Zintl materials were summarized. In addition, thermoelectric properties of *n*-type Mg$_3$Sb$_2$-based materials were also discussed, and the relationship between Mg vacancies, charge carriers scattering mechanism, and *n*-type or *p*-type conduction was identified. Finally, theoretical calculation on the thermoelectric properties of other *n*-type 1-2-2 Zintl materials was also presented. There are good reasons to believe that the wide variety of compositions in this same type structure makes it possible for further manipulating the structures and optimizing thermoelectric properties.

References

1. E. Zintl, W. Dullenkopf, Z. Phys. Chem. **16B**, 183–194 (1932)
2. E. Zintl, A. Woltersdorf, Ber. Der Bunsenges. Phys. Chem. Chem. Phys. **41**, 876–879 (1935)
3. E. Zintl, Angew. Chem. Int. Ed. **52**, 1–6 (1939)
4. H. Schäfer, B. Eisenmann, W. Müller, Angew. Chem. Int. Ed. **12**, 694–712 (1973)
5. H. Schäfer, B. Eisenmann, W. Müller, Angew. Chem. Int. Ed. **85**, 742–760 (1973)
6. H. Schäfer, Annu. Rev. Mater. Res. **15**, 1–42 (1985)
7. F. Laves, Naturwissenschaften **29**, 244–255 (1941)
8. S.M. Kauzlarich, S.R. Brown, G.J. Snyder, Dalton Trans. **21**, 2099–2107 (2007)
9. F. Gascoin, S. Ottensmann, D. Stark, S.M. Haïle, G.J. Snyder, Adv. Funct. Mater. **15**, 1860–1864 (2005)
10. J. Shuai, H. Geng, Y. Lan, Z. Zhu, C. Wang, Z. Liu, J. Bao, C.W. Chu, J. Sui, Z. Ren, Proc. Natl. Acad. Sci. U. S. A. **113**, E4125–E4132 (2016)

11. E.S. Toberer, C.A. Cox, S.R. Brown, T. Ikeda, A.F. May, S.M. Kauzlarich, G.J. Snyder, Adv. Funct. Mater. **18**, 2795–2800 (2008)
12. G. Rogl, A. Grytsiv, P. Rogl, E. Bauer, M. Zehetbauer, Intermetallics **19**, 546–555 (2011)
13. C. Fu, S. Bai, Y. Liu, Y. Tang, L. Chen, X. Zhao, T. Zhu, Nat. Commun. **6**, 8144 (2015)
14. H. Tamaki, H.K. Sato, T. Kanno, Adv. Mater. **28**, 10182–10187 (2016)
15. J. Zhang, L. Song, S.H. Pedersen, Y. Hao, T.H. Le, B.I. Bo, Nat. Commun. **8**, 13901 (2017)
16. C. Zheng, R. Hoffmann, R. Nesper, H.G.V. Schnering, J. Am. Chem.l Soc. **108**, 1876–1884 (1986)
17. J.K. Burdett, G.J. Miller, Chem. Mater. **2**, 12–26 (1990)
18. E.S. Toberer, A.F. May, B.C. Melot, E. Flage-Larsen, G.J. Snyder, Dalton Trans. **39**, 1046–1054 (2010)
19. J. Zhang, L. Song, G.K.H. Madsen, K.F.F. Fischer, W. Zhang, S. Xun, B.I. Bo, Nat. Commun. **7**, 10892 (2016)
20. J. Shuai, J. Mao, S. Song, Q. Zhang, G. Chen, Z. Ren, Mater. Today Phys. **1**, 74–95 (2017)
21. X.J. Wang, M.B. Tang, H.H. Chen, X.X. Yang, J.T. Zhao, U. Burkhardt, Y. Grin, Appl. Phys. Lett. **94**, 092106 (2009)
22. H. Zhang, M. Baitinger, M.B. Tang, Z.Y. Man, H.H. Chen, X.X. Yang, Y. Liu, L. Chen, Y. Grin, J.T. Zhao, Dalton Trans. **39**, 1101–1104 (2010)
23. J. Shuai, Z. Liu, H.S. Kim, Y. Wang, J. Mao, R. He, J. Sui, Z. Ren, J. Mater.Chem. A **4**, 4312–4320 (2016)
24. C.L. Condron, S.M. Kauzlarich, F. Gascoin, G.J. Snyder, J. Solid State Chem. **179**, 2252–2257 (2006)
25. A. Bhardwaj, D.K. Misra, RSC Adv. **4**, 34552–34560 (2014)
26. A. Zevalkink, W.G. Zeier, E. Cheng, J. Snyder, J.P. Fleurial, S. Bux, Chem. Mater.s **26**, 5710–5717 (2014)
27. J. Shuai, Y. Wang, Z. Liu, H.S. Kim, J. Mao, J. Sui, Z. Ren, Nano Energy **25**, 136–144 (2016)
28. X.J. Wang, M.B. Tang, J.T. Zhao, H.H. Chen, X.X. Yang, Appl. Phys. Lett. **90**, 232107 (2007)
29. C. Yu, T.J. Zhu, S.N. Zhang, X.B. Zhao, J. He, Z. Su, T.M. Tritt, J. Appl. Phys. **104**, 013705 (2008)
30. H. Zhang, J.T. Zhao, Y. Grin, X.J. Wang, M.B. Tang, Z.Y. Man, H.H. Chen, X.X. Yang, J. Chem. Phys. **129**, 164713 (2008)
31. Q.G. Cao, H. Zhang, M.B. Tang, H.H. Chen, J. Chem. Phys. **107**, 053714 (2010)
32. K. Guo, Q.G. Cao, X.J. Feng, M.B. Tang, H.H. Chen, X. Guo, L. Chen, Y. Grin, J.T. Zhao, Eur. J. Inorg. Chem. **2011**, 4043–4048 (2011)
33. A.F. May, M.A. Mcguire, J. Ma, O. Delaire, A. Huq, D.J. Singh, W. Cai, H. Wang, Phys. Rev. B Cond. Matter. **85**, 035202 (2012)
34. A. Bhardwaj, A. Rajput, A.K. Shukla, J.J. Pulikkotil, A.K. Srivastava, A. Dhar, G. Gupta, S. Auluck, D.K. Misra, R.C. Budhani, RSC Adv. **3**, 8504–8516 (2013)
35. J. Shuai, Y. Wang, H.S. Kim, Z. Liu, J. Sun, S. Chen, J. Sui, Z. Ren, Acta Mater. **93**, 187–193 (2015)
36. Q. Cao, J. Zheng, K. Zhang, G. Ma, J. Alloy Compd. **680**, 278–282 (2016)
37. J. Shuai, H.S. Kim, Z. Liu, R. He, J. Sui, Z. Ren, Appl. Phys. Lett. **108**, 183901 (2016)
38. L. Song, J. Zhang, B.I. Bo, J. Mater. Chem. A **5** (2017)
39. J. Shuai, J. Mao, S. Song, Q. Zhu, J. Sun, Y. Wang, R. He, J. Zhou, G. Chen, D.J. Singh, Energy Environ. Sci. **10**, 799–807 (2017)
40. G.S. Pomrehn, A. Zevalkink, W.G. Zeier, A. Walle, G.J. Snyder, Angew. Chem. Int. Ed. **126**, 3490–3494 (2014)
41. H.X. Xin, J.H. Jia, C.J. Song, K.X. Zhang, J. Zhang, X.Y. Qin, J. Phys. D. Appl. Phys. **42**, 165403 (2009)
42. K.X. Zhang, X.Y. Qin, H.X. Xin, H.J. Li, J. Zhang, J. Alloy Compd. **484**, 498–504 (2009)
43. F. Ahmadpour, T. Kolodiazhnyi, Y. Mozharivskyj, J. Solid State Chem. **180**, 2420–2428 (2007)
44. J. Shuai, H.S. Kim, Y. Lan, S. Chen, Y. Liu, H. Zhao, J. Sui, Z. Ren, Nano Energy **11**, 640–646 (2015)
45. Z. Ren, J. Shuai, J. Mao, Q. Zhu, S. Song, Y. Ni, S. Chen, Acta Mater. **143**, 265–271 (2018)

46. S. Kim, C. Kim, Y.-K. Hong, T. Onimaru, K. Suekuni, T. Takabatake, M.H. Jung, J. Mater. Chem. A **2**, 12311–12316 (2014)
47. J. Zhang, L. Song, S.H. Pedersen, H. Yin, L.T. Hung, B.B. Iversen, Nat. Commun. **8**, 13901 (2017)
48. S. Ohno, K. Imasato, S. Anand, H. Tamaki, S.D. Kang, P. Gorai, H.K. Sato, E.S. Toberer, T. Kanno, G.J. Snyder, Joule **2**, 141–154 (2018)
49. J. Shuai, J. Mao, S.W. Song, Q. Zhu, J.F. Sun, Y.M. Wang, R. He, J.W. Zhou, G. Chen, D.J. Singh, Z.F. Ren, Energy Environ. Sci. **10**, 799–807 (2017)
50. J. Mao, J. Shuai, S.W. Song, Y.X. Wu, J.W. Zhou, Z.H. Liu, J.F. Sun, Y.Z. Pei, Q.Y. Zhang, D.J. Singh, G. Chen, Z.F. Ren, Proc. Natl. Acad. Sci. U. S. A. **114**, 10548–10553 (2017)
51. J. Mao, Y.X. Wu, S.W. Song, J. Shuai, Z.H. Liu, Y.Z. Pei, Z.F. Ren, Mater. Today Phys. **3**, 1–6 (2017)
52. J. Mao, Y. Wu, S. Song, Q. Zhu, J. Shuai, Z. Liu, Y. Pei, Z. Ren, ACS Energy Lett. **2**, 2245–2250 (2017)
53. J. Shuai, B. Ge, J. Mao, S. Song, Y. Wang, Z. Ren, J. Am. Chem. Soc. **140**, 1910–1915 (2018)
54. W. Liu, X. Tang, H. Li, J. Sharp, X. Zhou, C. Uher, Chem. Mater. **23**, 5256–5263 (2011)
55. Z. Du, T. Zhu, Y. Chen, J. He, H. Gao, G. Jiang, T.M. Tritt, X. Zhao, J. Mater. Chem. **22**, 6838–6844 (2012)
56. K. Imasato, S. Ohno, S.D. Kang, G.J. Snyder, APL Mater. **6**, 016106 (2018)
57. E.G. Seebauer, M.C. Kratzer, Mater. Sci. Eng. R. Rep. **55**, 57–149 (2006)
58. T. Kanno, H. Tamaki, H.K. Sato, S.D. Kang, S. Ohno, K. Imasato, J.J. Kuo, G.J. Snyder, Y. Miyazaki, Appl. Phys. Lett. **112**, 033903 (2018)
59. J.J. Kuo, S.D. Kang, K. Imasato, H. Tamaki, S. Ohno, T. Kanno, G.J. Snyder, Energy Environ. Sci. **11**, 429–434 (2018)
60. F. Gascoin, S. Ottensmann, D. Stark, S.M. Haile, G.J. Snyder, Adv. Funct. Mater. **15**, 1860–1864 (2005)
61. H. Zhang, J.-T. Zhao, Y. Grin, X.-J. Wang, M.B. Tang, Z.Y. Man, H.H. Chen, X.X. Yang, J. Phys. Chem. **129**, 164713–164713 (2008)
62. E. Flage-Larsen, S. Diplas, Ø. Prytzr, E.S. Toberer, A.F. May, Phys. Rev. B **81**, 205204 (2010)
63. A.F. May, M.A. McGuire, D.J. Singh, R. Custelcean, G.E. Jellison Jr., Inorg. Chem. **50**, 11127–11133 (2011)
64. A.F. May, M.A. McGuire, J. Ma, O. Delaire, A. Huq, R. Custelcean, J. Appl. Phys. **111**, 033708 (2012)
65. A.F. May, M.A. McGuire, D.J. Singh, J. Ma, O. Delaire, A. Huq, W. Cai, H. Wang, Phys. Rev. B **85**, 035202 (2012)
66. D.J. Singh, D. Parker, J. Appl. Phys. **114**, 143703 (2013)
67. G.S. Pomrehn, A. Zevalkink, W.G. Zeier, A. van de Walle, G.J. Snyder, Angew. Chem. Int. Ed. **53**, 3422–3426 (2014)
68. A. Zevalkink, W.G. Zeier, E. Cheng, J. Snyder, J.-P. Fleurial, S. Bux, Chem. Mater. **26**, 5710–5717 (2014)
69. J.W. Zhang, L.R. Song, G.K.H. Madsen, K.F.F. Fischer, W.Q. Zhang, X. Shi, B.B. Iversen, Nat. Commun. **7**, 10892 (2016)
70. J. Sun, D.J. Singh, J. Mater. Chem. A **5**, 8499–8509 (2017)
71. J. Sun, J. Shuai, Z. Ren, D.J. Singh, Mater. Today Phys. **2**, 40–45 (2017)
72. X. Wang, W. Li, C. Wang, J. Lin, X. Zhang, B. Zhou, Y. Chen, Y. Pei, J. Mater. Chem. A **5**, 24185–24192 (2017)
73. Y. Takagiwa, Y. Sato, A. Zevalkink, I. Kanazawa, K. Kimura, Y. Isoda, Y. Shinohara, J. Alloy Compd. **703**, 73–79 (2017)
74. D.S. Parker, A.F. May, D.J. Singh, Phys. Rev. Appl. **3**, 064003 (2015)
75. G. Xing, J. Sun, Y. Li, X. Fan, W. Zheng, D.J. Singh, Phys. Rev. Mater. **1**, 065405 (2017)

Chapter 9
Skutterudites: Progress and Challenges

Gerda Rogl and Peter Rogl

Abstract Since the discovery of skutterudite (an arsenide mineral (Co, Fe, Ni)As$_3$) in 1845 in Norway, valuable research on isotypic Sb-based skutterudites achieved high-quality thermoelectric (TE) *p*- and *n*-type leg materials with remarkably high figure-of-merit ZT, forming the basis for large-scale production and application in thermoelectric power generation coupled with waste heat recovery.

We will cover the most important steps within these almost 200 years and discuss the development of TE skutterudites and the enhancement of those figures-of-merit. Besides unfilled CoSb$_3$-based grades, we will focus on filled *p*- and *n*-type skutterudites (La$_{1-x}$Fe$_4$Sb$_{12}$), to show improvement via tuning of the electronic band structure, to consider phonon engineering in order to reduce thermal conductivity, to discuss the nanoeffect (ball milling, high-energy ball milling, severe plastic deformation), and to highlight the importance of density and impact of nanoprecipitates. Issues concerning the stability of skutterudites and an overview of mechanical properties including thermal expansion will follow. Although laboratory records in the figure-of-merit ZT equal to 1.5 for *p*-type and 1.9 for *n*-type skutterudites were reached, large-scale production (about 50 kg batches of powders) has already achieved ZT values of 1.3 (*p*-type) and 1.5 (*n*-type) with thermoelectric conversion leg efficiencies of 13.4 and 14.5%. With these attractive values, one can conceive TE skutterudite materials being at the brink of a breakthrough in technological applications for thermoelectric power generation, given a large-scale thermoelectric (TE)-generator module producer.

9.1 Introduction

Thermoelectric (TE) materials, i.e., *n*- and *p*-type heavily doped semiconductors, have the ability to convert a heat flow from an external source (thermal radiation energy flow) into a flow of charge carriers within the material (i.e., electric current

G. Rogl (✉) · P. Rogl
Christian Doppler Laboratory for Thermoelectricity, Institute of Materials Chemistry, University of Vienna, Wien, Austria

Institute of Solid State Physics, TU Wien, Wien, Austria
e-mail: gerda.rogl@univie.ac.at

© Springer Nature Switzerland AG 2019
S. Skipidarov, M. Nikitin (eds.), *Novel Thermoelectric Materials and Device Design Concepts*, https://doi.org/10.1007/978-3-030-12057-3_9

and, hence, electric power). Many classes of materials like tellurides, silicides, copper selenides, oxides, zinc tellurides, antimonides, Zintl phases, clathrates, half-Heusler alloys, or skutterudites qualify as TE materials.

The most important features in favor to use skutterudites as TE materials are as follows:

1. Starting and raw materials are abundant and cheap in comparison to other TE materials.
2. Wide temperature range of operation (from room temperature to 900 K).
3. High TE quality.
4. Long-term stability of thermoelectric performance.
5. Reasonably good mechanical properties.

The quality of TE material depends usually on two properties. The first one is the figure-of-merit, $ZT = S^2 \sigma T/\kappa$, where S, σ, and κ are Seebeck coefficient, electrical conductivity, and thermal conductivity, respectively, κ consisting of electron κ_{el} and lattice κ_L parts. Therefore, a good TE material should have a large absolute value $|S|$, high σ, but low κ. The second property is the conversion efficiency η (for electric power generation):

$$\eta = \frac{T_h - T_c}{T_h} \times \frac{\sqrt{1 + ZT_{avg}} - 1}{\sqrt{1 + ZT_{avg}} + \frac{T_c}{T_h}}, \qquad (9.1)$$

where T_h and T_c are temperatures on hot and cold side and ZT_{avg} is the average ZT value in the temperature gradient between T_c and T_h. ZT_{avg} has to be evaluated integrating all measured data, as the curvature of ZT vs. T curve determines the result. Kim et al. [1] use a similar approach with detailed calculation, resulting in ZT_{eng} instead.

In various review articles [2–13], TE materials and/or skutterudites have been widely discussed. In this review we want to guide the reader through the development of skutterudites and those high ZT and efficiencies. We will show how easily skutterudites can be produced in the laboratory, as well as, in large quantities, discuss the problem of long-term stability and production of TE modules and generators. All that will be done step by step.

9.2   A Short Review of the History of Skutterudite Thermoelectrics

In 1828 Oftedal [14] identified the crystal structure of a new mineral, MAs_3 ($M =$ Co or Ni), which was found in Norway near the small town of Skutterud. He identified the crystal structure as body-centered cubic, containing 32 atoms per unit cell, belonging to the space group Im-3 No. 204. The general formula of binary skutterudites is MX_3 where M is a transition metal and X a pnictogen atom (P, As, Sb). The unit

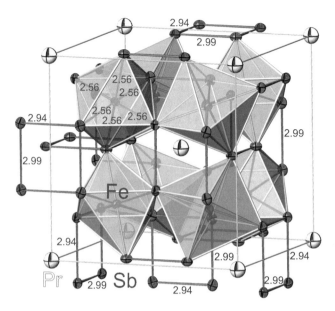

Fig. 9.1 Unit cell (in light blue) of filled skutterudite PrFeSb$_{12}$, highlighting Fe-centered, tilted Sb octahedra and red Sb rectangles. Numbers are interatomic distances in Å

cell is built of eight tilted but corner-connected octahedra formed by pnictogen atoms (Fig. 9.1). The transition metal atoms occupy 8c sites ($\frac{1}{4}$, $\frac{1}{4}$, $\frac{1}{4}$) at octahedral centers. Six of the eight sub-cubes of the unit cell are filled with planar rings of pnictogens, occupying 24g sites (0, y, z). Two icosahedral voids at 2a sites ((0, 0, 0) and ($\frac{1}{2}$, $\frac{1}{2}$, $\frac{1}{2}$)) can be stuffed with filler atoms, which should have ionic radii smaller than the cage volume in order to be loosely bound developing so-called rattling vibrations (for further details see below). In the 1970s Kjekshus et al. [15] showed that pnictogen atoms in skutterudites do not fulfill Oftedal relation for 24g sites: 2($y + z$) = 1. Therefore, he replaced the square pnictogen rings by rectangular ones.

In parallel to structure studies, already in the early 1950s of the twentieth century, Dudkin and Abrikosov [16] found that CoSb$_3$ had high electrical resistivity ρ combined with very high thermal conductivity κ. In addition, they studied the influence of Ni as dopant on ρ and hardness, both properties decreasing with increasing in Ni content. Also, other groups investigated the influence of dopants (Ni, Fe, Cu, Al, Ti, Zr), but all these attempts failed to produce useful TE materials because even though Fe and Ni could reduce κ, it still remained very high. Because of this high κ, ZT of CoSb$_3$ remained below 0.2 in the following years [17–19]. Furthermore, TE properties were found to change drastically dependent on the preparation method, and even a change from p- to n-type can occur as a function of temperature. Consequently, a breakthrough for CoSb$_3$ was not possible so far.

However, a first breakthrough in the development of skutterudites as TE materials arose with the appearance of filled skutterudites $E_yM_4X_{12}$ in 1977 [20], where E can be any electropositive element (lanthanoid, actinoid, alkaline, or alkaline earth) partially or fully occupying the voids in $2a$ sites. Fig. 9.1 reveals details on interatomic distances in $PrFeSb_{12}$. It should be mentioned that the transition metal atom exhibits uniform bonds to the vertices of non-regular Sb octahedra.

Concerning the ideal fillers, theory and experiment differ. Theoretically, optimal void fillers would be neutral atoms (e.g., Xe) with large atomic displacement parameters that would ensure enhanced scattering of heat-carrying phonons, thereby inferring a large reduction in the lattice thermal conductivity κ_L but only a minimum perturbation of electron transport properties. However, in praxis, so far, it was not possible to keep such rare gas species locked in the skutterudite cage [3].

Various review papers, e.g., Uher [3] and Rull-Bravo et al. [6], provide a detailed overview of the development of filled skutterudites; therefore, here only some benchmarks will be discussed.

In 1994 Slack and his group [21, 22] proposed the "phonon-glass electron-crystal" (PGEC) concept for a compound thermally conducting like a glass (low κ, low mobility, low effective mass) and at the same time conducting charge like an electron-crystal (high charge mobility, high S and high σ). For good TE materials, such a combination can indeed be found in filled skutterudites, as the rattling motion of the fillers interferes with the normal modes of the structure, reducing κ_L. Due to weak coupling of the filler with the rest of the structure and the fact that pnictogen orbitals are hardly affected by the rattling ions, the mobility of charge carriers remains large, and, therefore, the structure can keep its crystalline character and good electronic properties. The smaller and heavier the ion in the void is, the higher is atom disorder, and the lower is the phonon part of the thermal conductivity. This "PGEC model" can be regarded as a benchmark in the development of filled skutterudites as TE materials.

At about the same time, in year 1993, Hicks and Dresselhaus [23] showed theoretically that by using two-, one-, or even zero-dimensional structures, it is possible to decouple TE parameters, S, σ, and κ. Venkatasubramanian [24] and Chen [25] later backed up these ideas, showing that a significant reduction in κ_L is possible via nanostructuring, so a new approach to get higher ZT values was born.

In the 1990s and at the beginning of the twenty-first century, researchers all over the world experimented with various fillers for single, double, and multi-filled skutterudites, as well as with Fe/Co, Fe/Ni, and Co/Ni substitution and even doping at Sb site (see Sect. 9.5.1, band structure engineering) and in parallel with nanostructuring and nanocomposites (see Sect. 9.5.2, phonon engineering), in order to obtain TE materials with highly optimized TE power factors and ZT.

9.3 Why Do We Focus on Sb-Based Skutterudites?

Generally, for unfilled MX_3 or filled $E_yM_4X_{12}$ skutterudites, all elements of the fourth, fifth, and sixth main groups of the periodic table would qualify for X; however, skutterudites cannot be formed with most of these elements, whereas for $X = P$, As, Sb, it works well. As- and P-based skutterudites have been investigated but in most cases not for thermoelectric purposes. With As being toxic and P forming rather small cages suppressing rattling modes, these elements do not qualify for TE applications. Sb-based skutterudites have been studied most, as attractive ZT values could be reached due to high mobility of charge carriers, high atomic masses, high σ, and large S.

9.4 Preparation Methods on Lab Scale

In order to study physical and mechanical properties of any material, sufficiently large and homogeneous samples are a prerequisite. Although single crystals might have various advantages, these crystals are not necessary as skutterudites have a cubic structure and anisotropy is not a main concern.

A huge variety of preparation methods have been proposed for filled skutterudites, such as high-pressure-high-temperature synthesis, high-pressure synthesis, high-pressure torsion, induction melting, pulse plasma sintering, self-propagation synthesis, gas atomization, chemical synthesis routes, and melt-spinning followed by spark plasma sintering (for details see Ref. [10] and refs. therein). Although some of these methods were propagated as "fast" or even "super rapid," in most cases, results were rather poor, e.g., ZT = 0.9 (~670 K) for $Yb_yCo_4Sb_{12}$ and ZT < 0.2 (~580 K) for $Mm_{0.3}Fe_3CoSb_{12}$, where mischmetal, Mm, is a multifiller (consisting of 50.8% Ce + 28.1% La + 16.1% Nd + 5% Pr), both melt-spun and spark plasma sintered [26], ZT = 0.6 at 673 K for $In_{0.5}Co_4Sb_{11.5}Te_{0.5}$ with high-pressure-high-temperature synthesis [27] or ZT = 1 at 760 K for $Nd_{0.9}Fe_3CoSb_{12}$ via induction melting [28].

The most successful method to produce homogeneous polycrystalline dense specimens is still based on the preparation method already used by Dudkin in the late 1950s [16] (melting-annealing method). A master alloy is prepared in an evacuated and sealed quartz ampoule from Co(Fe,Ni) and Sb by melting and quenching on air. The appropriate amount of fillers is added to this bulk, again sealed in a quartz ampoule under vacuum, annealed for a couple of hours after which the temperature is slowly increased to the melting temperature. It turned out that additional annealing is not necessary. Therefore, after melting (still inside the quartz ampoule), the sample can be quenched on air and then crushed into small pieces prior to ball milling or high-energy ball milling followed by hot pressing or spark plasma sintering for densification. If well performed, this melting-annealing technique takes not more time than 48 h including ball milling and hot pressing

[29]. Various groups use the basics of this route; however, it is, of course, necessary to optimize ball milling and hot pressing conditions [30] in order to get dense samples with fine grains, as these properties influence severely the TE performance [31].

Sesselmann et al. [32] compared TE properties of $Ce_{0.6}Fe_2Co_2Sb_{12}$ (nominal composition) from various synthesis routes: (1) gas atomization, (2) conventional melting-annealing method, and (3) antimony master alloys $Fe_xCo_{1-x}Sb_2$ and $CeSb_2$ by ball milling and hot pressing. It turned out that these three methods lead to different final compositions of the skutterudite phase and secondary phases, which influence significantly all TE properties of the material. For method (3) the highest $ZT = 0.7$ at 700 K was reached.

9.5 Enhancement of ZT for *p*- and *n*-Type Skutterudites

9.5.1 Band Structure Engineering

Fillers

In the late 1990s, as soon as researchers had found out that filling the icosahedral voids in the crystal structure of the skutterudite with so-called rattlers increases in the number of phonon scattering centers and this way indeed reduces κ_L, the search for the best filler (single-filled skutterudites) or fillers (double- and multi-filled skutterudites) had started.

Slack and Tsoukala [21], e.g., investigated unfilled $IrSb_3$ and predicted that inserting impurities into the lattice voids could reduce κ and lead to better TE properties. Anisotropic interactions arising from the nonmetal (X) sublattice were already suggested by Kjekshus et al. [15] as the principal source of the deviation from Oftedal relation, i.e., a rectangular distortion of X_4 groups. Although a series of skutterudites may follow Oftedal relation, the bulk of skutterudites and, particularly, thermoelectric, filled skutterudites reveal a rectangular deviation from square X_4 groups. Large atom thermal displacement parameters, as derived from X-ray/neutron diffraction data for filler atoms, are consistent with low-frequency vibrational modes (Einstein oscillations) seen in heat capacity measurements with energies around 6 meV (~70 K) and 18 meV (200 K). Ab initio, lattice dynamics calculations [33] suggest that filler atom and Sb vibrations are coherently coupled in good agreement with experimental evidence from high-resolution inelastic neutron scattering data [34, 35]. It was concluded that the strong reduction in thermal conductivity results from the interaction between the vibrational dynamics of electropositive filler atoms with the acoustic phonons of M_4Sb_{12} matrix [36].

The mostly used fillers are alkaline (Li, Na, K, Rb), alkaline earth (Ca, Sr, Ba), and rare earth (La, Ce, Pr, Nd, Sm, Eu, Yb) elements, but also In [37–40], Sn [41–44], Ti [45, 46], and I [47, 48] have been tried successfully. In the beginning it was assumed that all voids are fully occupied; however, Chen et al. [49]

compared the amount of Ce in $Ce_yFe_4Sb_{12}$ and $Ce_yFe_{4-x}Co_xSb_{12}$, and they found that y decreases with increasing Co content and that for pure Co_4Sb_{12}, only 10% of the voids can be filled. Still, it could be shown that already a small amount of filling fraction can reduce thermal conductivity significantly [4, 22, 34, 50].

Calculating the limit of filling level and optimal filling fraction [51–53], one can tune TE properties of filled skutterudites via the adjustable filling fraction. Theoretical and experimental filling fraction limits were summarized by Shi et al. [51, 54]. Furthermore, for thermodynamic stability of filled skutterudites, the electronegativity of the filler minus the electronegativity of Sb must be bigger than 0.8 [55]. Yang et al. [56] showed theoretically that the more the resonance frequencies of two guest atoms differ from each other, the higher is the phonon scattering efficiency. These findings were the kickoff for experimentalists to produce double-, triple- and multi-filled skutterudites keeping always in mind that the fillers should have different ionic radii, masses, and atom vibration frequencies.

Lu et al. [57] were the first to introduce two fillers, and with $Ce_{0.1}La_{0.2}FeCo_3Sb_{12}$, ZT = 0.6 was reached at 773 K. For a long time, triple-filled $Ba_{0.08}La_{0.05}Yb_{0.04}Co_4Sb_{12}$ had the highest ZT = 1.7 at 850 K [54] till this value could be topped by ZT = 1.8 at 800 K for $Sr_{0.07}Ba_{0.07}Yb_{0.07}Co_4Sb_{12}$ before and ZT = 1.9 at 835 K after high-pressure torsion processing.

The natural double filler didymium, DD (consisting of 4.76% Pr and 95.24% Nd), and the multifiller, mischmetal, Mm, (consisting of 50.8% Ce + 28.1% La + 16.1% Nd + 5% Pr) have proven to be ideal fillers for p-type skutterudites, reaching ZT higher than 1.2 [58–62].

Substitutions

As already shown at the example of $CoSb_3$ [16], another route to enhance TE properties is by substitution at Co site, usually by similar elements with a higher number of electrons, which act as donors and thereby increase σ. For n-type skutterudites Ni, but also Pd and Pt are used to substitute for Co [63, 64]. Expensive Pt and Pd showed good results, but do not qualify for mass production. Substituting Ni for Co increases σ and thermopower because each Ni atom acts as electron donor. Dudkin and Abrikosov [16], however, found that if the Ni content rises above 1 at. %, then the trend reverses rapidly and Ni doped skutterudites show rather poor ZTs.

For p-type skutterudites, elements with a lower number of electrons are chosen in order to create holes and strengthen p-type conduction; therefore, Fe/Co substitution is a successful route. Unfilled p-type skutterudites are unstable compounds; therefore, it is necessary to stabilize the structure by introducing fillers. Morelli et al. [17] in 1995 demonstrated that effect with $Ce_yFe_xCo_{4-x}Sb_{12}$ which was confirmed by Fleurial et al. [65]. Tang et al. [66] combined Fe/Co substitution with Ba, Ce, and Y as fillers; however, ZT did not surpass 1.1 at that time. With a well-evaluated Fe/Co ratio (for $E_yFe_xCo_{4-x}Sb_{12}$, $x \sim 3$) [60] in combination with double and multi-filling [59–61], ZT reached 1.2.

Another approach is doping at Sb site, which influences the electronic structure; deforms the pnictogen ring, which dominates over the spectrum of κ; and introduces defects, although resulting in lower κ. Single doping with Te [27, 67–73], Ge [29, 74, 75], Bi [76], and Sn [29, 75, 77, 78], as well as double doping Te + Ge [79] or Te + Sn [80–82] at Sb site, proved to be successful. Liu et al. [80], in addition to double doping with Te/Sn and Te/Ge, tried also Te/Si, and Te/Pb; however, Si and Pb did not embed into the skutterudite lattice. Khan et al. [83] created a porous microstructure for double-doped Te/Si $CoSb_{2.75}Si_{0.075}Te_{0.175}$, containing nano- to microsized irregularly shaped and randomly oriented pores, which created rattling features and reduced tremendously κ_L, yielding the highest ZT = 1.7, ever reported for an unfilled skutterudite.

In many cases, doping at Sb site was combined with Fe/Co and Ni/Co doping. An overview is presented in [6]; the highest ZT = 1.3 at 820 K was reached for $Fe_{0.2}Co_{3.8}Sb_{11.5}Te_{0.5}$ [76] and ZT = 1.1 at 725 K for triple-doped $Co_{3.9}Ni_{1.1}Sb_{11.5}Te_{0.4}Se_{0.1}$ [84].

The combination of doping and filling can, of course, further increase in ZT. Zhang et al. [75] presented ZT = 1.1 at 700 K for $Nd_{0.6}Fe_2Co_2Sb_{11.7}Ge_{0.3}$, an enhancement of almost 60% in comparison to the starting alloy without Ge. Rogl et al. enhanced ZT = 1.1 of $DD_{0.60}Fe_{2.8}Co_{1.2}Sb_{12}$ to ZT = 1.3 at 775 K in $DD_{0.59}Fe_{2.7}Co_{1.3}Sb_{11.8}Sn_{0.2}$ [29].

9.5.2 Phonon Engineering

The Nano-effect (HM, BM, HBM, HPT)

Based on the findings that nanostructuring can enhance ZT [23–25], researchers tried to realize these ideas experimentally. In polycrystalline materials, micro- and nanostructures were found to affect electrical and thermal properties [85–87] by lowering κ due to grain boundary scattering, however, simultaneously increasing ρ. It could be shown, e.g., by Nakagawa et al [88], Yang et al. [89], or Jie et al. [90], that the net effect was positive, which means that ball-milled samples exhibit higher ZTs than simply hand ground ones, although the post consolidation via hot pressing or spark plasma sintering induces grain growth.

In this context, it is important to mention that Short et al. [91], comparing hand ground and ball-milled p-type skutterudite samples with extended X-ray absorption Fi(EXAFS) analysis, could prove that via ball milling, no damage to the crystal lattice of nanoparticles is done and, therefore, even smaller particles could be used to further improve ZT.

Rogl et al. [31] compared physical properties of p- and n-type skutterudites, fabricated from the same p- and n-type skutterudite powders and hot-pressed in exactly the same way but using three different particle sizes (150, 100, 50 μm) as starting material and for the smallest particle size three different ball milling conditions. It turned out that the grain and crystallite size had no influence on lattice vibrations as shown in Raman spectra; however, it was confirmed that the smaller the grain sizes, the higher is the TE performance due to a reduction of κ_L. ZT of p-type $DD_y(Fe_{1-x}Co_x)_4Sb_{12}$ was enhanced from ZT ~ 1.1 to ZT ~ 1.3 at 775 K and for n-type $(Mm,Sm)_yCo_4Sb_{12}$ from ZT ~ 1.0 to ZT ~ 1.6 at 825 K.

Another approach to reduce crystallite size and at the same time to increase in quantity of lattice defects and dislocations and, thereby, increase in the scattering of heat-carrying phonons is to apply severe plastic deformation (SPD) via high-pressure torsion (HPT) [92–101]. HPT-treated samples show reduced σ, almost unchanged S, and significantly reduced κ, which as net effect increases in ZT. Repeated heating or annealing during measurement cycles did not remove completely the beneficial microstructure achieved via HPT. The best results are ZT = 1.6 at 800 K for p-type $DD_{0.6}Fe_3CoSb_{12}$ (after temperature cycling ZT = 1.5) [96, 99] and ZT = 1.9 at 835 K for $Sr_{0.09}Ba_{0.11}Yb_{0.05}Co_4Sb_{12}$ (after temperature cycling ZT = 1.8) [98]. It must be noted that temperature-cycled samples are stable in respect to further cycling or annealing processes. The grain refinement due to SPD in addition infers a mechanical strengthening of the material as will be discussed in section 9.6.

Influence of Nanoprecipitates

A successful route to enhance ZT is to incorporate nanoparticles within bulk materials forming nanocomposites. By this approach, interfaces are built within bulk material, which should act in two ways: it should reduce κ more than σ by interface scattering and should increase S by energy filtering of charge carriers or by quantum confinement more than decreasing σ. The outcome is not only an increase in power factor, $PF = S^2\sigma$, but in ZT as well. As these nanocomposites might influence the charge carriers, nanoparticles must be selected, which enhance phonon scattering but do practically not disturb the charge carriers. In addition, with introduction of nanoparticles, the number of interfaces increases, and, eventually, impurity phases are introduced, reducing κ_L due to acting as scattering centers for phonons.

Various methods were employed to introduce the additive into the matrix, like solid-state or chemical-thermal reaction, chemical alloying methods, hydrothermal synthesis, in-situ reaction, sol-gel method, freeze-drying, melt milling, thermal diffusion, solvo-thermal method, and cryogenic grinding technique. However, in most cases the respective amount of additive was mixed (manually, via ball milling, BM, or high-energy ball milling, HBM) with the powder of the host material prior to hot pressing or spark plasma sintering. Also, chosen additives varied. A large group of additives comprises oxides (Al_2O_3, TiO_2, Cu_2O, ZnO, ZrO_2, MoO_2, WO_2, WO_3, Yb_2O_3, $La_{1.85}Sr_{0.15}CuO_4$, $BaFe_{12}O_{19}$) and the other one tellurides ($PbTe$, WTe_2, Bi_2Te_3, $AgSbTe_2$, $(Ag_2Te)_{0.40}(Sb_2Te_3)_{0.60}$, $(Ag_2Te)_{0.42}(Sb_2Te_3)_{0.58}$, $Co_4Sb_{11.5}Te_{0.5}$) or antimonides ($InSb$, $NiSb$, $GaSb$, $CoSb_3$ or simply excess of Sb). Some researchers dealt with metals (Ni, Ag, Co), borides ($Fe_{0.25}Co_{0.75}B$, $Ta_{0.8}Zr_{0.2}B$), nitrides (TiN), or silicides (Fe_3Si) as additives. Also, carbon-related materials (carbon fibers, multi-

or single-wall carbon nanotubes, reduced graphene oxide nanolayers, graphene, fullerenes) were introduced in the skutterudite matrix. A detailed description of the aforementioned methods and additives is given in a recent review ([102] and refs. therein).

Generally, it turned out that, independent of the chosen additive and preparation method, small amounts of additives, e.g., less than 2 wt. %, could enhance ZT more than bigger amounts. It can be noted that in almost 90% of all investigations, at least one sample with nanoprecipitates resulted in ZT enhancement, although in many cases the enhancement is within the error bar. However, the influence on mechanical properties was remarkable, as will be discussed in Sect. 9.6.

Most investigations were performed on n-type skutterudites, but here, selected examples of outstanding results will be discussed only. For many investigations $Yb_yCo_4Sb_{12}$ or $Ba_yCo_4Sb_{12}$ was used as matrix. ZT of $Yb_yCo_4Sb_{12}$ could be enhanced by 5–44% by adding either reduced graphene oxide layers [103] or multiwall carbon nanotubes [104] or Yb_2O_3 [105–107] reaching ZT values of ZT = 1.51, 1.43, 1.18, 1.16, 1.6, respectively. Rogl et al. [102], e.g., added excess Yb to $Sr_{0.09}Ba_{0.11}Yb_{0.05}Co_4Sb_{12}$ and found that Yb_2O_3 particles were formed and located at the grain boundaries of the matrix, as well as distributed within the grains as nanoscale inclusions: disturbing the phonon scattering, which reduced κ_L and enhanced ZT from 1.4 to 1.6. ZT was more than doubled with adding Ni as core shell particles [108] and almost tripled with $AgSbTe_2$, reaching ZT = 1.27 at 800 K [109]. The increased ZT was achieved by (a) enhanced mobility of charge carriers, which led to a threefold increase in σ, and (b) by enhanced S due to the energy filtering effect at interfaces between matrix and nanoinclusions, as well as (c) by reduction in κ. Adding Ag to $Ba_{0.3}Co_4Sb_{12}$ revealed 40% enhancement of ZT (Zhou et al. [110] and Peng et al. [111]). Here, a uniform dispersion of Ag nanoparticles was observed between grain boundaries and on grains' surfaces, leading to a large reduction in κ and a high ZT = 1.4 at 823 K. An even higher ZT enhancement, 44% and ZT = 1.3, was achieved for $Ba_yCo_4Sb_{12}$ with addition of fullerene [112]. Although Battabyal et al. [113] could triple ZT of $Ba_{0.4}Co_4Sb_{12}$, it still ended up with a rather low value of ZT = 0.9.

Much less was reported for p-type skutterudides as matrix; however, it is worth mentioning that for $CeFe_3CoSb_{12}$, ZT was raised from 1.05 to 1.22 (16%) with MoO_3 as additive [114] and even more, 50%, when Zhou et al. [115] added magnetic $BaFe_{12}O_{19}$. A quite meager ZT enhancement was reported for $CeFe_4Sb_{12}$ on addition of carbon fibers [116] or for $Ce_{0.85}Fe_3CoSb_{12}$ on addition of reduced graphene oxide layers [117] (8%) or for $DD_yFe_3CoSb_{12}$ with addition of $Ta_{0.8}Zr_{0.2}B$ (7%), in the latter case reaching ZT = 1.31 at 823 K [118].

Figure 9.2 shows a transmission electron microscopy (TEM) image as example. It shows how besides grains with sizes of about 200 nm two additional types of particles are present, bigger ones at the grain boundaries and smaller ones (smaller than 10 nm) inside the grains. These small grains provide more intensive phonon scattering and reduce κ_L, which results in ZT enhancement.

ZT values of $CoSb_3$ could be enhanced up to 60% with various additives, e.g., from ZT = 0.26 to 0.61 (graphene) [119], from 0.14 to 0.23 (fullerene) [51], from

Fig. 9.2 TEM image of $DD_yFe_3CoSb_{12}$ with addition of 1 wt. % of $Ta_{0.8}Zr_{0.2}B$

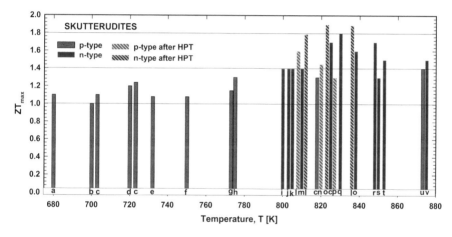

Fig. 9.3 Dependence of maximum ZT (ZT_{max}) of skutterudites on temperature. **a** - [48], **b** - [123], **c** - [102], **d** - [59], **e** - [90], **f** - [66], **g** - [124], **h** - [29], **i** - [93], **j** - [125], **k** - [126], **l** -[97], **m** - [127], **n** - [29], **o** - [98], **p** - [76], **q** - [128], **r** - [54], **s** - [129], **t** - [130], **u** - [65], **v** - [131]

0.25 to 0.4 (WO_2) [114], or from 0.45 to 0.71 by simply mixing micropowder of $CoSb_3$ with nanopowder of $CoSb_3$ [120].

Zhao et al. [121] and Tan et al. [122] used as matrix unfilled skutterudites $Co_4Sb_{11.7}Te_{0.3}$ and $FeSb_{2.2}Te_{0.8}$, respectively. Zhao added WO_3 and could raise ZT from 0.44 to 0.65 (48%), whereas Tan added InSb, but the rise was only 17%.

Fig. 9.3 shows an overview of the highest ZT values in dependence on the peak temperature. Generally seen, high ZT values of *p*-type skutterudites (higher than 1, reaching 1.3) are lower than ZT values of *n*-type skutterudites (higher than 1.3, reaching 1.8); in addition, maxima of *p*-type skutterudites occur at lower temperatures than those of *n*-type skutterudites. There are two exceptions: the maximum ZT with additives can be shifted to higher temperatures (818 K) and ZT of $CeFe_4Sb_{12}$, published by Fleurial in 1996, which for p-type skutterudite had a very high ZT = 1.4 at 900 K; however, this value could, so far, not be reproduced.

After SPD via HPT, p- and n-type skutterudites show enhanced ZT values as discussed in Sect. 9.5.2.1. above.

9.6   Mechanical Properties

As the authors provided a summary on mechanical properties of skutterudites in two articles [132, 133] and a review article [134], as well as an update [13], we will give here only a short overview but will explain the changes of mechanical properties under various influences in more details.

Generally, mechanical properties are dependent on various parameters such as (1) the sample's synthesis method and (2) the sample's composition, (3) surface preparation, as well as (4) grain size. Mechanical properties, furthermore, are strongly dependent on density/porosity and temperature. In addition, the used equipment can influence the values as, e.g., with resonant ultrasound spectroscopy, the whole sample is excited, whereas, with micro- or nanoindenter, only a very small area of the sample is measured. Usually, a difference exists between static hardness (measured with microhardness tester, where the imprint is measured after the force is released and the imprint has contracted) and dynamic hardness (measured with micro- or nanoindenter directly, while the load is applied). In addition, one has to consider that DFT calculated data refer always to samples with ideal structure at a temperature of zero Kelvin.

Average values of the crystallite size of p- and n-type skutterudites after hot pressing are about 150 nm. After HPT these sizes are decreased to about one third (~40–50 nm) and after measurement-induced heating or after annealing, crystallites grow slightly but never reach the size of the starting material. These important changes are accompanied by changes of (1) dislocation density (starting material, ~3 × 10^{13} m^{-2}; HPT: ~2 × 10^{14} m^{-2}; after annealing, ~1.5 × 10^{14} m^{-2}), (2) lattice parameter (slightly enhanced after HPT and a bit reduced after annealing), and (3) relative density (slightly lower after HPT and growing almost back to the starting values after annealing) [97, 133, 134].

Vickers hardness (HV) values for hand ground samples equal to HV0.1 = 240–300 for p-type and HV0.1 = 300–400 for n-type skutterudites are, due to bigger grain sizes and lower densities, lower than the values of ball-milled p- and n-type skutterudites with relative densities in the range of 89–99%, which, with one or two outliers, are HV0.1 = 430–530 and HV0.1 = 500–600, respectively. These values document that, generally, n-type skutterudites are harder than p-type counterparts.

Although after SPS via HPT, the relative density of samples is slightly decreased, following Hall-Petch relation (the smaller the crystallite size, the harder the alloy), hardness is enhanced [97]. HPT processed samples are not homogeneous with respect to shear strain, $\gamma = (2\pi Nr)/t$ (where N is number of revolutions, r is radius, and t is thickness of the sample) with $\gamma \sim 39$ at the rim and $\gamma \sim 8$ at the center of the sample (for $r=10$ mm, $t=1$ mm). Therefore, hardness values decrease from the rim to the center area, e.g., for $DD_{0.6}Fe_3CoSb_{12}$ from $HV0.1 = 580$ to $HV0.1 = 440$ for ball-milled sample, with hardness of $HV0.1 = 410$ before HPT [100]. Although samples, annealed for 1 h, show no changes in density or hardness, those annealed for 24 h at 673 K slightly soften (about 1% decrease), but do not show changes after further annealing.

Nanoparticles distributed evenly at the grain boundaries may influence mechanical properties significantly. With addition of borides and oxides, density of the samples increases slightly. Concerning hardness, the additives can strengthen or soften the material. As can be seen in Fig. 9.4, p- and, especially, n-type skutterudites have much higher hardness values (up to 50%) with addition of borides. The more boride is added, the harder the material, but ZT decreases as reported in ref. [102]. Addition of oxides, however, softens the skutterudites.

Young's modulus, E, of ball-milled skutterudites at room temperature and for relative densities above 95% ranges from 120 GPa to 135 GPa for p-type and is a little bit higher, 130 GPa to 145 GPa, for n-type skutterudites [133]. Fig. 9.4 shows the relation between hardness and Young's modulus. A linear fit to all ball-milled samples (Fig. 9.4) reveals the relation $HV = 3.75(16)E$. Only some data are beyond the error bar, and, of course, the data for hand ground samples are much lower. It is obvious that all HV - E data for samples with added borides are enhanced and lay

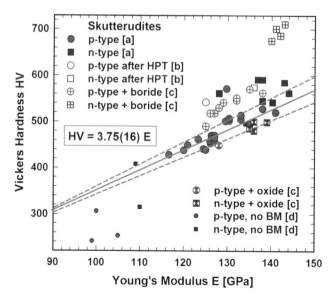

Fig. 9.4 Young's modulus vs. Vickers hardness of p- and n-type hand ground (small symbols) and ball-milled (large symbols) skutterudites before and after HPT or with addition of nanoprecipitates. [a] - [134] and refs. therein, [b] - [97], [c] - [102] and refs. therein, [d] - [134]

outside of the fit region, for n-type much more than for p-type. The values decrease after an oxide was added. For HPT processed samples, exemplarily, one value for p- and n-type is displayed and shows enhancement as already discussed.

As HPT processed samples are brittle and available in small disks only, it is not easy to study other mechanical properties except hardness and elastic moduli.

Compressive strength and flexural strength of $CoSb_3$ and of Sb-substituted unfilled and filled skutterudites were investigated [102] and appear enhanced with addition of oxides [135, 136], carbon fibers [116], or titanium nitride [71, 137] but reduced with addition of multiwall carbon nanotubes [138]. The authors saw an increase in compressive strength of p-type $DD_yFe_3CoSb_{12}$ from 450 MPa to 520 MPa after 5 wt. % of borides was added.

Fracture toughness, K_C, (or fracture resistance, in case it was determined from crack lines propagating out of hardness imprints) follows the behavior of compressive and flexural strength. With the addition of borides, K_C is enhanced [71, 116, 136], e.g., for $DD_yFe_3CoSb_{12}$ from 1.6 to 1.9 MPa $\times$ m$^{1/2}$ or for $(Mm,Sm)_yCo_4Sb_{12}$ from 1.8 to 2.0 MPa $\times$ m$^{1/2}$. However, multiwall carbon nanotubes decrease flexural strength and K_C as well [138].

According to Ashby plot [139] of elastic modulus vs. hardness, as shown in Fig. 9.5, skutterudites are located between TAGS (Te - Ag - Ge - Sb), Bi_2Te_3, and LAST (Pb - Ag - Sb - Te) TE materials and ceramics. The ratio of hardness to elastic moduli is higher for skutterudites than for metals and steels.

In order to avoid stresses on p and n legs of TE modules during repeated heating and cooling cycles, parameters of mechanical properties, especially, of thermal

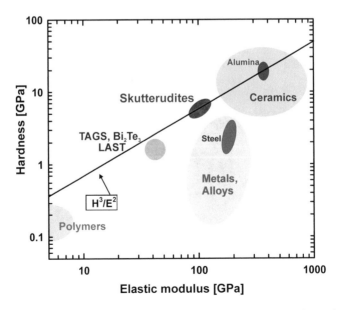

Fig. 9.5 Ashby plot of elastic modulus vs. hardness including Te-based thermoelectrics and skutterudites

expansion coefficients α, should not differ much. Generally, Sb-based filled p-type skutterudites ($\alpha_{av} = 11.2 \times 10^{-6}$ K^{-1}) have higher coefficients of thermal expansion than n-type counterparts ($\alpha_{av} = 9.6 \times 10^{-6}$ K^{-1}) (see detailed investigation of the expansion behavior of skutterudites [140]). For Fe/Ni-substituted multi-filled skutterudites, it was possible to get almost the same α for p-type, $Ba_{0.15}DD_{0.28}Yb_{0.05}Fe_3NiSb_{12}$, $\alpha(300–800$ K$) = 11.8 \times 10^{-6}$ K^{-1} and n-type, $Ba_{0.09}Sr_{0.02}DD_{0.22}Yb_{0.02}Fe_{2.4}Ni_{1.6}Sb_{12}$, $\alpha(300–800$ K$) = 11.9 \times 10^{-6}$ K^{-1}) [141]; however, both materials have ZT $\approx$ 1 only.

Using another approach, Rogl et al. [118] have demonstrated that borides dispersed in both p-type $DD_{0.7}Fe_3CoSb_{12}$ and n-type $(Mm,Sm)_yCo_4Sb_{12}$ can level out the discrepancy in α. By adding either 1.5 or 20 wt. % $Ta_{0.8}Zr_{0.2}B$ or adding 5 wt. % $Fe_{2.25}Co_{0.75}B$ to $DD_{0.7}Fe_3CoSb_{12}$, α could be reduced from $\alpha(300–800$ K$) = 11.3 \times 10^{-6}$ K^{-1} to $\alpha(300–800$ K$) = 10.6 \times 10^{-6}$ K^{-1}, $\alpha(300–800$ K$) = 10.4 \times 10^{-6}$ K^{-1} or $\alpha(300–800$ K$) = 9.7 \times 10^{-6}$ K^{-1}, respectively. On the other hand, $Ta_{0.8}Zr_{0.2}B$ (1 wt. %) dispersed in $(Mm,Sm)_yCo_4Sb_{12}$ enhanced α from α (300–800 K) $= 9.65 \times 10^{-6}$ K^{-1} to $\alpha(300–800$ K$) = 10.5 \times 10^{-6}$ K^{-1}. Considering ZT changes accompanying nanodispersions, luckily for n-type, ZT is higher with the addition of $Ta_{0.8}Zr_{0.2}B$. For p-type with 1.5 wt. % $Ta_{0.8}Zr_{0.2}B$, ZT is, within the error bar, almost the same (ZT $=$ 1.23 and ZT $=$ 1.26, respectively).

It should be mentioned that $La_{1.85}Sr_{0.5}CuO_4$ has the reverse effect, as it was also the case for hardness and Young's modulus; it enhances α of p-type and lowers α of n-type material. Ashby plot [139] of α vs. κ (Fig. 9.6) shows that skutterudites have

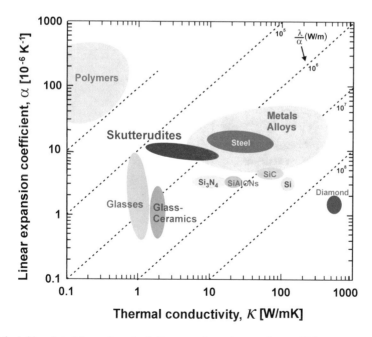

Fig. 9.6 Ashby plot of thermal conductivity, κ, vs. thermal expansion coefficient, α

thermal expansion coefficients in the range of metals, lower than polymers but higher than glasses; however, the thermal conductivity is mainly lower than that of metals.

The length change of HPT processed p- and n-type skutterudites is strange and could not be explained completely so far. Fact is that the measurements from 4.2 K to room temperature performed with a capacitance dilatometer, using the tilted plate principle, do not show, within the error bar, any discrepancies compared to unprocessed reference material. Above room temperature, using the zero-force method with a thermomechanical analyzer, a strange behavior appears. In the temperature range of 300–400 K, the graph of the length change vs. temperature is prolonged, but with increasing temperature, it becomes slightly less steep, as shown for $DD_{0.44}Fe_{2.1}Co_{1.9}Sb_{12}$ (Fig. 9.7) as an example. In case of $DD_{0.44}Fe_{2.1}Co_{1.9}Sb_{12}$, at about 450 K, a kink appears, and the sample contracts and afterward expands again but much more than prior to the compression and almost in the range of glass-ceramics, which is in favor of a high ZT.

This behavior was observed for all HPT processed skutterudites; however, the kink varies in temperature. It seems that during measurement-induced annealing, defects heal out, and/or internal microcracks fuse together, and stresses die down. After the first heating, the curve for decreasing temperature, as well as cycling, shows a linear expansion behavior without any kink. The value of α, before and after HPT processing in the low temperature range, is the same (for $DD_{0.44}Fe_{2.1}Co_{1.9}Sb_{12}$ $\alpha(115\text{–}300 \text{ K}) = 11.7 \times 10^{-6} \text{ K}^{-1}$). For the sample before HPT, this value does not change in the high temperature range and is, within the error bar, the same as for the HPT processed sample after heat treatment (for $DD_{0.44}Fe_{2.1}Co_{1.9}Sb_{12}$ α

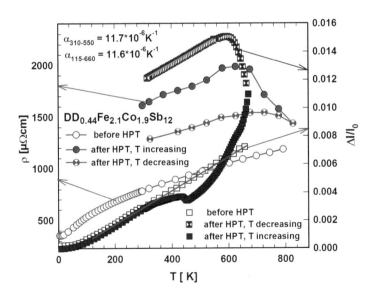

Fig. 9.7 Temperature-dependent electrical resistivity ρ (left scale) and thermal expansion $\Delta|l|_0$ of $DD_{0.44}Fe_{2.1}Co_{1.9}Sb_{12}$ before and after HPT. Pink lines display the linear fit to $\Delta|l|_0 - T$ curve

$(115–300 \text{ K}) = 11.7 \times 10^{-6} \text{ K}^{-1}$). It is interesting to note that ρ (Fig. 9.7) of HPT processed sample, although much higher than that of the reference sample, starts to decrease at about the same temperature where the kink in thermal expansion appears. In parallel to the thermal expansion, the resistivity measurement with decreasing temperature and further measurements do not reveal any kink.

All these changes in mechanical properties and thermal expansion show how flexible skutterudites are, as one can adjust those to the needs of any application, although in some cases these changes may be connected with a damage of ZT.

9.7 Thermal Stability, Sublimation, and Oxidation

As thermal stability and oxidation of any thermoelectric material is a crucial point for its application, it is necessary to investigate it, also for Sb-based skutterudites. In 2017 Rogl et al. [13] published a table with all known oxidation onset temperatures and activation energies.

Caillat et al. [142] in 2004 were first to investigate long-term stability of $CoSb_3$ as bulk in dynamic vacuum (10^{-4} Pa) and found a decomposition into $CoSb_2$ and $CoSb$ with the sublimation of Sb at 575 °C. They saw that time-dependent weight losses follow a parabolic-like behavior. Later (2011) Zhao et al. [143] confirmed such behavior when analyzing time-dependent mass losses of spark plasma sintered $CoSb_3$ in vacuum in the temperature range of 600–750 °C. They evaluated vaporization kinetics of Sb and took it as decomposition of $CoSb_3 \rightarrow CoSb_2 + Sb$. In addition, they saw a decrease in ZT from 0.24 to 0.16 at 327 °C after a thermal test duration of 16 days. At the same time, Leszczynski et al. [144], using powder samples and employing non-isothermal differential analysis and thermography, reported 420 °C as start of decomposition. In the same temperature range, oxidation of $CoSb_3$ thin films (380 °C) was reported by Savchuk et al. [145]. However, for Fe/Co-substituted alloys, oxidation starts earlier, with 300 °C. Wojciechowski et al. [146] observed the same oxidation temperature (380 °C) for $CoSb_3$ as well as for $CoSb_3$ with 1.5 at. % Te. Zhao et al. [147] reported surface antimony oxide layers, as soon as skutterudites are exposed to air at temperatures above 400 °C. At 600 °C, they even found two layers, which continue to grow. Park et al. [148] reported that no oxide layers formed in vacuum even after 120 h annealing at 550 °C.

Peddle et al. [149] observed the onset of oxidation of $EuFe_4Sb_{12}$ at 360 °C, whereas of $EuRu_4Sb_{12}$ at 460 °C. The same tendency was observed by Sigrist et al. [150] with an oxidation temperature of $CeRu_4Sb_{12}$ (470 °C), being more than 100 °C higher than that of $CeFe_4Sb_{12}$ and 70 °C higher than that of $CeRu_3CoSb_{12}$. The conclusion of these observations is that oxidation resistance increases with the content of noble elements.

Generally, activation energies for the oxidation of Co-based skutterudites (105–270 kJ/mol) are higher than those for rare earth-filled Fe-based skutterudites (14–40.5 kJ/mol), which can be due to the high oxygen affinity of rare earth atoms resulting in metastable Fe_4Sb_{12} that decomposes into $FeSb_2$ and Sb, accelerating the oxidation rate [13, 151]. In addition, for $CoSb_3$-based n-type skutterudites, a reduction in oxidation resistance was observed with the increase in the rare earth content by Xia et al. [152] for $Yb_yCo_4Sb_{12}$; however, Park et al. [148, 153] saw an opposite influence of the filler for $In_{0.25}Co_3MnSb_{12}$ and $In_{0.25}Co_3FeSb_{12}$, blaming it on the formation of InSb.

In any case protecting the skutterudite surface with an inert material can prevent oxidation processes.

Sklad et al. [154] and Xia et al. [151] studied the oxidation behavior of the filled p-type skutterudite $CeFe_4Sb_{12}$ and reported oxidation temperatures of 300 and 327 °C, respectively. In addition, Xia et al. found that below 800 K, an oxide layer of Sb_2O_3, Sb_2O_4, Ce_2O_2Sb, and amorphous Fe^{3+} oxide forms. At temperatures above 527 °C, an additional layer comprising amorphous Ce, $FeSb_2$, and Sb evolves beneath the first one, proving that oxidation is accompanied by decomposition of $CeFe_4Sb_{12}$. They calculated the activation energy of the oxidation of 4.0 kJ/mol (427–527 °C). Qiu et al. [155] observed about the same for $Ce_{0.9}Fe_3CoSb_{12}$, which has an oxidation temperature of 377 °C. Shin et al. [156] reported for $La_{0.3}Fe_3CoSb_{12}$ an only slightly higher oxidation temperature 400 °C.

Broz et al. [157, 158] investigated recently the thermal and phase stability of ball-milled and hot-pressed bulk $CoSb_3$ skutterudite by means of differential thermal analysis (DTA) and Knudsen effusion mass spectrometry (KEMS) accompanied by scanning electron microscopy. KEMS measurements (at 589, 618, and 637 °C) (Fig. 9.8) agree well with the diffusion profiles showing that Sb evaporation from $CoSb_3$ is a complex kinetic process including initial evaporation of free or weakly bound Sb (obviously excess Sb, from the material) followed by temperature-dependent Sb depletion resulting in phase transformations $CoSb_3 \rightarrow CoSb_2 \rightarrow CoSb \rightarrow \alpha(Co)$. However, structure stability up to ~620 °C was confirmed. Above 640 °C, $CoSb_2$ forms on the sample's surface (see insert in

Fig. 9.8 Temperature-dependent mass loss in mg per 24 h for BM (ball milled) and HP (hot-pressed) bulk $CoSb_3$ in vacuum. Insert: SEM micrograph of the surface layer at 637 °C

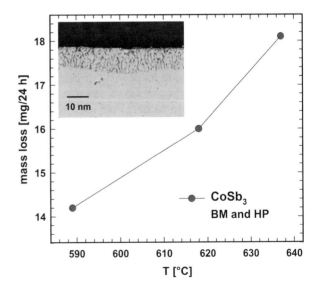

Fig. 9.8), which might grow with further increase of temperature. With low mass losses of Sb, as found from KEMS measurements (Fig. 9.8), $CoSb_3$ can be considered as stable basis for TE materials as long as it is used in protective atmosphere and below 600 °C.

In our recent review [13], we described a lot of effort in thermal stability tests, using commercially hot-pressed samples, as well as hot-pressed samples under laboratory conditions, in both cases from commercial p- and n-type skutterudite powders ($DD_yFe_3CoSb_{12}$ and $(Mm,Sm)_yCo_4Sb_{12}$), produced in Austria. From hot-pressed disks, samples were cut of various shapes and sizes, in parallel and perpendicular to the pressing direction, and room temperature electrical resistivity, as well as weight, was measured. Then the couples of p- and n-type material, as well as individual p- and n-type materials, were sealed in evacuated and Ar-filled quartz ampoules (length 40 cm). TE material was placed at the hot end of the ampoule, which was subjected to a temperature gradient $T_{hot} = 600$ °C and $T_{cold} = 80$ °C. About every 30 days, weight and electrical resistivity of the samples were controlled. After 504, 768, 2475, 4363, and 8800 h, the samples were quenched and investigated by SEM, EPMA (electron probe microanalysis), and XRD (X-ray diffraction), and, particularly, TE properties and those of the reference samples were measured in the temperature range from 300 K to 850 K. It turned out that for n-type reference samples, no differences in TE properties occurred; samples had ZT = 1.2 at 800 K. However, for the commercial hot-pressed p-type sample, secondary phases were detected, resulting in a lower ZT of 0.9 in comparison to laboratory hot-pressed samples with ZT = 1.1 at 800 K.

Summing up the long-term stability study [13], we can state the following: (1) long-term thermal stability tests at 600 °C show that Ar pressure had a beneficial effect on the suppression of Sb evaporation as compared to experiments in vacuum; (2) p-type materials show stronger Sb evaporation, which occurs in the surface zone, resulting in weight loss; (3) unusual weight exchange between p and n legs occurred for commercial hot-pressed samples, attributed to partial oxidation of p-type material; and (4) ZT at 800 K remains in the range of 0.9 for p-type but decreases from 1.2 to 1.0 for n-type, which was attributed to a decrease in the filling level from y = 0.17 to y = 0.11.

Generally, one can conclude that coating or encapsulation under argon is a necessity for long-term stability in order to prevent oxidation and/or sublimation. The optimal coating should have a low thermal and electrical conductivity, should be chemically inert, and should have good adhesion to the substrate and a thermal expansion coefficient as close as possible to that of the skutterudites.

9.8 Large-Scale Production of Skutterudite Material: Modules and TE Generators

The goal to develop high ZT skutterudites in laboratory has been very well achieved. Combining (1) the right fillers in (2) the right proportion with (3) substituting at Co site and/or Sb site together with (4) a well-developed preparation method to achieve dense samples with small grains leads already to spectacular ZT values and respective leg efficiencies: for p-type $DD_{0.59}Fe_{2.7}Co_{1.3}Sb_{11.8}Sn_{0.2}$ ZT = 1.3 with $\eta(300–800\ K) = 14.3\%$ [29] and for n-type $(Sr,\ Ba,\ Yb)_yCo_4Sb_{12} + 9.1$ wt. % $In_{0.4}Co_4Sb_{12}$ ZT = 1.8 with $\eta(300–800\ K) = 17.5\%$ [127]. To our knowledge, these are currently the highest values. As described in Sect. 9.5.2.1., these values can be further enhanced applying severe plastic deformation via high-pressure torsion.

All these high ZT values with high leg efficiencies, thermal stability, and proper mechanical properties are useless, as long as samples can only be produced in grams. Therefore, it is a big step forward that the commercial production of skutterudites powders has started. After many tests, industrial manufacturers have developed p- and n-type skutterudite powders, which have, after hot pressing, respectable ZT values at 800 K, i.e., ZT = 1.1 ± 0.1 for p-type and ZT = 1.3 ± 0.1 for the n-type. These values can be enhanced to ZT = 1.3 and ZT = 1.6, respectively, via ball milling under well-defined conditions [31, 32] prior to hot pressing. Nong and Hung [159], as well as Pryds [160], could confirm these outstanding results.

In a review article, Aswal et al. [161] present various highly efficient TE modules based also on skutterudites. It is well known that segmented modules, using skutterudites and other TE compounds, can reach high efficiencies like $\eta = 15\%$ (with temperature gradients of 1000 to 152 °C and 700 to 27 °C) [162] or calculated values of even 18% [163]; however, for mass production these segmented modules are not suitable. Therefore, it is notable that non-segmented, skutterudite-based modules reach already efficiencies of $\eta = 7\%$ (gradient of 500 to 40 °C) [164], $\eta = 9\%$ (gradient of 550 to 70 °C) [165], and even $\eta = 10\%$ (gradient of 550 to 90 °C) [166]. Durability tests performed by Ochi et al. [167] under thermal cycling from 200 °C to 600 °C for 8000 h on modules have shown that the power output decreases by 9.5% with simultaneous increase in module's resistance by 6.6%.

Pryds [160] built a TE generator based on commercial skutterudite powders, ball milled and hot-pressed according to evaluations presented in [31], and reached an efficiency of 11.2% for a temperature gradient of 533 K. Kober [168] used the same industrial starting material (hot-pressed only) and built a TE generator with an efficiency of almost 8% (temperature gradient equals to 480 K) with small dimensions (smaller than 3 dm^3) and a mass below 8 kg.

9.9 Conclusion

We guided the reader through the various steps of developments of skutterudites concerning thermoelectric behavior, stability, mechanical properties, and thermal expansion. With high ZT values for p- and n-type skutterudites, synthesized in the laboratory, as well as produced in large quantities, with sufficiently strong mechanical parameters and almost the same coefficient of thermal expansion for p- and n-type materials, we dare to say: yes, it is a breakthrough. Technological applications of skutterudite thermoelectrics for thermoelectric power generation are ready to start, given a large-scale module producer.

References

1. H.S. Kim, W. Liu, G. Chen, C.-W. Chu, Z. Ren, PNAS **112**(27), 8205 (2015)
2. G.S. Nolas, D.T. Morelli, T.M. Tritt, Ann. Rev. Mater. Sci. **29**, 89 (1999)
3. C. Uher, Sem. Ther. **69**(5), 139 (2001)
4. G.J. Snyder, E.S. Toberer, Nat. Mater. **7**, 105 (2008)
5. A.J. Minnich, M.S. Dresselhaus, Z.F. Ren, G. Chen, Energ. Environ. Sci. **2**, 466 (2009)
6. M. Rull-Bravo, A. Moure, J.F. Fernández, S. Martín-González, RSC Adv. **5**, 41653 (2015)
7. G. Schierning, R. Chavez, R. Schmechel, B. Balke, G. Rogl, P. Rogl, Transl. Mater. Res. **2**, 025001 (2015)
8. R. Fitriani, B.D. Ovik, M.C. Long, M. Barma, M.F.M. Riaz, S.M. Sabri, S.R. Saidur, Renew. Sustain. Energy Rev. **64**(635), 659 (2016)
9. G. Rogl, P. Rogl, in: *Current Opinion in Green and Sustainable Chemistry* , ed. by S. Schorr, (IOP Publishing Bristol 2017) p. 50
10. D.M. Rowe, *Thermoelectrics Handbook Macro to Nano* (CRC Taylor&Francis, Boca Raton, FL, 2006)
11. D.M. Rowe, C.M. Bhandari, *Modern Thermoelectrics* (Reston Publishing Company, Reston, VA, 1983)
12. H.J. Goldsmid, *Introduction to Thermoelectricity*, vol 121, 2nd edn. (Springer, Berlin, Heidelberg, 2016)
13. G. Rogl, A. Grytsiv, E. Bauer, P. Rogl, *Advanced Thermoelectrics: Materials, Contacts, Devices and Systems*, ed. by Z. Ren, Y. Lan, Q. Zhang, (CRC Press, Boca Raton, FL, 2018) p.193
14. I. Oftedal, Z. für Kristallogr. **66**, 517 (1928)
15. A. Kjekshus, T. Rakke, Acta Chem. Scand. A **28**(1), 99 (1974)
16. L.D. Dudkin, N. Kh, Z. Abrikosov, Neorg. Khim. **2**(1), 12 (1957)
17. D.T. Morelli, T. Caillat, J.P. Fleurial, A. Borshchevsky, J. Vandersande, B. Chen, C. Uher, Phys. Rev. B Cond. Matter **51**(15), 9622 (1995)
18. T. Caillat, A. Borshchevsky, J.-P. Fleurial, J. Appl. Phys. **80**(8), 4442 (1996)
19. L. Bertini, K. Billquist, M. Christensen, C. Gatti, L. Holmgreen, B. Iversen, E. Mueller, M. Muhammed, G. Noriega, A. Palmqvist, D. Platzek, D.M. Rowe, A. Saramat, C. Stiewe, M. Toprak, S.G. Williams, Y. Zhang, in *Proceedings of the Twenty-Second International Conference on Thermoelectrics, ICT2003, IEEE Cat. No.03TH8726* (IEEE, Piscataway, NJ, 2003), pp. 117–120
20. W. Jeitschko, D.J. Braun, Acta Cryst. B **33**, 401 (1977)
21. G.A. Slack, V.G. Tsoukala, J. Appl. Phys. **76**, 665 (1994)
22. C. Uher, B. Chen, S. Hu, D.T. Morelli, G.P. Meisner, Symp. Mater. Res. Soc. **348**, 315 (1997)
23. L.D. Hicks, M.S. Dresselhaus, Phys. Rev. B **47**, 2727 (1993)
24. R. Venkatasubramanian, In: *Semiconductors and Semimetals*, ed. by T. Tritt (Academic Press, 2001, Nashua, NH vol. 71) p. 175
25. G. Chen, In: *Semiconductors and Semimetals*, ed. by T. Tritt (Academic Press, 2001, Nashua, NH vol. 71) p. 203
26. I. Kogut, S. Nickalov, V. Ohorodnichuk, A. Dauscher, C. Candolfi, P. Masschelein, A. Jackot, B. Lenoir, Acta Phys. Pol. A **133**(4), 879 (2018)
27. L. Deng, L.B. Wang, J.M. Qin, T. Zheng, J. Ni, X. Peng Jia, H. An Ma, Mod. Phys. Lett. B **29**, 1550095 (2015)
28. L. Guo, X. Xu, J.R. Salvador, Appl. Phys. Lett. **106**, 231902 (2015)
29. G. Rogl, A. Grytsiv, P. Heinrich, E. Bauer, P. Kumar, N. Peranio, O. Eibl, J. Horky, M. Zehetbauer, P. Rogl, Acta Mater. **91**, 227 (2015)
30. L. Zhang, A. Grytsiv, M. Kerber, P. Rogl, E. Bauer, M.J. Zehetbauer, J. Wosik, G.E. Nauer, J. Alloys Compd. **481**, 106 (2009)
31. G. Rogl, A. Grytsiv, P. Rogl, E. Bauer, M. Hochenhofer, R. Anbalagan, R.C. Mallik, E. Schafler, Acta Mater. **76**, 434 (2014)

32. A. Sesselmann, G. Skomedal, H. Middleton, E. Müller, J. Electron. Mater. **45**(3), 1397 (2016)
33. J.L. Feldman, D.J. Singh, N. Bernstein, Phys. Rev. B: Cond. Matter Mater. Phys. **89**(22), 224304 (2014)
34. M.M. Koza, M.R. Johnson, R. Viennois, H. Mutka, L. Girard, D. Ravot, Nat. Mater. **7**(10), 805 (2008)
35. M. Rotter, P. Rogl, A. Grytsiv, W. Wolf, M. Krisch, A. Mirone, Phys. Rev. B **77**(14), 144301 (2008)
36. W. Li, N. Mingo, Phys. Rev. B **89**, 184304 (2014)
37. T. He, J. Chen, H.D. Rosenfeld, M.A. Subramanian, Chem. Mater. **18**, 759 (2006)
38. L. Wang, K. Cai, Y. Wang, H. Li, H. Wang, Appl. Phys. A Mater. Sci. Process. **97**, 841 (2009)
39. Y. Zhang, C. Li, Z. Du, C. Guo, J.C. Tedenac, CALPHAD 33(2), 405 (2009)
40. V.V. Khovaylo, T.A. Korolkov, A.I. Voronin, M.V. Gorshenkov, A.T. Burkov, J. Mater. Chem. A **5**(7), 3541 (2017)
41. G.S. Nolas, H. Takizawa, T. Endo, H. Sellinschegg, D.C. Johnson, Appl. Phys. Lett. **77**(1), 52 (2000)
42. R.C. Mallik, J.-Y. Jung, V.D. Das, S.-C. Urband, I.-H. Kim, Solid State Commun. **141**(4), 233 (2007)
43. R.C. Mallik, J.-Y. Jung, V. Damodara Das, S.-C. Ur, I.-H. Kim, in *Proceedings of the Twenty-Fifth International Conference on Thermoelectrics* (ICT, Vienna, Austria, 2006), p. 431
44. C. Godart, E.B. Lopes, A.P. Goncalves, Acta Phys. Pol. A **113**, 403 (2008)
45. B.C. Sales, B.C. Chakoumakos, D. Mandrus, Phys. Rev. B **61**(4), 2475 (2000)
46. S. Choi, K. Kurosaki, Y. Ohishi, H. Muta, S. Yamanaka, J. Appl. Phys. **115**(2), 023702 (2014)
47. X. Li, B. Xu, L. Zhang, F. Duan, X. Yan, J. Yang, Y. Tian, J. Alloys Compd. **615**, 177 (2014)
48. L. Zhang, B. Xu, X. Li, F. Duan, X. Yan, Y. Tian, Mater. Lett. **139**, 249 (2015)
49. B. Chen, J.-H. Xu, C. Uher, D.T. Morelli, G.P. Meisner, J.-P. Fleurial, T. Caillat, A. Borshchevsky, Phys. Rev. B Cond. Matter **55**(3), 1476 (1997)
50. G.S. Nollas, J.L. Cohn, G.A. Slack, Phys. Rev. B **58**(1), 164 (1998)
51. X. Shi, L.D. Chen, J. Yang, G.P. Meisner, Appl. Phys. Lett. **84**, 2301 (2004)
52. Z.G. Mei, J. Yang, Y.Z. Pei, W. Zhang, L.D. Chen, J. Yang, Phys. Rev. B **77**, 045202 (2008)
53. L. Xi, J. Yang, W. Zhang, L. Chen, J. Yang, Am. Chem. Soc. **131**(15), 5563 (2009)
54. X. Shi, J. Yang, J.R. Salvador, M. Chi, J.Y. Cho, H. Wang, S. Bai, J. Yang, W. Zhang, L. Chen, J. Am. Chem. Soc. **133**, 7837 (2011)
55. X. Shi, W. Zhang, L.D. Chen, J. Yang, Phys. Rev. Lett. **95**, 185503 (2005)
56. J. Yang, W. Zhang, S.Q. Bai, Z. Mei, L.D. Chen, Appl. Phys. Lett. **90**(19), 192111 (2007)
57. Q.M. Lu, J.X. Zhang, X. Zhang, Y.Q. Liu, D.M. Liu, M.L. Zhou, J. Appl. Phys. **98**(10), 106107 (2005)
58. G. Rogl, A. Grytsiv, E. Bauer, P. Rogl, M. Zehetbauer, Intermetallics **18**, 57 (2010)
59. G. Rogl, A. Grytsiv, P. Rogl, E. Bauer, M. Zehetbauer, Intermetallics **19**, 546 (2011)
60. G. Rogl, A. Grytsiv, P. Rogl, E. Bauer, M. Zehetbauer, Solid State Phenom. **170**, 2435 (2011)
61. L. Zhang, A. Grytsiv, M. Kerber, P. Rogl, E. Bauer, M. Zehetbauer, J. Alloys Compd. **490**, 19 (2010)
62. J. Yang, G.P. Meisner, C.J. Rawn, H. Wang, B.C. Chakoumakos, J. Martin, G.S. Nolas, P.L. Pedersen, J.K. Stalick, J. Appl. Phys. **102**, 083702 (2007)
63. H. Anno, H. Tashiro, K. Matsubara, in *Proceedings of the Eighteenth International Conference on Thermoelectrics. ICT 1999 Baltimore, Maryland 1999 (Cat. No.99^{TH}8407)* (IEEE, Piscataway, NJ, 1999), p. 169
64. L. Bertini, C. Stiewe, M. Toprak, J. Appl. Phys. **93**(1), 438 (2003)
65. J.P. Fleurial, A. Borshchevsky, T. Caillat, D.T. Morelli, G.P. Meisner, in *Proceedings of the Fifteenth International Conference on Thermoelectrics, Pasadena, California, 1996. (Cat. No.96^{TH}8169)*, vol. 507 (IEEE, New York, NY, 1996), p. 91
66. X. Tang, Q. Zhang, L. Chen, T. Goto, T. Hirai, J. Appl. Phys. **97**(9), 93712 (2005)
67. K.T. Wojciechowski, J. Tobola, J. Leszczynski, J. Compd. **361**, 19 (2003)
68. X.Y. Li, L.D. Chen, J.F. Fan, W.B. Zhang, T. Kawahara, T. Hirai, J. Appl. Phys. **98**, 083702 (2005)

69. J. Y. Jung, M. J. Kim, S. W. You, S. C. Ur, I. H. Kim, in *Proceedings of the Twenty-Fifth International Conference on Thermoelectrics* (ICT, Vienna, Austria, 2006), p. 443
70. S. Liu, B.P. Zhang, J.F. Li, H.L. Zhang, L.D. Zhao, J. Appl. Phys. **102**, 103717 (2007)
71. B. Duan, P. Zhai, P. Wen, S. Zhang, L. Liua, Q. Zhang, Scr. Mater. **67**, 372 (2012)
72. L. Fu, Q. Jiang, J. Yang, J. Peng, Y. Xiao, Y. Luo, Z. Zhou, D. Zhang, J. Mater. Chem. A **4**, 16499 (2016)
73. L. Fu, J. Yang, Q. Jiang, Y. Xiao, Y. Luo, D. Zhang, Z. Zhou, J. Electron. Mater. **45**(3), 1240 (2016)
74. N. Shaheen, X. Shen, M.S. Javed, H. Zhan, L. Guo, R. Alsharafi, T. Huang, X. Lu, G. Wang, X. Zhou, J. Electron. Mater. (2016). https://doi.org/10.1007/s11664-016-5079-z
75. L. Zhang, F. Duan, X. Li, X. Yan, W. Hu, L. Wang, Z. Liu, Y. Tian, B. Xu, J. Appl. Phys. **114**, 083715 (2013)
76. R.C. Mallik, R. Anbalagan, G. Rogl, E. Royanian, P. Heinrich, E. Bauer, P. Rogl, S. Suwas, Acta Mater. **61**, 6698 (2013)
77. K.H. Park, S.C. Ur, I.H. Kim, S.M. Choi, W.S. Seo, J. Korean Phys. Soc. **57**, 1000 (2010)
78. J. Mackey, F. Dynys, B.M. Hudak, B.S. Guiton, A. Sehirlioglu, J. Mater. Sci. **51**, 6117 (2016)
79. X. Su, H. Li, G. Wang, H. Chi, X. Zhou, X. Tang, Q. Zhang, C. Uher, Chem. Mater. **23**, 2948 (2011)
80. W.S. Liu, B.P. Zhang, L.D. Zhao, J.F. Li, Chem. Mater. **20**, 7526 (2008)
81. R.C. Mallik, J.-Y. Jung, S.-C. Ur, I.-H. Kim, Metals Mater. Int. **14**, 615 (2008)
82. J. Navratil, T. Plecha, C. Drasar, V. Kucek, F. Laufek, E. Cernoskova, L. Benes, M. Vlcek, J. Electron. Mater. **45**(6), 2905 (2016)
83. A.U. Khan, K. Kobayashi, D.-M. Tang, Y. Yamauchi, K. Hasegawa, M. Mitome, Y. Xue, B. Jiang, K. Tsuchiya, D. Goldberg, Y. Bando, T. Mori, Nano Energy **31**, 152 (2017)
84. C. Xu, B. Duan, S. Ding, P. Zhai, Q. Zhang, J. Electron. Mater. **43**, 2224 (2014)
85. B. Poudel, Q. Hao, Y. Ma, Y. Lan, A. Minnich, B. Yu, X. Yan, D. Wang, A. Muto, D. Vashaee, X. Chen, J. Liu, M.S. Dresselhaus, G. Chen, Z. Ren, Science **320**, 634 (2008)
86. G.H. Zhu, H. Lee, Y.C. Lan, X.W. Wang, G. Joshi, D.Z. Wang, J. Yang, D. Vashaee, H. Gilbert, A. Pillitteri, M.S. Dresselhaus, G. Chen, Z.F. Ren, Phys. Rev. Lett. **102**(196803) (2009)
87. Y.C. Lan, A.J. Minnich, G. Chen, Z.F. Ren, Adv. Funct. Mater. **20**, 357 (2010)
88. H. Nakagawa, H. Tanaka, A. Kasama, H. Anno, K. Matsubara, in *Proceedings of the Sixteenth International Conference on Thermoelectric* (ICT, Dresden, Germany, 1997), p. 97
89. L. Yang, J.S. Wu, L.T. Zhang, J. Alloys Compd. **375**, 114 (2004)
90. Q. Jie, H. Wang, W. Liu, H. Wang, G. Chen, Z. Ren, Phys. Chem. Chem. Phys. **15**, 6809 (2013)
91. M. Short, F. Bridges, T. Keiber, G. Rogl, P. Rogl, Intermetallics **63**, 80 (2015)
92. L. Zhang, A. Grytsiv, B. Bonarski, M. Kerber, D. Setman, E. Schafler, P. Rogl, E. Bauer, G. Hilscher, M. Zehetbauer, J. Alloys Compd. **494**, 78 (2010)
93. G. Rogl, M. Zehetbauer, M. Kerber, P. Rogl, E. Bauer, Mater. Sci. Forum **1089**, 667–669 (2011)
94. G. Rogl, D. Setman, E. Schafler, J. Horky, M. Kerber, M. Zehetbauer, M. Falmbigl, P. Rogl, E. Royanian, E. Bauer, Acta Mater. **60**, 2146 (2012)
95. G. Rogl, Z. Aabdin, E. Schafler, J. Horky, D. Setman, M. Zehetbauer, M. Kriegisch, O. Eibl, A. Grytsiv, E. Bauer, M. Reinecker, W. Schranz, P. Rogl, J. Alloys Compd. **537**, 183 (2012)
96. G. Rogl, A. Grytsiv, P. Rogl, E. Royanian, E. Bauer, J. Horky, D. Setman, E. Schafler, M. Zehetbauer, Acta Mater. **61**, 6778 (2013)
97. G. Rogl, P. Rogl, in: *Thermoelectric Nanomaterials*, ed. by K. Kuomoto, T. Mori Springer-Verlag, Berlin, Heidelberg, 2013) p.193
98. G. Rogl, A. Grytsiv, P. Rogl, N. Peranio, E. Bauer, M. Zehetbauer, O. Eibl, Acta Mater. **63**, 30 (2014)
99. G. Rogl, A. Grytsiv, J. Horky, R. Anbalagan, E. Bauer, R.C. Mallik, P. Rogl, M. Zehetbauer, Phys. Chem. Chem. Phys. **17**, 3715 (2015)

100. G. Rogl, D. Setman, E. Schafler, J. Horky, M. Kerber, M. Zehetbauer, M. Falmbigl, P. Rogl, E. Bauer, in: *The NATO Science for Peace and Security Programme, Series B: Physics and Biophysics*, ed. by V. Zlatic and A. Hewson, (Springer, Dordrecht, 2012) p. 81

101. R. Anbalagan, G. Rogl, M. Zehetbauer, A. Sharma, P. Rogl, S. Suwas, R.C. Mallik, J. Electron. Mater. **43**(10), 3817 (2014)

102. G. Rogl, P. Rogl, Mater. Today Phys. **3**, 48 (2017)

103. P. Zong, X. Chen, Y. Zhu, Z. Liu, Y. Zeng, L. Chen, J. Mater. Chem. A **3**, 8643 (2015)

104. Q. Zhang, Z. Zhou, M. Dylla, M.T. Agne, Y. Pei, L. Wang, Y. Tang, J. Liao, S. Bai, W. Jiang, L. Chen, G.J. Snyder, Nano Energy **41**, 501 (2017)

105. J. Ding, H. Gu, P. Qiu, X. Chen, Z. Xiong, Q. Zheng, X. Shi, L. Chen, J. Electron. Mater. **42**(3), 382 (2013)

106. J. Ding, R.-H. Liu, H. Gu, L.-D. Chen, J. Inorg. Mater. **29**(2), 209 (2014)

107. X.Y. Zhao, X. Shi, L.D. Chen, W.Q. Zhang, S.Q. Bai, Y.Z. Pei, X.Y. Li, T. Goto, Appl. Phys. Lett. **89**, 092121 (2006)

108. L. Fu, J. Yang, J. Peng, Q. Jiang, Y. Xiao, Y. Luo, D. Zhang, Z. Zhou, M. Zhang, Y. Cheng, F. Cheng, J. Mater. Chem. A **3**(3), 1010 (2015)

109. L. Fu, J. Yang, Y. Xiao, J. Peng, M. Liu, Y. Luo, G. Li, Intermetallics **43**, 79 (2013)

110. X. Zhou, G. Wang, L. Zhang, H. Chi, X. Su, J. Sakamoto, C. Uher, J. Mater. Chem. **22**, 2958 (2012)

111. K. Peng, L. Guo, G. Wang, X. Su, X. Zhou, X. Tang, C. Uher, Sci. Adv. Mater. **9**(3-4), 682 (2017)

112. X. Shi, L.D. Chen, S.Q. Bai, X.Y. Huang, X.Y. Zhao, Q. Yao, J. Appl. Phys. **102**, 103709 (2007)

113. M. Battabyal, B. Priyadarshini, D. Sivaprahasam, N.S. Karthiselva, R. Gopalan, J. Phys. D Appl. Phys. **48**, 455309 (2015)

114. S. Katsuyama, H. Okada, J. Jpn. Soc. Powder Powder Metall. **54**(5), 375 (2007)

115. H.-Y. Zhou, W.-Y. Zhao, W.-T. Zhu, J. Yu, P. Wei, D.-G. Tang, Q.-J. Zhang, J. Electron. Mater. **43**(6), 1498 (2014)

116. S. Wan, X. Huang, P. Qiu, S. Bai, L. Chen, Mater. Des. **67**, 379 (2015)

117. P. Zong, R. Hanus, M. Dylla, Y. Tang, J. Liao, Q. Zhang, G.J. Snyder, L. Chen, Energy Environ. Sci. **10**, 183 (2017)

118. G. Rogl, J. Bursik, A. Grytsiv, S. Puchegger, V. Soprunyuk, W. Schranz, X. Yan, E. Bauer, P. Rogl, Acta Mater. **145**, 359 (2018)

119. B. Feng, J. Xie, G. Cao, T. Zhu, X. Zhao, J. Mater. Chem. A **1**, 13111 (2013)

120. J.L. Mi, X.B. Zhao, T.J. Zhu, J.P. Tu, Appl. Phys. Lett. **91**, 172116 (2007)

121. D. Zhao, M. Zuo, Z. Wang, X. Teng, H. Geng, J. Nanosci. Nanotechnol. **15**(4), 3076 (2015)

122. G. Tan, H. Chi, W. Liu, Y. Zheng, X. Tang, J. He, C. Uher, J. Mater. Chem. C **3**, 8372 (2015)

123. L. Zhou, P. Qiu, C. Uher, X. Shi, L. Chen, Intermetallics **32**, 209 (2013)

124. T. Dahal, Q. Jie, W. Liu, K. Dahal, C. Guo, Y. Lan, Z. Ren, J. Alloys Compd. **623**, 104 (2015)

125. H. Li, X. Tang, Q. Zhang, C. Uher, Appl. Phys. Lett. **94**, 102114 (2009)

126. S. Ballikaya, C. Uher, J. Alloys Compd. **585**, 168 (2014)

127. S. Ballikaya, N. Uzar, S. Yildirim, J.R. Salvador, C. Uher, J. Solid State Chem. **193**, 31 (2012)

128. G. Rogl, A. Grytsiv, K. Yubuta, S. Puchegger, E. Bauer, C. Raju, R.C. Mallik, P. Rogl, Acta Mater. **95**, 201 (2015)

129. B. Duan, J. Yang, J.R. Salvador, Y. He, B. Zhao, S. Wang, P. Wei, F.S. Ohuchi, W. Zhang, R.P. Hermann, O. Gourdon, S.X. Mao, Y. Cheng, C. Wang, J. Liu, P. Zhai, X. Tang, Q. Zhang, J. Yang, Energ. Environ. Sci. **9**, 2090 (2016)

130. S. Wang, J.R. Salvador, J. Yang, P. Wei, B. Duan, J. Yang, NPG Asia Mater. **8**, 285 (2016)

131. M. Matsubara, R. Asahi, J. Electron. Mater. **45**(3), 1669 (2016)

132. L. Zhang, G. Rogl, A. Grytsiv, S. Puchegger, J. Koppensteiner, F. Spieckermann, H. Kabelka, M. Renecker, P. Rogl, M. Zehetbauer, M.A. Carpenter, Mat. Sci. Eng. B **170**, 26 (2010)

133. G. Rogl, S. Puchegger, M. Zehetbauer, A. Grytsiv, in *Proceedings of the Materials Research Society*, 2011, 1325, 845, mrss 11–1325-e03. doi https://doi.org/10.1557/opl.2011.845

134. G. Rogl, P. Rogl, Sc. Adv. Mater. **3**, 1 (2011)
135. P. Wen, H. Mei, P. Zhai, B. Duan, J. Mater. Eng. Perform. **22**, 3561 (2013)
136. P.-A. Zong, L.-D. Chen, J. Inorg. Mater. **32**(1), 33 (2017)
137. P. Wen, P. Zhai, S. Ding, B. Duan, L. Yao, J. Electron. Mater. **46**(5), 2807 (2017)
138. A. Schmitz, C. Schmid, J. de Boor, E. Müller, J. Nanosci. Nanotechnol. **17**, 1547 (2017)
139. M.F. Ashby, *Materials Selection in Mechanical Design*, 3rd edn. (Elsevier, Oxford, 2005)
140. G. Rogl, L. Zhang, P. Rogl, A. Grytsiv, D. Rajs, H. Müller, E. Bauer, J. Koppensteiner, W. Schranz, M. Zehetbauer, J. Appl. Phys. **107**, 043507 (2010)
141. G. Rogl, A. Grytsiv, E. Royanian, P. Heinrich, E. Bauer, P. Rogl, M. Zehetbauer, S. Puchegger, M. Reinecker, W. Schranz, Acta Mater. **61**, 4066 (2013)
142. T. Caillat, J. Sakamoto, L. Lara, A. Jewell, A. Kisor, *Oral Presentation at the Twenty-Third International Conference on Thermoelectrics* (Adelaide, Australia, 2004)
143. D. Zhao, C. Tian, Y. Liu, C. Zhan, L. Chen, J. Alloys Compd. **509**, 3166 (2011)
144. J. Leszczynski, K.T. Wojciechowski, A.L. Malecki, J. Therm. Anal. Calorim. **105**(1), 211 (2011)
145. V. Savchuk, J. Schumann, B. Schupp, G. Behr, N. Mattern, D. Souptel, J. Alloys Compd. **351**(1–2), 248 (2003)
146. K.T. Wojciechowski, J. Leszczynski, R. Gajerski, *Seventh European Workshop on Thermoelectrics*, (Pamplona, Spain, 2002)
147. D. Zhao, C. Tian, S. Tang, Y. Liu, L. Chen, J. Alloys Compd. **504**(2), 552 (2010)
148. K.H. Park, W.S. Seo, S.-M. Choi, I.-H. Kim, J. Korean Phys. Soc. **64**(1), 79 (2014)
149. J.M. Peddle, W. Gaultois, P. Michael, A. Grosvenor, Inorg. Chem. **50**(13), 6263 (2011)
150. J.A. Sigrist, J.D.S. Walker, J.R. Hayes, M.W. Gaultois, A.P. Grosvenor, Solid State Sci. **13**(11), 2041 (2011)
151. X. Xia, P. Qiu, X. Huang, S. Wan, Y. Qiu, X. Li, J. Electron. Mater. **43**(6), 1639 (2014)
152. X. Xia, P. Qiu, X. Shi, X. Li, X. Huang, L. Chen, J. Electron. Mater. **41**(8), 2225 (2012)
153. K.H. Park, S.W. You, S.C. Ur, I.-H. Kim, S.-M. Choi, W.-S. Seo, J. Electron. Mater. **41**(6), 1051 (2012)
154. L.C. Sklad, M.W. Gaultois, A.P. Grosvenor, J. Alloys Compd. **505**(1), 6 (2010)
155. P. Qiu, X. Xia, X. Huang, M. Gu, Y. Qiu, L. Chen, J. Alloys Compd. **612**, 365 (2014)
156. D.K. Shin, I.H. Kim, K.H. Park, S. Lee, W.-S. Seo, Electron. Mater. **44**(6), 1858 (2015)
157. P. Brož, F. Zelenka, Intern. J. Mass Spectronom. **383–384**, 13 (2015)
158. P. Brož, F. Zelenka, Z. Kohoutek, J. Vřešťál, V. Vykoukal, J. Buršík, A. Zemanová, G. Rogl, P. Rogl, Calphad **65**, 1 (2019)
159. N.V. Nong, L.T. Hung, *Oral Presentation at EMN Meeting on Thermoelectric Materials* (Orlando, FL, 2016)
160. N. Pryds, *Oral Presentation at the Autumn School (Thermoelectrics)* (Duisburg, 2015)
161. D.K. Aswal, R. Basu, A. Singh, Energ. Conver. Manage. **114**, 50 (2016)
162. T. Caillat, A. Borshchevsky, J.G. Snyder, in *Proceedings of the AIP Conf. Albuquerque* (New Mexico, 2001), 552, 1107
163. P.H. Ngan, D.V. Christensens, G.J. Snyder, L.T. Hung, S. Linderoth, N.V. Nong, N. Pryds, Phys. Status Solidi A **211**, 9 (2014)
164. J.R. Salvador, J.Y. Cho, Z. Ye, J.E. Moczygemba, A.J. Thompson, J.W. Sharp, J.D. Koenig, R. Maloney, T. Thompson, J. Sakamoto, H. Wang, A.A. Wereszczak, Phys. Chem. Chem. Phys. **16**(24), 12510 (2014)
165. A. Muto, J. Yang, J. Poudel, Z. Ren, Adv. Energy Mater. **3**, 245 (2013)
166. J.Q. Guo, H.Y. Geng, T. Ochi, S. Suzuki, M. Kikuchi, Y. Yamaguchi, S. Ito, J. Electron. Mater. **41**, 1036 (2012)
167. T. Ochi, G. Nie, S. Suzuki, M. Kikuchi, S. Ito, J.Q. Guo, J. Electron. Mater. **43**(6), 2344 (2014)
168. M. Kober, *Oral presentation at EMN Meeting on Thermoelectric Materials* (Orlando, FL, 2016)

Chapter 10
Half-Heusler Thermoelectrics

Ran He, Hangtian Zhu, and Shuo Chen

Abstract Half-Heusler (HH) compounds are superb thermoelectric (TE) materials due to combined features of excellent TE performances, high thermal stability, mechanical robustness, simple contact, and non-toxicity. Over the past two decades, significant improvements in TE properties have been realized in a variety of compounds. Here we summarize the recent development of TE properties of HH compounds with special emphasis on the experimentally demonstrated strategies such as alloying effects, energy filter effects, nanostructuring, band convergence, etc. that were beneficial to improve TE performances.

10.1 Introduction

Burning nonrenewable fossil fuels satisfies ~80% of global energy consumptions. In this process, however, over 50% of produced energy is not used for its intended purpose and rejected as waste heat. Techniques that are capable of recovering even a small portion of waste heat could generate great benefits economically and environmentally. Thermoelectric generators (TEGs) operating on the basis of thermoelectric effects can convert some of waste heat into electricity, improve the efficiency, and reduce consumption of fossil fuel. TEGs are special due to solid-state nature. TEGs are reliable, noiseless, scalable, and maintenance free. The drawback of TEG is relatively low efficiency due to limited performance of TE materials and high cost of generated power. Therefore, it is crucially important to boost TE performances of TE materials and/or to reduce significantly those costs.

For ideal TEG, where heat can be supplied and removed effectively, and with negligible contact losses, heat-to-power conversion efficiency (η) can be expressed as:

R. He (✉)
Leibniz Institut für Festkörper- und Werkstoffforschung Dresden e.V., Institut für Metallische Werkstoffe, Dresden, Germany
e-mail: r.he@ifw-dresden.de

H. Zhu · S. Chen
Department of Physics and TcSUH, University of Houston, Houston, TX, USA

© Springer Nature Switzerland AG 2019
S. Skipidarov, M. Nikitin (eds.), *Novel Thermoelectric Materials and Device Design Concepts*, https://doi.org/10.1007/978-3-030-12057-3_10

$$\eta = \frac{T_{\mathrm{h}} - T_{\mathrm{c}}}{T_{\mathrm{h}}} \times \frac{\sqrt{1 + \mathrm{ZT}_{\mathrm{avg}}} - 1}{\sqrt{1 + \mathrm{ZT}_{\mathrm{avg}}} + \frac{T_{\mathrm{c}}}{T_{\mathrm{h}}}}, \tag{10.1}$$

where T_{h} and T_{c} are temperatures on hot and cold side and $\mathrm{ZT}_{\mathrm{avg}}$ is average value of TE figure-of-merit ZT of TE material in temperature gradient between T_{c} and T_{h}.

$$\mathrm{ZT} = \frac{\mathrm{PF}}{\kappa} T, \tag{10.2}$$

$$\mathrm{PF} = \frac{S^2}{\rho} = S^2 \sigma, \tag{10.3}$$

$$\kappa = \kappa_{\mathrm{e}} + \kappa_{\mathrm{L}}, \tag{10.4}$$

where PF, T, κ, S, ρ, σ, κ_{e}, and κ_{L} are power factor, absolute temperature, total thermal conductivity, Seebeck coefficient, electrical resistivity, electrical conductivity, electronic thermal conductivity, and lattice thermal conductivity, respectively. Consequently, higher ZT corresponds to higher conversion efficiency.

Among various TE materials, HH compounds are one of the most promising candidates for heat-to-power energy conversion applications. HH compounds are crystallized in the space group $F\bar{4}3m$ with three interpenetrating face-centered cubic (fcc) structures. HH compounds possess the general formula XYZ, where X, Y, and Z occupy Wyckoff positions $4b(\frac{1}{2}, \frac{1}{2}, \frac{1}{2})$, $4c(\frac{1}{4}, \frac{1}{4}, \frac{1}{4})$ and $4a(0, 0, 0)$, respectively, and $4d(\frac{3}{4}, \frac{3}{4}, \frac{3}{4})$ positions are voids. Among hundreds of HH compounds, HHs with valence electron count 8 and 18 are potential TE materials due to semiconductors' nature [1]. The robustness of HH compounds originates from the properties of decent figure-of-merit, ZT [2–12], high mechanical and thermal stability [5, 13, 14], low cost [15], low toxicity, and ultrahigh output power density [16, 17]. The only drawback of HHs is relativity high κ, especially κ_{L}. Therefore, suppressing the phonon transport is crucial for obtaining high performance TE properties in HH compounds.

Among the various HH compounds, MNiSn- and MCoSb-based compounds ($M = $ Hf, Zr, Ti) are the most widely studied n- and p-type HH materials. In recent years, TE properties of n-type MCoSb and p-type NbFeSb- and ZrCoBi-based compounds were also investigated, and promising ZT have been reported [6, 18–20]. For example, Fu et al. reported high ZT reaching 1.47 in Hf-doped NbFeSb at 1200 K [17]. Zhu et al. realized similar ZT $\sim$ 1.42 in ZrCoBi-based compounds at lower temperature (973 K) [20]. He et al. reported a record high PF $\sim$ 106 μW $\times$ cm^{-1} $\times$ K^{-2} in Ti-doped NbFeSb at room temperature, which, subsequently, yielded a record output power density of $\sim$22 W $\times$ cm^{-2} in between 293 K and 868 K [16]. These works boost greatly the applicability of HH compounds for TE power generation.

Successful improvement of TE properties in half-Heusler compounds was realized through either reducing in κ or boosting PF. For reducing κ, approaches such as nanoparticle inclusion, bulk nanostructuring, phase separations, alloying effects, etc.

have been studied, while strategies such as band engineering, tuning the scattering mechanism, and energy filter effects, etc. have been introduced for increasing in PF. In this chapter, we introduce these strategies in detail and present those effects in improving TE performances of HH compounds.

10.2   Suppressing the Thermal Conductivity

10.2.1   Suppressing Thermal Conductivity in Single Phase

Because of the small phonon group velocity of the optic branch, most of the heat transport in the material is attributed to acoustic branch [21]. Therefore, κ_L can be described by Callaway model [22]:

$$\kappa_L = \frac{1}{3} \int_0^{\omega_{max}} C_s(\omega) v_g(\omega)^2 \tau(\omega) d\omega, \tag{10.5}$$

where ω is frequency of phonon, C_s is spectral heat capacity, v_g is phonon group velocity, and τ is phonon relaxation time. Assuming that phonon distribution follows Boltzmann equation [21], Eq. (10.5) can be rewritten as:

$$\kappa_L = \frac{k_B}{2\pi^2 v_{ph}} \left(\frac{k_B T}{\hbar} \right)^3 \int_0^{\theta_D/T} \frac{x^4 e^x}{\tau_C^{-1}(e^x - 1)^2} dx, \tag{10.6}$$

where k_B is Boltzmann constant, v_{ph} is phonon velocity, $\hbar$ is reduced Planck constant, θ_D is Debye temperature, τ_C is combined (total) phonon relaxation time, and $x = \frac{\hbar \omega}{k_B T}$ is dimensionless variable for integration. Based on this equation, κ_L can be regarded as function of v_{ph}, θ_D, and total relaxation time τ_C. Total relaxation time τ_C is determined by combined effect of various phonon scattering processes. Assuming all scattering process is independent to each other, τ_C^{-1} can be estimated as the sum of reciprocals of relaxation times for different processes:

$$\tau_C^{-1} = \sum_i \tau_i^{-1} \tag{10.7}$$

Low κ of TE material is one of essential requirements for achieving high thermoelectric performance. Seeking new materials with intrinsically low κ_L and developing effective strategies to suppress κ of state-of-the-art TE materials become two most important approaches to improve ZT value.

The Strategies to Suppress Thermal Conductivity

Umklapp process, which is determined by intrinsic properties of the material, is the key process for phonon scattering in single crystal and dominates in high temperature thermal transport in all materials. In comparison to most of promising TE materials that have intrinsically low κ_L, half-Heusler compounds possess relatively high κ_L due to the strong bonding that facilitate phonon transport. Seeking HH compounds with intrinsically low κ_L is a good starting point on which further reduction can be achieved through manipulating the microstructure. Therefore, understanding the origins of low κ_L is beneficial to identify new promising TE materials.

Usually κ_L can be expressed as [21, 23]:

$$\kappa_L = \frac{1}{3} c_v v_{ph} l, \qquad (10.8)$$

where c_v is specific heat and v_{ph} is phonon velocity, while l is phonon mean free path. In single crystal, especially, at temperature much higher than θ_D, Umklapp scattering is dominant in phonon propagation. As will be shown later, extensive works have been introduced to reduce phonon mean free path. Meanwhile, v_{ph} as another important parameter determining κ_L can also be tailored for phonon engineering. Usually, v_{ph} is simply approximated by low-frequency sound velocity $v \propto \sqrt{\frac{B}{\delta}}$, where B is elastic modulus and δ is density of compound. Therefore, materials with low v usually tend to have low κ_L.

This was demonstrated in a comparative study of TiCoSb-, ZrCoSb-, and ZrCoBi-based HH compounds [20]. Figure 10.1 shows relationship between mean sound velocity v_m and Young's modulus E of several HH compounds. In comparison to Sn- and Sb-based half-Heusler compounds, ZrCoBi possesses the

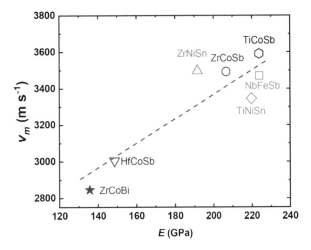

Fig. 10.1 Relationship between Young's modulus E and mean sound velocity v_m of several HH compounds [20]

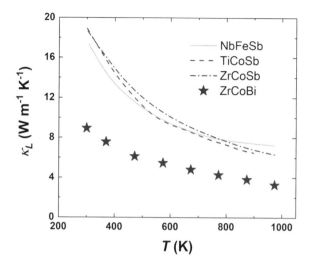

Fig. 10.2 Temperature-dependent κ_L of undoped NbFeSb, TiCoSb, ZrCoSb, and ZrCoBi [20]

lowest $v_m \sim 2850$ m $\times$ s^{-1} and Young's modulus. Low v and E originate from weaker chemical bonding and heavy atomic mass of Bi. For ZrCoBi-based compounds, strong relativistic effect of Bi contracts $6s$ shell and increases in its inertness for bonding. Therefore, low v_m and E will jointly contribute to intrinsically low κ_L of ZrCoBi.

Temperature-dependent thermal conductivities of several undoped HH compounds are plotted in Fig. 10.2 including NbFeSb, TiCoSb, ZrCoSb, and ZrCoBi [16, 19, 20, 24]. The pristine ZrCoBi shows much lower κ_L compared to the other HHs. Room temperature κ_L is equal to ~ 19 W $\times$ m^{-1} $\times$ K^{-1} for TiCoSb, $\sim$ 19 W $\times$ m^{-1} $\times$ K^{-1} for ZrCoSb, and ~ 17 W $\times$ m^{-1} $\times$ K^{-1} for NbFeSb, but $\sim$ 9 W $\times$ m^{-1} $\times$ K^{-1} for ZrCoBi, which is only half of the other p-type HHs. Significant suppressing of κ_L originates from reduced v_{ph} and low E, as suggested in Fig. 10.1 and Eq. (10.8). Based on intrinsic low κ_L, subsequent strategies were employed to further suppress κ_L, such as alloying effects and nano-bulk structuring. As a result, low $\kappa_L \sim 2.2$ W $\times$ m^{-1} $\times$ K^{-1} were realized in ZrCoBi-based compounds which further yields high ZT ~ 1.42 at 973 K, which is one of the highest value at this temperature among HH compounds [20].

However, Umklapp process is not the only source in single-phase material for phonon scattering. In real TE material, phonon has to pass through specific microstructure to transport heat and may be scattered by all kind of boundaries, point defects, dislocations, etc., that are strongly related to material processing and synthesis. Therefore, thermal conductivity reduction is the result of various scattering processes, mostly from grain boundary scattering and point defect scattering for HH compounds.

Nano-bulk structuring. Polycrystalline materials are composed of many grains with varying size and orientation. In recent years, the technique to controlling size and arrangement of the grain has been applied in TE material to manipulate of transport properties. Grain boundary can provide very effective low-frequency

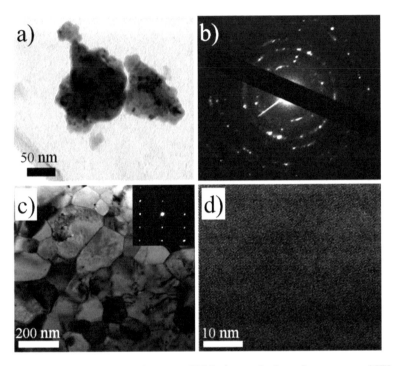

Fig. 10.3 Transmission electron microscopy (TEM) characterizations of nanostructured HH. (**a**) TEM image of the powders of $Hf_{0.5}Zr_{0.5}CoSb_{0.8}Sn_{0.2}$ after ball milling. (**b**) Electron diffraction pattern of selected area showing the polycrystalline nature of agglomerated cluster in **a**). (**c**) Low magnification TEM image of hot-pressed $Hf_{0.5}Zr_{0.5}CoSb_{0.8}Sn_{0.2}$, showing grain size of ~200 nm. Inset: Selected area electron diffraction (SAED) pattern of one grain showing single crystallization. (**d**) High-resolution TEM image showing crystallinity at atomic level [2]

phonon scattering $\tau_B^{-1} \sim \omega^0$ at low temperature. A significant performance improvement has been demonstrated in bulk $Hf_{0.5}Zr_{0.5}CoSb_{0.8}Sn_{0.2}$ with nano-bulk structuring [2]. Nanocrystalline bulk material is prepared by ball milling of arc-melted ingot to obtain nanopowders, which were sintered into a dense sample using spark plasma sintering (SPS) or hot pressing at elevated temperatures (from ~1100 to 1150 °C) and external pressure (from 50 MPa to 80 MPa). Figure 10.3a, b shows that grain sizes after ball milling are usually of ~10 nm; meanwhile, SPS samples are well crystallized with average grain size of ~200 nm, as shown in Fig. 10.3c, d. The obtained grain size is much smaller comparing with the samples prepared by using other approaches, where grain sizes are usually in the order of 10–100 μm.

Figure 10.4 shows compared TE properties of bulk and nanostructured $Hf_{0.5}Zr_{0.5}CoSb_{0.8}Sn_{0.2}$. Nanostructured compounds possess higher S and lower σ, as shown in Fig. 10.4a, b. PF of nanostructured compounds are slightly higher than bulk counterpart, as shown in Fig. 10.4c. On the other hand, a significant drop of κ occurs in nanostructured sample, as shown in Fig. 10.4d. Figure 10.4e shows similar

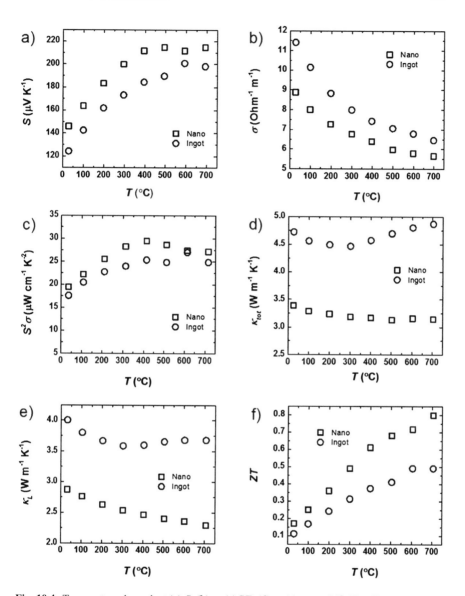

Fig. 10.4 Temperature-dependent (**a**) S, (**b**) σ, (**c**) PF, (**d**) κ, (**e**) κ_L, and (**f**) ZT of ball milled and hot-pressed $Hf_{0.5}Zr_{0.5}CoSb_{0.8}Sn_{0.2}$ sample in comparison with that of the ingot [2]

decreasing in κ_L. As a result, peak ZT ~ 0.8 has been achieved. In comparison, bulk ingot possesses of peak ZT ~ 0.5 with identical nominal compositions. Ball milling and hot-pressing techniques have also been proved very effective in preparing HH compounds with different composition, and remarkable improvements have been demonstrated [2, 4].

Point defect scattering. Fluctuations of mass and strain in the atomic scale have been employed to scatter phonon effectively. The intrinsic nature of ternary HH compounds provides great opportunity to introduce point defects. Specifically, partial substitution of isoelectric element is widely studied due to its minimum effects on transport of charge carriers. For example, M site in MCoSb- and MNiSn-based compounds could be arbitrary combinations of Hf, Zr, and Ti, albeit phase separations might occur. Due to κ sensitivity to composition ZT, optimization is possible. As an example, Fig. 10.5 shows TE properties of compounds $Hf_{1-x}Zr_xCoSb_{0.8}Sn_{0.2}$ and $Hf_{1-y}Ti_yCoSb_{0.8}Sn_{0.2}$ with diatomic and $Hf_{1-x-y}Zr_xTi_yCoSb_{0.8}Sn_{0.2}$ with triatomic combination on M site [2–4, 15]. Specifically noted from Fig. 10.5e, $Hf_{1-y}Ti_yCoSb_{0.8}Sn_{0.2}$ with diatomic combination on M site possesses much lower κ than the other two groups of compounds (diatomic $Hf_{1-x}Zr_xCoSb0.8Sn0.2$ and triatomic $Hf_{1-x-y}Zr_xTi_yCoSb_{0.8}Sn_{0.2}$) as a result of stronger point defect scattering because of larger difference between atomic mass Hf/Ti. The variation trends of κ show the effects of point defect scattering in TE performances of HH compounds.

10.2.2 Thermal Conductivity Suppression with Mixed Phases

Nanoinclusions

The reduction in κ could also be realized through a mixture of multiple phases. The lattice mismatch across phase boundaries serves as scattering center in phonon propagation. Note that the inclusion of secondary phases also scatters charge carrier, therefore, this approach is particularly useful for compounds with high κ, such as HHs, where the phonon transport could be more sensitive than electron transport to presence of secondary phases. Besides, tunable parameters, such as size and Fermi level of the inclusion, can be employed to filter low-energy electrons and thus benefits S. Experimentally, such effects were realized in HH compounds either through introduction of secondary phases, such as oxides or with phase decompositions from nominal compositions, such as phase separation and half-Heusler/full-Heusler (HH/FH) mixtures.

It's straightforward to decrease κ by incorporating dispersed secondary phases as phonon scattering centers. Experimentally reported are oxides include Al_2O_3, C_{60}, WO_3, NiO, and MO_2 in MCoSb- and MNiSn-based compounds with M being Hf, Zr, or Ti [25–31]. These oxides could be introduced either in situ or ex situ. Hsu et al. studied the effects of HfO_2 inclusions in TE properties of $Zr_{0.5}Hf_{0.5}CoSb_{1-x}Sn_x$ [32]. The in situ grown HfO_2 powders were obtained through ball milling of arc-melted ingots under atmosphere pressure (1 atm, 25 °C and 40% RH). Obtained powders were sintered at 1373 K for 15 mins under 80 MPa by SPS process. The dispersed HfO_2 nanoparticles at interfaces between grains not only impede phonon transport, but also impede grain growth during SPS procedure, which could further enhance phonon scattering. κ of HH compounds with dispersed HfO_2 was suppressed greatly. Peak ZT ~ 0.8 was obtained at 873 K, which is similar to nano-bulk structured HH compounds.

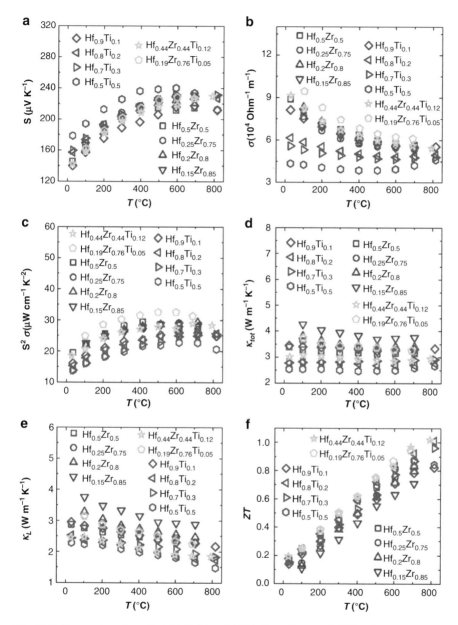

Fig. 10.5 Temperature-dependent (**a**) S, (**b**) σ, (**c**) PF, (**d**) κ, (**e**) κ_L, and (**f**) ZT of nanostructured MCoSb$_{0.8}$Sn$_{0.2}$

In another work by Poon et al., approximately 1–2 vol. % of ZrO$_2$ nanoparticles were mixed with Hf$_{0.3}$Zr$_{0.7}$CoSn$_{0.3}$Sb$_{0.7}$ powders [31]. TEM observation suggests ZrO$_2$ particles with grain size from ~50 to 80 nm at interfaces between grains, while typical grain sizes of main matrix are 5–25 μm. TE properties of p-type $Hf_{0.3}Zr_{0.7}CoSn_{0.3}Sb_{0.7}$ with and without 2 vol. % ZrO_2 nanoparticles additive were comparatively plotted in Fig. 10.6. Figure 10.6a shows that ρ increases upon addition of ZrO_2 since it is an insulator. Second-phase nanoparticles can also serve as potential barriers that prohibit low-energy charge carriers from passing through the grain boundaries, hence increasing S (Fig. 10.6b). Similar effects were also reported in HH compounds with InSb semiconductor nanoparticles [33]. Overall PF increases by up to ~13% at 970 K, as shown in Fig. 10.6c. On the other hand, κ is suppressed, especially, at elevated temperatures (Fig. 10.6d). Increased PF and suppressed κ yield increase in ZT of ~23% from 0.65 to 0.8 at 970 K (Fig. 10.6e).

Another type of inclusions, such as InSb compound and FH alloys, could not only enhance phonon scattering but also benefit charge carriers transport due to energy filter effect at the boundaries. Xie et al. studied TE properties of $TiCo_{0.85}Fe_{0.15}Sb$ with x at. % in situ formed InSb [34]. All TE parameters were simultaneously enhanced with up to 1.0 at. % InSb, i.e., σ and S become higher, and κ_L becomes lower. As shown in Fig. 10.7a, b, improved σ is due to enhanced mobility of charge carriers since mobility in TiCoSb is as low as ~0.5 $cm^2/(V \times s)$, while InSb has very high mobility of charge carriers on the order of ~10^4 $cm^2/(V \times s)$ for electrons and ~10^3 $cm^2/(V \times s)$ for holes. Furthermore, the presence of InSb yields energy filter effects, as suggested by decreased concentration of charge carriers (Fig. 10.7b) and increased S (Fig. 10.7c). Therefore, PF is greatly improved with up to 1 at. % InSb inclusions (Fig. 10.7d).

On the other hand, κ shows also descending trend with respect to increased InSb content up to 1.5 at. %, as shown in Fig. 10.7e; κ decreases not only because InSb is secondary phase but also because precipitated InSb on the grain boundaries impedes the growth of grains, which yields a stronger phonon scattering.

Figure 10.7f shows peak ZT ~ 0.3 at ~ 925 K with 1 at. % InSb inclusions as a result of simultaneously improvement of all three TE parameters. Note that with content of InSb inclusions more than 1.0 at. %, TE properties become worse, showing the existence of optimum InSb concentration. As seen from Fig. 10.7b, with InSb content exceeding 1.5 at. %, mobility of charge carriers dropped, and concentration increased. This suggests the disappearing of energy filtering effect. Similar effects were also observed in half-Heusler compounds with FH inclusions [35–37].

A small concentration of FH precipitations could effectively enhance S and σ and suppress κ [37]. In both cases, of InSb or FH inclusions, the energy filter effect originates from enlarged band gap due to the quantum confinement effects. Therefore, precipitates should be small enough; otherwise the energy filter effect vanishes. This is, indeed, the case, since there exists a maximum concentration of these precipitates, beyond which TE properties decays greatly. The details of energy filter effects will be discussed in Sect. 10.3.3.

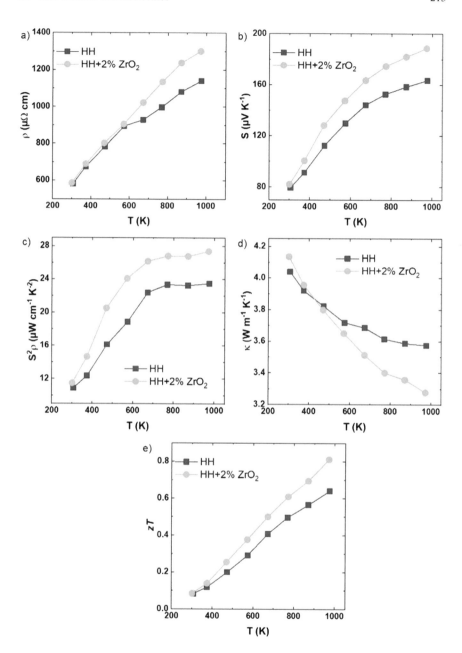

Fig. 10.6 Temperature-dependent (**a**) ρ, (**b**) S, (**c**) PF, (**d**) κ, and (**e**) ZT of p-type $Hf_{0.3}Zr_{0.7}CoSn_{0.3}Sb_{0.7}$ with and without 2 vol. % ZrO_2 additive of nanoparticles [31]

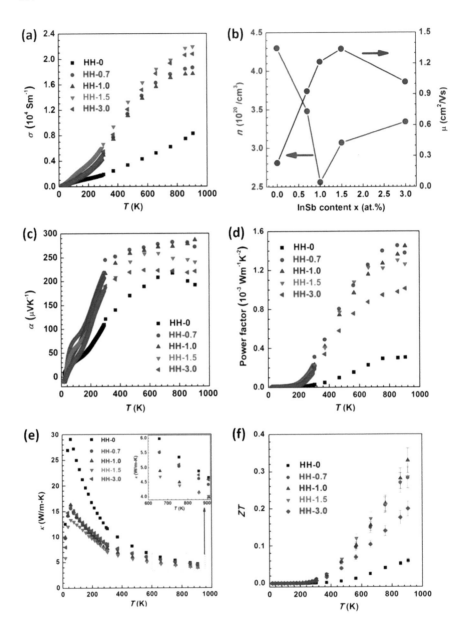

Fig. 10.7 (**a**) Temperature-dependent σ. (**b**) Dependences of concentration and mobility of charge carriers on InSb content. Temperature-dependent (**c**) S, (**d**) PF, (**e**) κ, and (**f**) ZT of $TiCo_{0.85}Fe_{0.15}Sb$ with different amount of InSb nanoinclusions [34]

Phase Separation

Another effective approach for reducing κ is phase separation. Rausch et al. studied systematically the effects of phase separation in TE performance of MCoSb-based HH compounds [38–40]. Typically, raw element materials were arc melted to form initial ingots. These ingots were annealed at 900 °C for 7 days and treated by ice water cooling. Figure 10.8a, b shows secondary electron image (SEM) with clearly visible multiple phases in the samples with nominal composition $Zr_{0.5}Hf_{0.5}CoSb_{0.8}Sn_{0.2}$ and $Ti_{0.5}Hf_{0.5}CoSb_{0.8}Sn_{0.2}$. Clearly, Hf − Ti diatomic system is more obvious in phase separation than Hf − Zr diatomic system due to similar atomic radii between Hf and Zr. This is further demonstrated in Table 10.1.

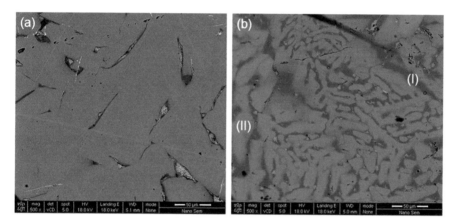

Fig. 10.8 SEM image of (**a**) $Zr_{0.5}Hf_{0.5}CoSb_{0.8}Sn_{0.2}$ and (**b**) $Ti_{0.5}Hf_{0.5}CoSb_{0.8}Sn_{0.2}$ [38]

Table 10.1 The nominal compositions; energy-dispersive X-ray spectroscopy (EDX) measured average compositions and separated phases of MCoSb$_{0.8}$Sn$_{0.2}$ [38]

M	EDX composition	Phases
Ti	$Ti_{0.97}Co_{0.99}Sb_{0.79}Sn_{0.25}$	$Ti_{1.01}Co_{0.97}Sb_{0.91}Sn_{0.01}$
		$Ti_{0.91}Co_{1.19}Sb_{0.23}Sn_{0.67}$
Zr	$Zr_{1.01}Co_{0.96}Sb_{0.77}Sn_{0.25}$	$Zr_{1.02}Co_{0.96}Sb_{0.87}Sn_{0.16}$
	$Zr_{1.01}Co_{0.96}Sb_{0.15}Sn_{0.88}$	$Zr_{1.01}Co_{0.96}Sb_{0.15}Sn_{0.88}$
Hf	$Hf_{1.07}Co_{0.93}Sb_{0.80}Sn_{0.20}$	$Hf_{1.11}Co_{0.92}Sb_{0.83}Sn_{0.14}$
		$Hf_{0.89}Co_{1.28}Sb_{0.07}Sn_{0.76}$
$Ti_{0.5}Zr_{0.5}$	$Ti_{0.49}Zr_{0.50}Co_{0.94}Sb_{0.89}Sn_{0.18}$	$Ti_{0.45}Zr_{0.56}Co_{0.96}Sb_{0.94}Sn_{0.09}$
		$Ti_{0.64}Zr_{0.38}Co_{0.98}Sb_{0.89}Sn_{0.13}$
$Zr_{0.5}Hf_{0.5}$	$Zr_{0.47}Hf_{0.56}Co_{0.94}Sb_{0.84}Sn_{0.19}$	$Zr_{0.47}Hf_{0.56}Co_{0.93}Sb_{0.88}Sn_{0.15}$
		$Zr_{0.52}Hf_{0.51}Co_{0.97}Sb_{0.72}Sn_{0.28}$
$Ti_{0.5}Hf_{0.5}$	$Ti_{0.52}Hf_{0.56}Co_{0.98}Sb_{0.65}Sn_{0.29}$	(I) $Ti_{0.65}Hf_{0.31}Co_{1.21}Sb_{0.23}Sn_{0.61}$[a]
		(II) $Ti_{0.38}Hf_{0.70}Co_{0.98}Sb_{0.84}Sn_{0.1}$[a]

[a]Roman numbers correspond to the areas in Fig. 10.8b

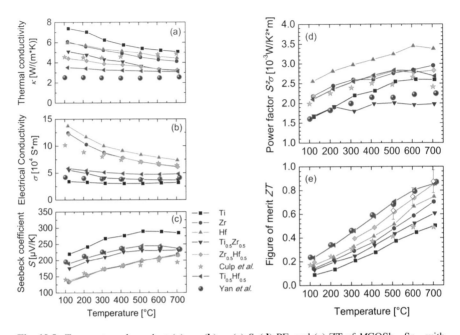

Fig. 10.9 Temperature-dependent (**a**) κ_L, (**b**) σ, (**c**) S, (**d**) PF, and (**e**) ZT of MCOSb$_{0.8}$Sn$_{0.2}$with phase separation. For comparison, TE properties of bulk ingots (star) and nanostructured (ball) Hf$_{0.5}$Zr$_{0.5}$CoSb$_{0.8}$Sn$_{0.2}$ are also plotted [38]

Figure 10.9 shows TE properties of phase-separated MCoSb$_{0.8}$Sn$_{0.2}$. For comparison, bulk (Culp et al. [41]) and nano-bulk (Yan et al. [2]) structured Hf$_{0.5}$Zr$_{0.5}$CoSb$_{0.8}$Sn$_{0.2}$ are plotted also in Fig. 10.9. Hf$_{0.5}$Zr$_{0.5}$CoSb$_{0.8}$Sn$_{0.2}$ with phase separation possesses much lower κ than that of the bulk ingot, as shown in Fig. 10.9a. On the other hand, both σ and S of Hf$_{0.5}$Zr$_{0.5}$CoSb$_{0.8}$Sn$_{0.2}$ with phase separation are similar to the bulk ingots, as shown in Fig. 10.9b, c. Accordingly, PF and ZT are plotted in Fig. 10.9d, e, respectively. Maximal ZT equals to approximately 0.9 at 700 °C for Ti$_{0.5}$Hf$_{0.5}$ and Zr$_{0.5}$Hf$_{0.5}$, which is comparable to the nano-bulk structured compound.

10.3 Enhancing Power Factor

10.3.1 Band Engineering

The thermoelectric effect in the solid is the result of electrical and thermal transport under electric field and temperature gradient. Electrical properties are determined mainly by the band structure and doping concentration of the materials. For HH compounds, the band structure could be considered as rigid and behavior of charge carriers usually fits well to single parabolic band (SPB) model within reasonable doping levels. These features facilitate the manipulation and analysis of electronic properties of HH compounds.

One important strategy to enhance PF is band engineering. Schmitt et al. and Xie et al. analyzed the band structure of ZrNiSn using different approaches and revealed the existence of certain amount of Ni at $4d$ vacancy site $(\frac{3}{4},\ \frac{3}{4},\ \frac{3}{4})$ [42–44]. The interstitial Ni yields an impurity band in the gap of ordered ZrNiSn, which is beneficial to n-type properties but undermines p-type performance of ZrNiSn. Importantly, this work reveals the sensitivity of band structures at atomic scale.

Subsequently, Fu et al. compared TE performances of p-type $Nb_{0.6}V_{0.4}FeSb$ and NbFeSb [7, 45]. The substitution of V at Nb site broadens of band edge and enhances of effective mass. The enhanced effective mass decreases in mobility of charge carriers and yields low PF. Therefore, higher TE performances were obtained by eliminating V. Peak PF $\sim$ 47 $\mu W \times cm^{-1} \times K^{-2}$ at 900 K was reported in p-type NbFeSb with concentration of charge carriers of $\sim 1.9 \times 10^{21}$ cm^{-3}, while for $Nb_{0.6}V_{0.4}FeSb$, PF is equal to only $\sim$30 $\mu W \times cm^{-1} \times K^{-2}$ at similar doping levels.

In another work by Zhu et al., the band structures of ZrCoBi, ZrCoSb, and TiCoSb were comparatively investigated [20]. As shown in Fig. 10.10a, the valence band maxima (VBM) locates at Γ point (marked by blue color), while the valence band at L point (marked by red color) show a negligible energy difference (ΔE) of $\sim$0.001 eV lower than that of Γ point. Due to the negligible energy difference, all the valence bands that converge at L and Γ points will contribute jointly to the hole transport, therefore, in total yielding a high band degeneracy of 10 for p-type ZrCoBi (Fig. 10.10b). To highlight such a high band degeneracy in ZrCoBi, band structures for isostructural HHs ZrCoSb (Fig. 10.10c) and TiCoSb (Fig. 10.10e) were also calculated for comparison. In contrast, there is an appreciable energy difference of L and Γ points for ZrCoSb ($\Delta E \sim 0.13$ eV) and TiCoSb ($\Delta E \sim 0.11$ eV), which means only the valence bands at one of the points will contribute to the hole transport. To better illustrate the differences in band degeneracy among the three compounds, iso-energy surface at 0.1 eV below VBM is plotted (Fig. 10.10b, d, f). The band degeneracy equals to 8 for ZrCoSb and only 3 for TiCoSb, both of which are noticeably lower than that of ZrCoBi.

Due to the band convergence, higher PF is expected [46]. The electronic properties of doped ZrCoBi are shown in Fig. 10.11. Figure 10.11a, b shows σ and S of Sn-doped ZrCoBi. It is noteworthy that pristine ZrCoBi shows an intrinsic n-type transport characteristic, and Sn doping (Sn content as low as $\sim$ 5 at. %) converts it successfully into fully p-type. The combined conductivity and S yield PF, as shown in Fig. 10.11c. For comparison, PF of ZrCoSb and TiCoSb with similar concentration of charge carriers are also plotted. Clearly, Sn-doped ZrCoBi shows noticeably higher PF than that of Sn-doped ZrCoSb and TiCoSb. Furthermore, band degeneracy-dependent PF at different temperatures is further plotted for the three compounds, as shown in Fig. 10.11d. PF increases monotonically with the band degeneracy at all of the temperatures. This unambiguously demonstrates that band degeneracy plays a pivotal role in PF. In other words, the high PF achieved in ZrCoBi-based compounds should be mainly ascribed to high band degeneracy for this compound as indicated by the theoretical calculations (Fig. 10.11b).

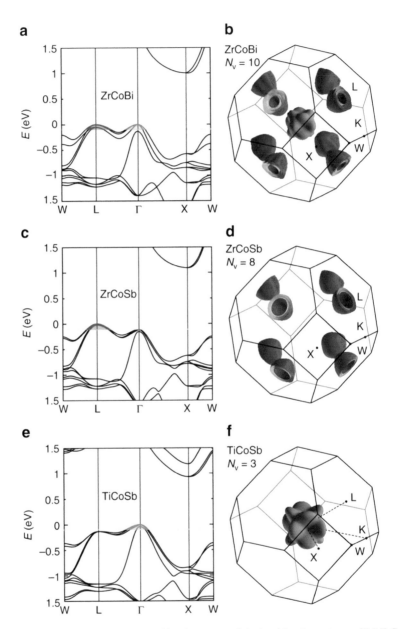

Fig. 10.10 First principle calculation of band structure. Calculated band structures of (**a**) ZrCoBi, (**c**) ZrCoSb, and (**e**) TiCoSb. The blue lines represent energy level of 0.1 eV below VBM. The corresponding iso-energy surfaces at 0.1 eV below VBM in Brillouin zone of (**b**) ZrCoBi, (**d**) ZrCoSb, and (**f**) TiCoSb [20]

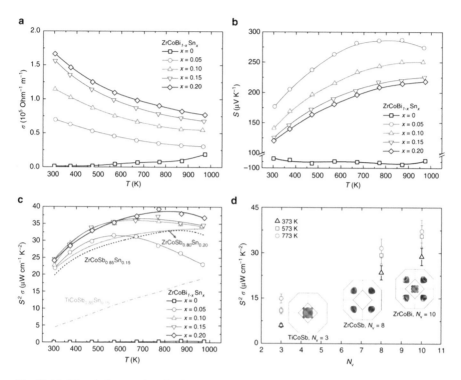

Fig. 10.11 Electrical properties of ZrCoBi$_{1-x}$Sn$_x$. Temperature-dependent (**a**) σ, (**b**) S, and (**c**) PF of ZrCoBi$_{1-x}$Sn$_x$ (x = 0, 0.05, 0.10, 0.15, and 0.20). (**d**) Band degeneracy-dependent PF for 15 at. % Sn-doped TiCoSb, ZrCoSb, and ZrCoBi at different temperatures with concentration of charge carriers of 1.62×10^{21}, 1.47×10^{21}, and 2.2×10^{21} cm^{-3}, respectively

10.3.2 Tuning the Scattering Mechanisms

Free charge carriers may interact with multiple scattering centers such as acoustic and optical phonons, ionized and neutral impurities, structural defects (dislocations, vacancies, et al.) and other obstacles. In most good TE materials, charge carriers scatter mostly by acoustic phonons and ionized impurities. For acoustic phonon scattering, the lattice vibrations are represented by longitudinal and transverse waves. When these waves are propagated along a crystal, an additional periodic potential is superimposed on the internal periodic potential. This alters the energy structure of the crystal, which is equivalent to an alternating potential energy of charge carriers, known as the deformation potential. On the other hand, an ionized impurity center in the lattice of a semiconductor produces long-range Coulomb field, which scatters charge carrier. The ionized impurity scattering is highly related to the amount of guest atoms and defects. Problems of mixed scattering have been considered with Boltzmann type electron gas in calculations of σ, Hall coefficient R_H, S, and Nernst-Ettingshausen effect [47]. A simpler consideration involves power-law dependence of mobility of charge carriers with respect to temperature. For heavily doped semiconductor, mobility of charge carriers follows $\sim T^{-3/2}$ and $\sim T^{3/2}$ variation trends if the dominate scattering originates from acoustic phonons and ionized impurities, respectively. Tuning scatter mechanisms from ionized impurities to acoustic phonons was shown effective in improving TE performance in Mg_3Sb_2-based compounds [48]. Similar effects were also observed in HH compounds $Nb_{0.95}Ti_{0.05}FeSb$ with different sintering temperatures. By increasing sintering temperature, variation of σ with respect to temperature gradually approaches to $\sim T^{-3/2}$ law in the vicinity of room temperature (Fig. 10.12a) [16]. As a result, the value of PF at room temperature increases from ~ 50 $\mu W \times cm^{-1} \times K^{-2}$ to ~ 106 $\mu W \times cm^{-1} \times K^{-2}$, as shown in Fig. 10.12b. The obtained PF is not only the highest among HH compounds but also much higher than most TE materials. This work validates the effectiveness in improve TE performances by tuning scattering mechanisms of charge carriers.

Fig. 10.12 Temperature dependences (**a**) σ and (**b**) PF of $Nb_{0.95}Ti_{0.05}FeSb$ sintered at temperatures 1073 K, 1173 K, 1273 K and 1373 K [16]

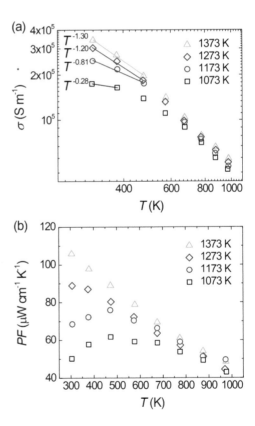

10.3.3 Energy Filter Effect

Based on Mott formula, Seebeck coefficient can be written as:

$$S = \frac{\pi^2}{3} \times \frac{k_B}{q} \times k_B T \left\{ \frac{d[\ln(\sigma(E))]}{dE} \right\}_{E=E_F}$$

$$= \frac{\pi^2}{3} \times \frac{k_B}{q} \times k_B T \left\{ \frac{1}{n} \frac{dn(E)}{dE} + \frac{1}{\mu} \frac{d\mu(E)}{dE} \right\}_{E=E_F}, \quad (10.9)$$

where q, E_F, $n(E)$, and $\mu(E)$ are electron charge, Fermi energy, concentration, and mobility of charge carriers at energy E, respectively [49]. Enhancing both $\frac{dn(E)}{dE}$ and $\frac{d\mu(E)}{dE}$ will lead to high S, where the latter case can be realized through selective scattering of low-energy charge carriers near E_F, also named energy filtering effect. To enable scattering, potential barriers near E_F are necessary. In principle, any types of interfaces can produce potential barriers, but effective barriers require sophisticate designs. For example, single-phase boundaries, such as defective grain boundaries, may induce potential barriers. However, Bachmann et al. point out theoretically that double Schottky barriers due merely to the surface states are determined by chemical potential and concentrations of charge carriers as well [50]. When the matrix is heavily doped with concentration of charge carriers above 10^{20} cm^{-3}, potential barriers may become screened and ineffective. The paper suggests also other selective scattering centers, such as structural disorders at the grain boundaries, can still result in energy filtering.

Heterojunctions, where electronic band mismatch of different compositions takes place, can also create barriers together with the structural disorders. For instance, Chen et al. sintered 0, 3, 6, and 9 vol. % ZrO$_2$ nanoparticles with n-type Zr$_{0.5}$Hf$_{0.5}$Ni$_{0.8}$Pd$_{0.2}$Sn$_{0.99}$Sb$_{0.01}$ [29]. The resulting grain size of ZrO$_2$ is around 20 nm. The samples with 6 and 9 vol. % ZrO$_2$ both show increased in absolute value peak S of ~ -170 μV/K, compared to ~ -150 μV/K of ZrO$_2$ free sample. However, mobilities of charge carriers are reduced with increasing in vol. % of ZrO$_2$. Correspondingly, the sample with 9 vol. % ZrO$_2$ exhibits the highest resistivity and the lowest PF among all samples. The sample with 6 vol. % ZrO$_2$ shows ~15% enhanced PF compared with ZrO$_2$ free sample (~30 μW $\times$ cm^{-1} $\times$ K^{-2} vs. ~26 μW $\times$ cm^{-1} $\times$ K^{-2}). In addition to physical mixing of two compounds to create heterojunctions, the interfaces can also be produced by in situ precipitation. For instance, Xie et al. produced (0, 1, 3, and 7 at. %) InSb nanoinclusions (10–300 nm) at the grain boundaries of Ti$_{0.5}$Zr$_{0.25}$Hf$_{0.25}$Co$_{0.95}$Ni$_{0.05}$Sb via induction melting, pulverization, and spark plasma sintering [33]. Bulk InSb has higher mobility of charge carriers $\sim 10^4$ cm^2/(V $\times$ s) compared with HH matrix 1–10 cm^2/(V $\times$ s). However, in this work, mobilities of charge carriers reduced after introducing InSb, which is probably due to enhanced boundary scatterings. Increased S in the samples with InSb suggest that the scattering is stronger toward low-energy electrons. In addition, these boundaries serve as dopants to improve σ. Among different samples, the one with 1 at. % InSb shows the highest S and σ with PF improved by nearly twofold at 570 K ($\sim$15 $\mu W \times cm^{-1} \times K^{-2}$) compared with that of HH matrix. Interestingly, Xie et al. observed also enhanced S in p-type HH with 0.7, 1.0, 1.5, and 3.0 at. % InSb nanoinclusions [34]. The sample preparation follows the same procedure as that of n-type work. However, electron mobility increases with InSb concentration and is higher than that of the sample without InSb. This trend is opposite to n-type HH case. In addition, the samples contain InSb all show higher S compared with InSb free sample. The authors again attribute it to the energy filtering effect at InSb-HH interfaces. Overall, the sample with 1.0 at. % InSb shows the highest peak $S \sim 285$ $\mu V \times K^{-1}$ at 900 K. It also exhibits the highest PF = 15 $\mu W \times cm^{-1} \times K^{-2}$ at 900 K, i.e., $\sim$360% improvement compared with InSb free sample.

Another type of heterojunction, coherent interfaces between in situ formed HH matrix and FH nanoinclusion (HH/FH), has been studied extensively, especially, by Poudeu's group [36, 37, 51–56]. The most interesting property of these interfaces is structural coherency, which leads to both energy filtering and preservation of charge mobilities, in contrast to reduced mobilities due to the incoherent boundaries introduced in previous paragraph. The effectiveness of such interfaces is reported in both n-type HH and p-type HH. In general, FH has composition of XY_2Z. The coherent HH/FH interfaces can be formed by adding excessive Y during the alloying of XYZ. The distinct crystal structures of FH (space group $Fm\overline{3}m$) and HH (space group $F\overline{4}3m$) lead to phase separation and formation of FH inclusions in HH matrix, while the similar atomic distributions of FH and HH result in coherent boundaries. Although bulk FH is metallic, at the dimension of 10 nm and below, quantum confinement can turn FH to semiconducting. Accordingly, the potential barriers near HH/FH vary significantly. In addition, Fermi level of HH matrix can be tuned by doping. Last, the distances between barriers should be greater than carrier momentum relaxation length and less than carrier energy relaxation length [57]. Therefore, size, number density, and distribution of FH, as well as, composition of HH matrix, all play important roles in energy filtering.

For example, in n-type HH $Zr_{0.25}Hf_{0.75}NiSn$ compound, extra Ni (2 at. % and 5 at. %, respectively) was added during solid-state reaction of elements in powder forms [37]. The mixture was held at $\sim$$10^{-4}$ Torr and 1173 K for 2 days. TEM observation indicates formation of $Zr_{0.25}Hf_{0.75}Ni_2Sn$ nanoinclusions in both samples. In the sample containing 2 at. % extra Ni, nanoinclusions' sizes range from <1 to 3 nm and HH/FH interfaces are coherent. The room temperature S is 67% higher than that of pure HH matrix. The authors attribute the enhancement to energy filtering by HH/FH interfaces, as they observed a threefold decrease in room temperature concentration of charge carriers (from 5.6×10^{19} cm^{-3} for HH matrix to 1.9×10^{19} cm^{-3} for the sample with FH). Interestingly, the authors also observe 80% increase in mobility of charge carriers at 300 K. They propose that the higher mobility is due to weakened electron-electron scattering when concentration of charge carriers decreases.

In the sample with 5 at. % extra Ni, sizes of FH nanoinclusions increase to 1–20 nm, and lamellar structures with 2–8 nm thick and up to 30 nm long show up. As a result, the large metallic FH nanoinclusions serve as dopants, and the room temperature concentration of charge carriers nearly doubles that of HH matrix, despite the energy filtering at HH/FH interfaces. This again indicates the importance of size control of FH. Besides using $Zr_{0.25}Hf_{0.75}NiSn$ matrix, energy filtering by HH/FH is also effective with $Zr_{0.25}Hf_{0.75}Ni_{1+x}Sn_{0.975}Bi_{0.025}$ [37] and $Ti_{0.1}Zr_{0.9}Ni_{1+x}Sn$ [52] as the matrix, which is evidenced by decrease in concentration of charge carriers, higher mobility, and improved S after introducing FH nanoinclusions. However, if HH matrix is heavily doped by Sb in solid-state reaction produced $Zr_{0.25}Hf_{0.75}Ni_{1+x}Sn_{1-y}Sb_y$ ($y = 0.01$, 0.025) [55] and $Ti_{0.1}Zr_{0.9}Ni_{1+x}Sn_{0.975}Sb_{0.025}$ [53], energy filtering at HH/FH is greatly suppressed, especially when large (>20 nm) FH inclusions exist. This can be explained by large population of Sb and metallic FH-introduced extrinsic charge carriers, whose energy is high enough to cross the potential barrier. Furthermore, ball milling of the ingots obtained by solid-state reactions will significantly increase in the number of small FH nanoinclusions (<10 nm) and boost the effect of energy filtering, even when HH matrix is heavily doped $Ti_{0.4}Hf_{0.6}Ni_{1+x}Sn_{0.975}Sb_{0.025}$ [54].

The stability of FH inclusion on HH matrix has been studied by Chai, et al. on n-type $ZrNi_{1.1}Sn$ [56]. The materials were prepared by arc melting and solidified by optical floating zone method. As-prepared material contains $ZrNi_2Sn$ FH nanoprecipitates in plate (semi-coherent, average size of 330 nm long, 20 nm thick, and 100 nm wide) and disk (coherent, average size of $\sim$ 110 nm in diameter and $\sim$ 7 nm thick) morphologies. These nanoprecipitates orient preferentially along $\langle 001 \rangle$ directions of ZrNiSn HH matrix, and the interfaces are sharp under TEM [56, 58, 59]. Room temperature S of as-prepared sample is equal to $-208\ \mu V \times K^{-1}$, which is 20% higher than that of ZrNiSn ($-170\ \mu V \times K^{-1}$), presumably due to the energy filtering effect from the precipitates. In contrast, after the sample is measured up to 1073 K and cooled down, room temperature S drops to $-180\ \mu V \times K^{-1}$ and does not change much after another heating-cooling cycle. TEM shows that size of precipitates decreases to <20 nm after the cycles. However, smaller precipitates do not lead to stronger energy filtering as expected based on Poudeu's papers. The authors attribute the suppressed energy filtering to a reduced sharpness of HH/FH interfaces, stronger effect than size reduction, after the sample being heated to above the dissolution temperature of FH precipitates (1067 K).

In p-type HH, Poudeu's group firstly reports that HH/FH interfaces enable also energy filtering in $Ti_{0.5}Hf_{0.5}Co_{1+x}Sb_{0.9}Sn_{0.1}$ ($x = 0$, 0.04, 0.05, 0.06, 0.07, 0.08) [35]. The samples were prepared by solid-state reactions of elements in powder form at 10^{-4} Torr and 1173 K for 14 days. The powders were then sintered at 900 °C and 100 MPa. At $x = 0.05$, 5–60 nm spherical FH nanoinclusions are formed with coherent to semi-coherent interfaces with HH matrix. At room temperature, hole concentration in sample with $x = 0.05$ is equal to $\sim 2.8 \times 10^{20}\ cm^{-3}$, which is much lower than that of HH matrix ($\sim 8.0 \times 10^{20}\ cm^{-3}$). The authors exclude electron doping from metallic FH as the reason of reduced hole concentration, because excitations of intrinsic charge carriers are negligible at room temperature. In addition, room temperature hole mobility increases from $\sim$0.9 to $\sim$1.5 $cm^2/(V \times s)$.

The increase may be due to the following reasons: (1) suppressed scattering of charge carriers with decreased hole concentration, (2) weakened scattering by ionized impurities due to spatial separation of ionized impurities in HH matrix and holes in the valence band of FH inclusion, and (3) higher mobility of charge carriers with higher energy (passing through the energy barriers at HH/FH interfaces). At 300 K, as x increases from 0.05 to 0.08, σ increases from 62 S $\times$ cm^{-1} to 135 S $\times$ cm^{-1}, while S drops from 300 to 140 μV $\times$ K^{-1}. The origin of such changes isn't discussed in the paper. Nevertheless, even at $x = 0.08$, S of HH/FH composite sample is still 40% higher than that of HH matrix, indicating effective energy filtering at HH/FH boundaries. Besides this work, Chauhan et al. report also about energy filtering due to HH/FH interfaces in another p-type compound, $ZrCo_{1+x}Sb_{0.9}Sn_{0.1}$ ($x = 0.01$, 0.03, 0.05) [36].

Overall, energy filtering takes place at varied types of boundaries in HH. At that, incoherent interfaces degrade mobility of charge carriers and are less favorable than coherent ones. However, the existence of interface potential barriers and energy filtering effects is speculated through transport measurements on the whole samples. More work on quantifying magnitude of individual potential barrier and efficient controls of dimension, distribution, microstructure, and electronic structures of interfaces are challenging but highly desired.

10.4 Summary

In this chapter, we introduced TE properties of half-Heusler compounds. Promising ZT and PF have been found in material systems including MCoSb, MNiSn, NbFeSb, ZrCoBi, etc. Peak ZT values of 1–1.5 were reported in a variety of compounds. A record high output power density of ~22 W $\times$ cm^{-2} was also realized in NbFeSb-based compounds. Specifically, we summarize the strategies for improving thermoelectric performance. Band engineering, tuning scattering mechanisms of charge carriers, and energy filtering effects are the common strategies for PF enhancement; meanwhile nano-bulk structuring, alloying effects, and phase separations, etc. have been experimentally demonstrated effective to reduce κ_L. The success of these strategies suggests also the great potentials of half-Heusler compounds for power generation applications.

References

1. T. Graf, C. Felser, S.S.P. Parkin, Prog. Solid State Chem. **39**(1), 1–50 (2011)
2. X. Yan, G. Joshi, W. Liu, Y. Lan, H. Wang, S. Lee, J.W. Simonson, S.J. Poon, T.M. Tritt, G. Chen, Z.F. Ren, Nano Lett. **11**(2), 556–560 (2011)
3. X. Yan, W. Liu, H. Wang, S. Chen, J. Shiomi, K. Esfarjani, H. Wang, D. Wang, G. Chen, Z. Ren, Energy Environ. Sci. **5**(6), 7543 (2012)

4. X. Yan, W. Liu, S. Chen, H. Wang, Q. Zhang, G. Chen, Z. Ren, Adv. Energy Mater. **3**(9), 1195–1200 (2013)
5. G. Joshi, R. He, M. Engber, G. Samsonidze, T. Pantha, E. Dahal, K. Dahal, J. Yang, Y. Lan, B. Kozinsky, Z. Ren, Energy Environ. Sci. **7**(12), 4070–4076 (2014)
6. J. Yu, C. Fu, Y. Liu, K. Xia, U. Aydemir, T.C. Chasapis, G.J. Snyder, X. Zhao, T. Zhu, Adv. Energy Mater. **8**(1), 1701313 (2018)
7. C. Fu, T. Zhu, Y. Liu, H. Xie, X. Zhao, Energy Environ. Sci. **8**(1), 216–220 (2015)
8. S. Chen, K.C. Lukas, W. Liu, C.P. Opeil, G. Chen, Z. Ren, Adv. Energy Mater. **3**(9), 1210–1214 (2013)
9. G. Joshi, T. Dahal, S. Chen, H. Wang, J. Shiomi, G. Chen, Z. Ren, Nano Energy **2**(1), 82–87 (2013)
10. C. Yu, T.-J. Zhu, R.-Z. Shi, Y. Zhang, X.-B. Zhao, J. He, Acta Mater. **57**(9), 2757–2764 (2009)
11. H. Xie, H. Wang, Y. Pei, C. Fu, X. Liu, G.J. Snyder, X. Zhao, T. Zhu, Adv. Funct. Mater. **23** (41), 5123–5130 (2013)
12. G. Rogl, P. Sauerschnig, Z. Rykavets, V.V. Romaka, P. Heinrich, B. Hinterleitner, A. Grytsiv, E. Bauer, P. Rogl, Acta Mater. **131**, 336–348 (2017)
13. R. He, S. Gahlawat, C. Guo, S. Chen, T. Dahal, H. Zhang, W. Liu, Q. Zhang, E. Chere, K. White, Z. Ren, Phys. Status Solidi A **212**(10), 2191–2195 (2015)
14. G. Rogl, A. Grytsiv, M. Gürth, A. Tavassoli, C. Ebner, A. Wünschek, S. Puchegger, V. Soprunyuk, W. Schranz, E. Bauer, H. Müller, M. Zehetbauer, P. Rogl, Acta Mater. **107**, 178–195 (2016)
15. R. He, H.S. Kim, Y. Lan, D. Wang, S. Chen, Z. Ren, RSC Adv. **4**(110), 64711–64716 (2014)
16. R. He, D. Kraemer, J. Mao, L. Zeng, Q. Jie, Y. Lan, C. Li, J. Shuai, H.S. Kim, Y. Liu, D. Broido, C.-W. Chu, G. Chen, Z. Ren, Proc. Natl. Acad. Sci. **113**(48), 13576–13581 (2016)
17. C. Fu, S. Bai, Y. Liu, Y. Tang, L. Chen, X. Zhao, T. Zhu, Nat. Commun. **6**, 8144 (2015)
18. Y. Liu, C. Fu, K. Xia, J. Yu, X. Zhao, H. Pan, C. Felser, T. Zhu, Adv. Mater. **30**, e1800881 (2018)
19. R. He, H. Zhu, J. Sun, J. Mao, H. Reith, S. Chen, G. Schierning, K. Nielsch, Z. Ren, Mater. Today Phys. **1**, 24–30 (2017)
20. H. Zhu, R. He, J. Mao, Q. Zhu, C. Li, J. Sun, W. Ren, Y. Wang, Z. Liu, Z. Tang, Nat. Commun. **9**(1), 2497 (2018)
21. D.T. Morelli, G.A. Slack, High lattice thermal conductivity solids, in *High Thermal Conductivity Materials*, ed. by S. L. Shindé, J. S. Goela, (Springer, New York, NY, 2006)
22. J. Callaway, Phys. Rev. **113**(4), 1046–1051 (1959)
23. E.S. Toberer, A. Zevalkink, G.J. Snyder, J. Mater. Chem. **21**(40), 15843–15852 (2011)
24. M. Zhou, L. Chen, C. Feng, D. Wang, J.-F. Li, J. Appl. Phys. **101**(11), 113714 (2007)
25. X.Y. Huang, Z. Xu, L.D. Chen, X.F. Tang, Key Eng. Mater. **249**, 79–82 (2003)
26. X.Y. Huang, Z. Xu, L.D. Chen, Solid State Commun. **130**(3–4), 181–185 (2004)
27. R. Yaqub, P. Sahoo, J.P.A. Makongo, N. Takas, P.F.P. Poudeu, K.L. Stokes, Sci. Adv. Mater. **3** (4), 633–638 (2011)
28. D.K. Misra, J.P.A. Makongo, P. Sahoo, M.R. Shabetai, P. Paudel, K.L. Stokes, P.F.P. Poudeu, Sci. Adv. Mater. **3**(4), 607–614 (2011)
29. L.D. Chen, X.Y. Huang, M. Zhou, X. Shi, W.B. Zhang, J. Appl. Phys. **99**(6), 064305 (2006)
30. X.Y. Huang, L.D. Chen, X. Shi, M. Zhou, Z. Xu, Key Eng. Mater. **280–283**, 385–388 (2007)
31. S.J. Poon, D. Wu, S. Zhu, W. Xie, T.M. Tritt, P. Thomas, R. Venkatasubramanian, J. Mater. Res. **26**(22), 2795–2802 (2011)
32. C.-C. Hsu, Y.-N. Liu, H.-K. Ma, J. Alloys Compd. **597**, 217–222 (2014)
33. W.J. Xie, J. He, S. Zhu, X.L. Su, S.Y. Wang, T. Holgate, J.W. Graff, V. Ponnambalam, S.J. Poon, X.F. Tang, Acta Mater. **58**(14), 4705–4713 (2010)
34. W.J. Xie, Y.G. Yan, S. Zhu, M. Zhou, S. Populoh, K. Gałązka, S.J. Poon, A. Weidenkaff, J. He, X.F. Tang, T.M. Tritt, Acta Mater. **61**(6), 2087–2094 (2013)
35. P. Sahoo, Y. Liu, J.P.A. Makongo, X.-L. Su, S.J. Kim, N. Takas, H. Chi, C. Uher, X. Pan, P.F.P. Poudeu, Nanoscale **5**(19), 9419 (2013)

36. N.S. Chauhan, A. Bhardwaj, T.D. Senguttuvan, R.P. Pant, R.C. Mallik, D.K. Misra, J. Mater. Chem. C **4**(24), 5766–5778 (2016)
37. J.P.A. Makongo, D.K. Misra, X. Zhou, A. Pant, M.R. Shabetai, X. Su, C. Uher, K.L. Stokes, P.F.P. Poudeu, J. Am. Chem. Soc. **133**(46), 18843–18852 (2011)
38. E. Rausch, B. Balke, S. Ouardi, C. Felser, Phys. Chem. Chem. Phys. **16**(46), 25258–25262 (2014)
39. E. Rausch, B. Balke, T. Deschauer, S. Ouardi, C. Felser, APL Mater. **3**(4), 041516 (2015)
40. E. Rausch, B. Balke, S. Ouardi, C. Felser, Energ. Technol. **3**(12), 1217–1224 (2015)
41. S.R. Culp, J.W. Simonson, S.J. Poon, V. Ponnambalam, J. Edwards, T.M. Tritt, Appl. Phys. Lett. **93**(2), 022105 (2008)
42. J. Schmitt, Z.M. Gibbs, G.J. Snyder, C. Felser, Mater. Horiz. **2**(1), 68–75 (2015)
43. H.-H. Xie, J.-L. Mi, L.-P. Hu, N. Lock, M. Chirstensen, C.-G. Fu, B.B. Iversen, X.-B. Zhao, T.-J. Zhu, Cryst. Eng. Comm. **14**(13), 4467 (2012)
44. H. Xie, H. Wang, C. Fu, Y. Liu, G.J. Snyder, X. Zhao, T. Zhu, Sci. Rep. **4**(1), 6888 (2014)
45. C. Fu, T. Zhu, Y. Pei, H. Xie, H. Wang, G.J. Snyder, Y. Liu, Y. Liu, X. Zhao, Adv. Energy Mater. **4**(18), 1400600 (2014)
46. Y. Pei, X. Shi, A. LaLonde, H. Wang, L. Chen, G.J. Snyder, Nature **473**(7345), 66–69 (2011)
47. J.P. Heremans, C.M. Thrush, D.T. Morelli, Phys. Rev. B **70**(11), 115334 (2004)
48. J. Shuai, J. Mao, S. Song, Q. Zhu, J. Sun, Y. Wang, R. He, J. Zhou, G. Chen, D.J. Singh, Z. Ren, Energy Environ. Sci. **10**(3), 799–807 (2017)
49. M. Cutler, N.F. Mott, Phys. Rev. **181**(3), 1336–1340 (1969)
50. M. Bachmann, M. Czerner, C. Heiliger, Phys. Rev. B **86**(11), 245115 (2012)
51. J.P.A. Makongo, D.K. Misra, J.R. Salvador, N.J. Takas, G. Wang, M.R. Shabetai, A. Pant, P. Paudel, C. Uher, K.L. Stokes, P.F.P. Poudeu, J. Solid State Chem. **184**(11), 2948–2960 (2011)
52. Y. Liu, P. Sahoo, J.P.A. Makongo, X. Zhou, S.-J. Kim, H. Chi, C. Uher, X. Pan, P.F.P. Poudeu, J. Am. Chem. Soc. **135**(20), 7486–7495 (2013)
53. Y. Liu, A. Page, P. Sahoo, H. Chi, C. Uher, P.F.P. Poudeu, Dalton Trans. **43**(21), 8094–8101 (2014)
54. P. Sahoo, Y. Liu, P.F.P. Poudeu, J. Mater. Chem. A **2**(24), 9298–9305 (2014)
55. Y. Liu, J.P.A. Makongo, A. Page, P. Sahoo, C. Uher, K. Stokes, P.F.P. Poudeu, J. Solid State Chem. **234**, 72–86 (2016)
56. Y.W. Chai, T. Oniki, Y. Kimura, Acta Mater. **85**, 290–300 (2015)
57. Y. Nishio, T. Hirano, Jpn. J. Appl. Phys. **36**(Part 1, 1A), 170–174 (1997)
58. Y.W. Chai, Y. Kimura, Acta Mater. **61**(18), 6684–6697 (2013)
59. Y. Wang Chai, Y. Kimura, Appl. Phys. Lett. **100**(3), 033114 (2012)

Part II
Novel Inorganic Materials

Chapter 11
Polymer-Derived Ceramics: A Novel Inorganic Thermoelectric Material System

Rakesh Krishnamoorthy Iyer, Adhimoolam Bakthavachalam Kousaalya, and Srikanth Pilla

Abstract Ever-increasing global energy consumption and its resultant impact on climate change have pushed researchers toward the development of multiple novel sustainable technologies. Among these technologies, utilization of waste heat energy via use of thermoelectric generators (TEGs) has gained credence over the past few years, especially given the vast improvement in TEGs conversion efficiency through innovative and unique thermoelectric (TE) materials. However, these improvements have been marked by concerns regarding poor thermal and mechanical stability at higher (operational) temperatures, as well as toxicity and high cost of specific elements used in such TEGs, thereby rendering TEGs unfit for commercial purposes. Polymer-derived ceramics (PDCs) offer a possible solution to this unsolved conundrum, given the excellent mechanical and thermal stability of PDCs (till as high as 2000 °C), along with the relatively low toxicity and abundance of elements used in such systems. PDCs also provide the opportunity for tuning of microstructure to achieve the desired properties via control of polymeric precursor and processing parameters. This chapter provides an overview of PDCs, focusing on TE properties of PDCs in detail, especially with regard to changes in operational temperature, as well as the underlying electrical conduction mechanism observed in these PDCs. Finally, a succinct summary of the existing literature, as well as the desired direction of work for truly harnessing potential as commercial thermoelectrics, has also been provided.

R. K. Iyer · A. B. Kousaalya
Department of Automotive Engineering, Clemson University, Greenville, SC, USA

Clemson Composites Center, Clemson University, Greenville, SC, USA

S. Pilla (✉)
Department of Automotive Engineering, Clemson University, Greenville, SC, USA

Clemson Composites Center, Clemson University, Greenville, SC, USA

Department of Materials Science and Engineering, Clemson University, Clemson, SC, USA

Department of Mechanical Engineering, Clemson University, Clemson, SC, USA
e-mail: spilla@clemson.edu

© Springer Nature Switzerland AG 2019
S. Skipidarov, M. Nikitin (eds.), *Novel Thermoelectric Materials and Device Design Concepts*, https://doi.org/10.1007/978-3-030-12057-3_11

11.1 Introduction

Ever-increasing global energy consumption and its associated impacts on anthropogenic climate change have led to enhanced attention on sustainable, clean, and green energy technologies. Among such technologies, efficient generation and utilization of energy by conversion of waste heat to electricity via TEGs remains an attractive possibility. TEGs can be highly beneficial for sectors and processes that generate enormous amounts of waste heat, such as home heating, nuclear and thermal power plants, automotive exhausts, industrial processes, and waste incineration, all of which together cause wastage of approximately two-thirds of total primary energy used [1]. Application of TEGs across all these sectors can help improve the efficiency of fossil fuel usage, reduce the associated consumer bills, and also reduce both pollutant and greenhouse gas emissions, thereby mitigating local pollution and climate change, as well as resultant impacts on our planet.

However, commercial application of TEGs necessitates high conversion efficiencies (from waste heat to electricity), exceeding 5% [2]. Efficiency (η) of TEG is determined by its TE figure of merit (ZT)—a dimensionless quantity that is desired to be high for a material to exhibit excellent TE performance (Eq. 11.1). ZT of a material in turn is a function of three properties: thermal conductivity (κ) which can be divided into electronic (κ_e) and lattice (κ_L) components, electrical conductivity (σ), and Seebeck coefficient S (Eqs.11.2–11.4). However, achieving high ZT remains a difficult and challenging objective, since electrical and thermal conductivities of any material have a direct correlation through proportionality between σ and κ_e [3].

$$\eta = \frac{T_h - T_c}{T_h} \times \frac{\sqrt{1 + ZT_{avg}} - 1}{\sqrt{1 + ZT_{avg}} + \frac{T_c}{T_h}}, \tag{11.1}$$

where T_h and T_c are temperatures on hot and cold side and ZT_{avg} is average value of TE figure of merit ZT of TE material in temperature gradient between T_c and T_h.

$$ZT = \frac{PF}{\kappa} T, \tag{11.2}$$

$$PF = S^2 \sigma, \tag{11.3}$$

$$\kappa = \kappa_e + \kappa_L, \tag{11.4}$$

where PF, T, κ, S, σ, κ_e, and κ_L are power factor, absolute temperature, total thermal conductivity, Seebeck coefficient, electrical conductivity, electronic thermal conductivity, and lattice thermal conductivity, respectively. Consequently, higher ZT corresponds to higher conversion efficiency.

Despite these limitations, numerous initiatives have been undertaken to reach higher ZT, resulting in the development of several novel TE materials over the past few decades which can be categorized into three groups (Fig. 11.1) [4]. The first

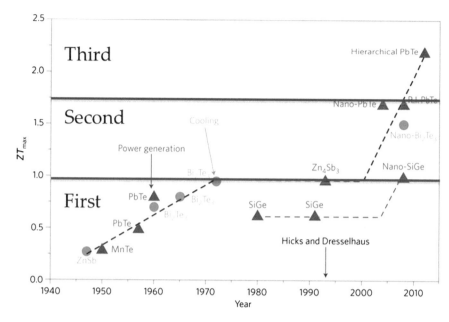

Fig. 11.1 Thermoelectric materials across different generations over the years (reproduced from [4] after modification)

group comprises mainly of four compounds, namely, ZnSb—the first compound that exhibited Seebeck effect [4]—and three naturally occurring semiconductors: Bi_2Te_3, PbTe, and SiGe. While this group achieved a maximum ZT of 1.0, it was believed that all three TE properties (κ, σ, and S) are intrinsic to a TE material and cannot be artificially manipulated. This belief led to a near-complete halt in research in the field of thermoelectrics until the 1990s, when pioneering work by Hicks and Dresselhaus [5, 6] on enhancing ZT via nanostructuring led to a renaissance in this field. This led to the development of second group of thermoelectrics that achieved significant increase in ZT (up to ~1.7) by significantly reducing κ via enhanced phonon scattering. In contrast, the third generation of thermoelectrics, which advent is fairly recent, has focused on the use of initiatives aimed at increase in either/both of σ and S in order to enhance ZT of TE materials beyond 1.8 [7, 8]. Table 11.1 gives a list of prominent TE materials (developed till date) including ZT values.

However, despite the availability of multiple bulk TE materials — both in research and in commercial applications (Table 11.1)—these suffer from a plethora of issues. Most TE materials are highly toxic and brittle and are characterized by poor mechanical properties at temperatures where these materials exhibit the highest ZT value (~0.75 of melting point). Furthermore, most bulk TE materials contain elements with high densities, making those unsuitable for specific uses such as in automobiles, where the focus is on lightweighting vehicles and improving fuel economy [24]. The listed TE materials are also environmentally unsustainable due to the use of highly energy-intensive processing methods for production and often

Table 11.1 Thermoelectric materials: ZT values and other properties

Material	M	T	T_m, °C	H	A	ZT p-type	n-type	d, g/cm³	Ref.
Bi_2Te_3	✗	✓	586	✗	✗	1.9 (320 K)	1.2 (370 K)		[9, 10]
LAST	!	!		✗	✗	1.7 (800 K)	2.2 (800 K)	8	[11, 12]
PbTe	!	!	924	✗	✗	2.2 (915 K)	1.6 (775 K)	8.16	[1, 13]
Clathrates	✓	✓		!	✗	0.61	1.35 (900 K)	5.8	[14]
$\beta - Zn_4Sb_3$	✗	!	567	!	!	1.3 (670 K)	–	6.2	[15]
Skutterudites	!	✓	1350	!	✗	0.93	1.7 (850 K)	7.4	[16]
SnSe	✗	✗	861	✗	✗	2.6 (900 K)	–	6.18	[17]
Mg_2Si	✗	✗	1102	✓	✓	0.65	1.3 (700 K)	2	[18]
Half-Heusler	✓	✓		!	!	1 (1073 K)	1 (1073 K)	10	[19]
SiGe	✓	✓		✓	✗	1	1.3 (1173 K)	4	[20]
PLEC	!	!		!	✓	1.5 (1000 K)	0.6	6.5	[21]
Zintl phase	!	!		✗	✗	1.3 (1223 K)	0.86	8.3	[22]
Oxides	!	✓		✓	✓	1.4 (923 K)	1.4 (923 K)	8	[23]

✓ safe, ✗ unsafe, ! OK, M mechanical stability, T thermal stability, H nontoxic, A abundance, T_m melting temperature, d material density

contain rare earth elements that make those very expensive [25]. Further, most TE materials can achieve ZT values only up to ~1.8, while higher conversion efficiency (desired for commercial applications) often necessitates ZT $\geq$ 3 [2].

In light of these issues, this chapter discusses and evaluates the potential of polymer-derived ceramics (PDCs) as an alternative to existing TE materials. The choice of PDCs stems from ability to address the aforementioned issues. PDCs are mainly constituted of low-density elements, are stable up to very high temperatures (~2000 °C), can be processed via energy-efficient methods, and are relatively inexpensive. This chapter presents a succinct introduction of PDCs, followed by a discussion of the existing literature on TE properties of PDCs. Finally, the chapter ends with a section that highlights the probable future direction of research on enhancing thermoelectric potential of PDCs.

11.2 Polymer-Derived Ceramics: A Brief Overview

Polymer-derived ceramics (PDCs), as the name suggests, refer to ceramics, mostly silicon-based, that are derived from polymers through a series of transformative steps [26]. While the advent of PDCs commenced with the use of molecular precursors to produce non-oxide ceramics [26–28], practical polymer-to-ceramic transformation of silicon-based polymers was first reported in the 1970s [26, 29]. Presence of strong bonds between different constituent elements in these precursors enables the synthesis of multinary ceramics [26, 30, 31] that cannot be obtained via conventional routes. Such multinary ceramics can be categorized as

Fig. 11.2 Steps associated with PDC route (based on [26, 30])

binary (SiC and BN), ternary (such as SiOC and SiCN), and quaternary (such as SiOCN, SiBCN, and SiAlOC) systems [26]. Recent times have also witnessed successful attempts at incorporating new elements, such as Zr and Hf in PDCs [32, 33], thereby expanding the range to pentanary ceramic systems.

An interesting facet of PDCs is the ability to synthesize ceramics via bottom-up route with precise control of microstructure, chemical composition, and phase distribution via careful selection of polymeric precursors and process parameters [26, 30]. Such precise control helps in obtaining tunable properties of the final ceramic, be it mechanical, electrical, thermal, or chemical. Moreover, ceramics processed via PDC route offer two significant advantages over ceramics produced using conventional route. First, it offers the scope for synthesizing ceramics at lower temperatures, such as SiC and Si_3N_4 — at up to 1000 °C via PDC route vis-à-vis at 1700–2000 °C via powder metallurgy route [26]. Second, ceramics synthesized via PDC route exhibit higher thermal stability and better mechanical properties (such as oxidation and creep resistance) compared to those via conventional routes [34, 35]. For instance, polymer-derived SiBCN has been reported to remain stable till 2200 °C in non-oxide atmosphere due to the dominance of kinetics over thermodynamics of phase transformation [26, 36].

In terms of processing, PDCs are produced from polymeric precursors through a series of steps as shown in Fig. 11.2.

After shaping, these precursors undergo cross-linking at lower temperatures (up to 200 °C) [26, 30]. Cross-linked polymer is subsequently pyrolyzed at higher temperatures (≥ 1000 °C) in inert (non-oxide) atmosphere, enabling its conversion to a ceramic (inorganic species) via loss of organic moieties [26, 30]. Often, fillers are added prior to ceramization (pyrolysis of polymer) in order to reduce both porosity and shrinkage caused by the loss of such moieties [26, 37]. Table 11.2 shows major polymeric precursors used to synthesize commonly processed PDCs.

In terms of microstructure, PDCs undergo phase evolution as a function of ceramization (pyrolysis) temperature [26, 30]. When pyrolyzed at lower temperature (800–1200 °C), PDCs consist of an amorphous matrix that is coupled with "free" carbon phase where carbon atoms are bonded to each other (but not to silicon) in graphite-like structure. With increase in ceramization temperature (>1200 °C), PDCs exhibit localized molecular rearrangement, resulting in phase separation in the amorphous matrix (such as the amorphous matrix in SiCN separating into SiC and Si_3N_4 phases), followed by the crystallization of separated phases [26, 30]. Further increase in crystallized content is observed with increase in ceramization temperature, which later decomposes to release gaseous by-products (such as CO_2) [26, 30].

Table 11.2 List of common polymeric precursors used for producing commonly processed PDCs (based on [26, 30, 32, 33])

PDC system	Polymeric precursors
SiOC	Polysiloxanes
SiCN	Polyorganosilazanes
	Polysilylcarbodiimides
SiBCN	Borazines
	Polyborosilazanes
	Polyborosilylcarbodiimides
SiBOC	Polyorganoborosiloxane
SiAlOC	Polyaluminosiloxane
SiHfCNO	Polysilazane and Hafnium tetra (*n*-butoxide)
SiOC/ZrO$_2$	Polymethylsilsesquioxane and Zirconium tetra (*n*-propoxide)

The combination of amorphous matrix and free carbon phase resembles the "phonon-glass electron-crystal" (PGEC) structure—the desired structure of an ideal TE material as hypothesized by Slack [38]. This hypothesis is based on the premise that while phonon glass ensures low κ (due to low κ_L as a result of its amorphous nature), the electron crystal exhibits high σ, typical of a single crystal. Thus, the thermoelectric potential of PDCs stems from its resemblance to PGEC structure, as "free" carbon phase could exhibit σ similar to graphite, while the bulk amorphous matrix could lead to low κ. Such potential of PDCs, coupled with other critical properties, such as excellent thermal and chemical stability, exceptional creep resistance till ~1500 °C, abundance of constituent elements, and relatively higher eco-friendliness compared to existing thermoelectrics, enhances the possibility of application for commercial purposes [26, 39–41]. This spawns interest in exploring and analyzing the possibility and means of developing PDCs as an alternative TE material.

Yet, barring one research work [25], no single study has been undertaken on the simultaneous analysis of all three TE properties (κ, σ, and S) of any single PDC system. However, individual studies have focused on measuring only one or two of these properties for the analyzed PDC. In the following section, a detailed discussion is presented on the literature published in this regard for different PDC systems.

11.3 Polymer-Derived Ceramics: Thermoelectric Properties

11.3.1 Electrical Conductivity (σ)

Among all the three thermoelectric properties, electrical conductivity (σ) remains the most highly explored property for PDCs. Table 11.3 shows the values of σ that have been measured for various PDC systems.

Table 11.3 Electrical conductivity (σ) of different PDC systems

PDC system	σ (S/cm)	Measurement temperature	Refs.
SiOC	10^{-12} to 10^{0}	20–1500 °C	[25, 37, 39–59]
Filled SiOC	10^{-5} to 10^{3}	20–1400 °C	
SiCN	10^{-10} to 10^{0}	20–1000 °C	[60–75]
Filled SiCN	10^{-8} to 10^{2}	20 °C	
SiOCN	10^{-5} to 10^{-1}	20–800 °C	[53–55]
SiBCN	10^{-13} to 10^{-1}	20–450 °C	[68–70, 76, 77]
SiAlCO	10^{-11} to 10^{-1}	20–300 °C	[78, 79]
SiBOC	10^{-5} to 10^{-3}	20 °C	[80]
SiCN(H)	10^{-7} to 10^{1}	50–400 °C	[81]

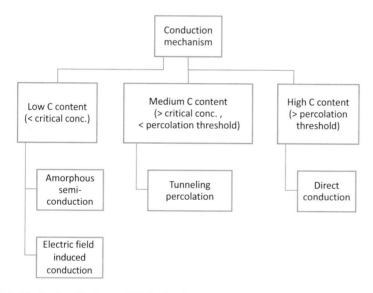

Fig. 11.3 Mechanisms/Regimes of PDCs electrical conduction: regimes

As is evident from data in Table 11.3, σ of any PDC varies over a wide range from the insulating regime ($<10^{-6}$ S/cm) to the semiconducting regime (10^{-6} to 10^{3} S/cm), indicating that it can be tuned to desired values. For all the aforementioned PDC systems that are listed in Table 11.3, σ values, as well as the underlying conduction mechanism, are both observed to vary primarily with the amount of graphitic-like free carbon content present. This variation, both in values and mechanism of electrical conduction, can be categorized into three groups or regimes that are shown in Fig. 11.3.

The first regime corresponds to high levels of free carbon content, where the graphitic-like phase is beyond percolation threshold, and is, therefore, continuous throughout the concerned PDC. Interestingly, such percolation threshold for free carbon phase can be as low as 5 vol. % [56], with the phase continuity enabling seamless movement of charge carriers across the specimen, thus resulting in higher symbol for electrical conductivity via direct conduction mechanism in the aforementioned PDC system [25, 26, 53, 63, 70, 82]. For instance, general effective media (GEM)-based modeling indicates higher σ for SiOC ($\geq 10^{-2}$ S/cm) with low percolation threshold (5 vol. %) when compared to that of SiOC ($\sigma = 10^{-10}$ to 10^{-8} S/cm) with higher percolation threshold (20 vol. %) [56]. This has also been observed for SiOC elsewhere, with lower electrical resistivity seen for SiOC ceramized at higher temperature (0.14 Ohm×m at 1400 °C vis-à-vis 0.35 Ohm×m at 1100 °C) on account of the formation of percolating network of turbostratic carbon [46]. Similar findings have been reported for polysilazane-derived SiCN that shows higher ($\sigma \sim 10^{-2}$ S/cm) due to the presence of percolating free carbon network [71].

On the other hand, the second regime focuses on PDCs where the graphitic-like carbon content is in the medium range, i.e., it is below percolation threshold and is, therefore, not entirely continuous, but it is not so low as to be fully discontinuous [26, 56, 71]. In this regime, free carbon phase is also surrounded by another (other) phase(s) present in the concerned PDC. Usually, this "other phases" refer to the amorphous matrix that exhibits lower σ values vis-à-vis free carbon network, thereby acting as a barrier to the movement of charge carriers [26, 30]. However, this amorphous barrier is quite thin, resulting in easy tunneling of charge carriers through it to hop across the specimen or move along the free carbon phase (in local regions where it is continuous) [60, 64, 71, 79, 83]. This mechanism is termed "tunneling percolation", which combines the "tunneling" of electrons across the amorphous barrier with the movement of charge carriers across the incomplete network of percolated free carbon.

Unlike the direct conduction mechanism, tunneling percolation corresponds to lower σ values. For example, SiOC shows lower ($\sigma = 10^{-12}$ to 10^{-3} S/cm), corresponding to the tunneling percolation regime when ceramized at lower temperatures, but exhibits higher ($\sigma \geq 10^{-2}$ S/cm) via direct conduction at higher temperatures [56]. Similarly, SiCN exhibits lower the symbol for electrical conductivity in the tunneling percolation regime (10^{-7} to 10^{-3} S/cm) vis-à-vis direct conduction regime ($\sigma = 10^{-3}$ to 10^{-2} S/cm) [56, 71, 72]. This can be easily understood as reduction in conductivity due to the emergence of discontinuity in free carbon network, which cannot be fully compensated by tunneling of electrons across the thin amorphous matrix.

Finally, the third regime is associated with PDCs that contain low levels of free carbon content, i.e., the amount of free carbon phase is below a critical concentration limit that enables tunneling of charge carriers across the amorphous matrix despite discontinuity in this phase [60, 79]. Hence, the barrier (amorphous matrix) between free carbon clusters is so thick that it is not possible for charge carriers to tunnel across it to other free carbon clusters across PDC specimen. As a result, PDCs in this regime exhibit much lower values of σ when compared to the aforementioned regimes. An example of this is SiCN that shows much lower ($\sigma = 10^{-9}$ to 10^{-6} S/cm) due to low amount of free carbon content—an outcome of lower ceramization temperature [60]—when compared to σ of SiCN in other studies mentioned earlier.

Unlike the first two regimes, electrical conduction in PDCs in this regime (i.e., low free carbon content) occurs via two mechanisms. The first mechanism is amorphous semiconduction, which arises because of the presence of dangling carbon atoms in the amorphous matrix [53, 63, 71, 81, 84]. These atoms are like defect dopants that both provide charge carriers (electrons) and also give rise to energy states between valence and conduction bands in the amorphous matrix, resulting in the bandgap transforming into mobility gap [53, 63]. Hence, charge carriers can easily occupy and hop across these energy states, allowing for electron movement across the specimen and giving rise to electrical conductivity [53, 63, 82, 85, 86]. Simultaneously, free carbon atoms are also hypothesized to generate electric field within PDC specimen at low concentrations [60]. This is expected to improve electrical conductivity in the thick barrier region (amorphous matrix) between free carbon clusters, thereby easing the movement of charge carriers and enabling electrical conduction in PDCs that have low free carbon content.

Thus, encompassing all three regimes, electrical conductivity, both in terms of its magnitude and mechanism, is strongly impacted by the amount of free carbon content present in the concerned PDC. However, free carbon content is itself dependent primarily on processing temperature. Additionally, σ is also dependent on the filler used, as well as to some extent on the measurement temperature. All these aspects are discussed in some detail in the corresponding subsections.

Effect of Processing Temperature

Of the aforementioned critical parameters, the most commonly analyzed parameter is processing temperature that refers to either of ceramization, sintering, hot-pressing, or annealing temperature, as the case may be. For almost all studies mentioned in Table 11.3, increase in processing temperature results in increase in σ, as can be seen in Fig. 11.4 for a variety of PDC systems [39, 60, 68, 79].

As can be seen in Fig. 11.4a-d, graphs between electrical conductivity and the reciprocal of processing temperature indicate Arrhenius relationship between these two parameters [39, 60, 68, 79]. This is explained by the relationship between processing temperature and free carbon content, which influences electrical conductivity. Initially, increase in processing temperature leads to enhancement in free carbon content due to the conversion of sp^3-hybridized carbon atoms (present in amorphous matrix) to sp^2-hybridized carbon atoms (that become a part of the graphitic-like free carbon phase) [39, 42, 55, 61, 63, 83]. This increase in free carbon content results in an increase in σ, since free carbon phase is similar to graphite, and is, therefore, a good electrical conductor. This hypothesis is backed by the similarity in activation energies of the Arrhenius relationship between σ and processing temperature with that for the conversion of sp^3-hybridized carbon to sp^2-hybridized carbon (~3.5 eV) [55, 60, 68, 87]. Thus, initial enhancement in processing temperature improves σ of PDCs via increase in free carbon content.

However, subsequent increase in processing temperature influences the nature of both phases in PDCs (i.e., amorphous matrix and free carbon). Free carbon phase undergoes transformation from amorphous to nanocrystalline state, followed by in-plane growth of these nanocrystal clusters [40, 60, 79]. This can be seen in Fig. 11.5, which shows the cluster size of free carbon as a function of processing temperature for SiAlCO [79] and SiCN [83].

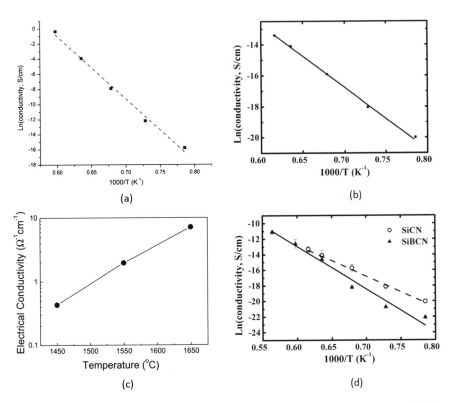

Fig. 11.4 Electrical conductivity as a function of (**a**) ceramization temperature for SiAlCO ceramics (reproduced from [79]); (**b**) ceramization temperature for SiCN ceramics (reproduced from [60]); (**c**) hot-pressing temperature for SiOC ceramics (reproduced from [39]); and (**d**) ceramization temperature for SiCN and SiBCN ceramics (reproduced from [68])

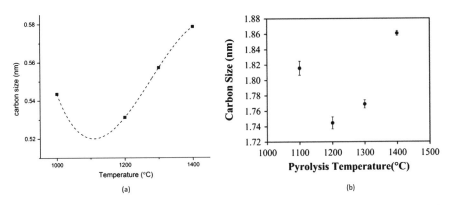

Fig. 11.5 Carbon size as function of ceramization temperature for (**a**) SiAlCO ceramic (reproduced from [79]) and (**b**) SiCN (reproduced from [83])

As can be seen for both PDCs, free carbon cluster size initially shows a decrease, which can be explained by the shift from amorphous state (i.e., distorted aromatic rings) to nanocrystalline state (i.e., regular six-membered rings) [88–90]. Subsequently, in-plane growth of these clusters results in an increase in cluster size, as well as the crystallite size of this phase, thereby reducing electron scattering and simultaneously enhancing electrical conductivity [79, 88, 89]. However, other studies highlight that it is not necessary for all PDCs to exhibit both the initial decrease and subsequent increase in cluster size, but instead show only one of these two phenomena [42, 43, 91]. This may be related to the nature of polymeric precursors used and/or processing conditions employed.

The increase in crystallinity of free carbon is also indicated by the emergence of a sharp peak at ~26.5° in X-ray diffraction (XRD) spectra of PDCs (Fig. 11.6b, e) that corresponds to crystalline graphite [42, 43, 46, 50]. However, this increase in crystallinity and cluster size of free carbon upon increase in processing temperature is simultaneously accompanied by the transformation of amorphous matrix to crystalline phases [25, 40, 42, 43, 46, 49, 80, 84, 92]. For instance, SiOC ceramics exhibit phase separation in amorphous matrix, whereby amorphous phases (such as SiC and SiO$_2$) separate from the matrix and crystallize subsequently with increase in processing temperature [40, 42, 43, 46, 93]. Such crystallization, especially in case

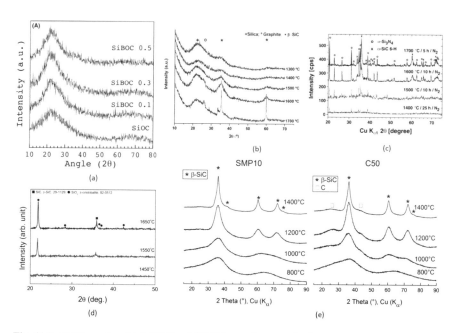

Fig. 11.6 XRD spectra of (**a**) SiOC and SiBOC samples ceramized at 1200 °C (reproduced from [80]); (**b**) SiOC sintered via spark plasma sintering at 1300, 1400, 1500, 1600, and 1700 °C (reproduced from [42]); (**c**) SiCN ceramics annealed at temperatures of 1400, 1500, 1600, and 1700 °C (reproduced from [84]); (**d**) SiOC + 1 mol.% Ba hot-pressed at 1450, 1550, and 1650 °C (reproduced from [40]); and (**e**) SiOC ceramized at temperatures of 800, 1000, 1200, and 1400 °C (reproduced from [43])

of SiC via reaction of Si with some amount of free carbon, leads to significant reduction in free carbon cluster size [25, 94]. However, its consequences on σ are not entirely clear or consistent, for while reduction in cluster size can lead to decrease in σ, formation of crystalline SiC, that possesses good thermal and electrical properties, is expected to enhance σ of PDCs.

The aforementioned increase in free carbon content with increase in processing temperature causes also change in the mechanism of electrical conduction, which can be understood by analyzing the behavior between σ and measurement temperature of the concerned PDC [25, 44, 54, 55, 69, 70, 81]. At lower processing temperatures, it is observed that σ of PDCs follows the relationship given by Eq. (11.5), where T_0 is constant, and T is measurement temperature[1] [56, 81]. Conversely, at higher processing temperatures, it is observed that $\sigma(T)$ of PDCs follows the relationship given by Eq. (11.6), where E_a is activation energy and k_B is Boltzmann constant [56, 81].

$$\sigma \sim e^{\left(-\frac{T_0}{T^{1/4}}\right)},$$ (11.5)

$$\sigma \sim e^{\left(-\frac{E_a}{k_B T}\right)},$$ (11.6)

For instance, methylvinyldichlorosilane-derived SiCN shows the relationship corresponding to Eq. (11.5) at processing temperatures of 775–1200 °C while exhibiting the relationship corresponding to Eq. (11.6) at higher processing temperatures [81] (as shown in Fig. 11.7).

The first relationship (Eq. (11.5)) was earlier attributed to Mott's variable range hopping (VRH) behavior [44, 69, 70, 81] but has been later ascribed to band range hopping (BRH) behavior in subsequent studies [53]. Both hopping mechanisms are characteristic of amorphous semiconductors as these mechanisms indicate localized transfer of charge carriers, thereby highlighting lower free carbon content in the given ceramic [95–97]. This is in sync with existing literature that shows increase in free carbon content with rise in processing temperature, as discussed earlier. Thus, methylvinyldichlorosilane-derived SiCN exhibits hopping conduction due to lower free carbon content at lower processing temperature, as can be seen in Fig. 11.7a [81], while similar findings are also reported elsewhere for SiCN in other studies [69, 70], SiOCN [53], and SiOC [25].

On the other hand, the second relationship (Eq. 11.6) can be ascribed to a combination of two factors: one, increased ability of charge carriers to tunnel across excited energy states, and two, greater probability of charge carriers getting enough excitation energy to occupy energy states in conduction band [56, 81]. While the first

---

[1]Measurement temperature is not the same as processing temperature – processing temperature refers to the temperature at which the polymer is converted or sintered or annealed to obtain the ceramic, while measurement temperature refers to the temperature at which the concerned property of the ceramic is measured.

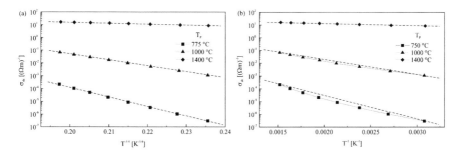

Fig. 11.7 Variation in σ as function of (**a**) $T^{1/4}$ and (**b**) T^{-1} for SiCN ceramics processed at three temperatures (775/750, 1000, and 1400 °C) (reproduced from [81])

factor is indicative of tunneling percolation mechanism, the second factor is associated with a fully percolated and continuous network of free carbon [56, 81]. Thus, increase in processing temperature can be said to significantly augment free carbon content levels, such that this leads to change in the conduction mechanism of final ceramic, and thereby, improve σ. This can be seen in case of methylvinyldichlorosilane-derived SiCN that shows much higher σ at processing temperature of 1400 °C (compared to 775/750 and 1000 °C), even at lower measurement temperatures (Fig. 11.7b) [81]. Again, similar observations have also been made for other PDC systems, such as SiOC [44] and SiOCN [54]. Thus, for a number of PDCs, increase in ceramization temperature leads to increase in σ (via augmenting free carbon content), as well as variation in the associated conduction mechanism. However, it is not necessary that all PDCs exhibit both these relationships (Eqs. (11.5)–(11.6)) with increase in processing temperature – some PDCs may show only one of these two relationships [44].

Effect of Polymeric Precursor

Polymeric precursors remain another key determining factor behind both the amount of free carbon content and the resultant σ of concerned PDC. Mostly, this can be ascribed to the amount of carbon in polymer and its subsequent effect on free carbon content in the final ceramic. For instance, SiOCN obtained using polysilazane shows lower σ compared to SiOCN obtained using trione (amine) [55]. This is partly explained by the higher amount of carbon content in trione (precursor) that in turn enhances free carbon content in the final ceramic. Similar variation is observed in σ of SiOC, with higher σ observed for SiOC derived from polymeric precursors with high carbon content [58].

Effect of Filler

In addition to processing temperature and polymeric precursor, the nature and amount of filler used play also a vital role in influencing σ of the final ceramic. Fillers are often used to reduce the porosity attained often during polymer-to-ceramic transformation, thereby enhancing the density of final ceramic [37]. Use of conductive fillers enhances σ of the ceramic mostly via enabling the formation of percolation paths that may also interact with the free carbon phase, even at low concentrations of these fillers [37, 52, 64, 83].

For instance, the inclusion of graphitic oxide (GO) as filler in (polysiloxane) SILRES62-derived SiOC (with the use of polyvinyl alcohol or PVA) is reported to form percolation networks even at very low filler concentration (<2 wt. %), thereby enhancing its σ [52]. This is attributed to the occurrence of GO-induced crystallization of graphite that improves σ of free carbon phase, and, thereby, of the final ceramic. Similarly, the formation of percolation network of $MoSi_2$ is responsible for significant rise in σ of polymethylsiloxane-derived SiOC, even when free carbon content is low (at temperatures below 1200 °C) [37]. However, unlike GO, $MoSi_2$ does not induce precipitation or crystallization of free carbon till 1200 °C [37]. Other PDC systems are also reported to exhibit significant rise in σ upon the incorporation of conductive fillers at low concentrations due to the formation of percolation network, such as CNT (<1 vol. %) [67] and reduced graphene oxide (0.2 wt. %) [72] in SiCN. As in case of $MoSi_2$, these fillers have not been observed to induce any significant extent of crystallization of free carbon phase in SiCN ceramics, and thus, increase in σ is solely due to the introduction of well-dispersed conductive filler in the ceramic. Interestingly, the use of higher concentrations of such conductive fillers (beyond the percolation threshold) does not cause any drastic improvement in σ of the ceramic, such as in case of CNT [67] and reduced graphene oxide [72] in SiCN. This is reported to be possibly due to the agglomeration or random dispersion of filler in the concerned ceramic.

11.3.2   Thermal Conductivity (κ)

Thermal conductivity (κ) of any material is strongly associated with two aspects: its nature (crystalline or amorphous material, along with grain/crystal size) and presence of defects (such as porosity) [98, 99]. With regard to PDCs, κ has not been studied to the extent to which σ has been explored. Yet, of the limited attempts undertaken in this regard, the first two aspects (nature of material and porosity) have been observed to be critical to κ of PDCs, both at room and high temperatures.

The importance of the nature of the concerned PDC can be understood via analyzing changes observed in its microstructure with increase in processing temperature. As described earlier, PDCs consist of two phases (amorphous matrix and free carbon phase) at lower processing temperatures [26, 30]. Of these phases, while

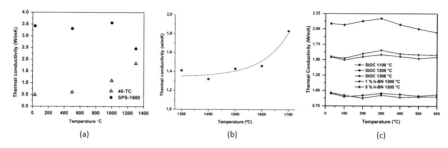

Fig. 11.8 Thermal conductivity (**a**) as a function of measurement temperature for SiBCN foam and SPS sample (reproduced from [98]); (**b**) as a function of sintering temperature for SPS SiOC (reproduced from [47]); (**c**) as a function of measurement temperature for SPS SiOC sintered at different temperatures (reproduced from [25])

free carbon phase is thermally conductive, the amorphous matrix is thermally insulating as it can scatter phonons and, thereby, reduce thermal conductivity of the final ceramic. However, subsequent increase in processing temperature causes crystallization of both amorphous matrix and free carbon phases, thereby transforming the nature of the ceramic (from amorphous to crystalline). Further, increase in ceramization temperature enables better flow of viscous matrix, thereby enabling better reduction in porosity of the material. This in turn is expected to improve κ of the concerned PDC system, and the same has been reported in several studies.

For instance, SiBCN foam shows lower κ compared to SiBCN processed using spark plasma sintering (SPS) (Fig. 11.8a), since the foam is completely amorphous and contains pores, while SPS sample is crystalline in nature and is highly dense [91]. This trend (of lower κ for foam vis-à-vis SPS sample) is observed even with increase in measurement temperature from RT to 1300 °C (Fig. 11.8a). On the other hand, another study highlights increase in κ of SiOC with increase in SPS (processing) temperature, with a significant amount of jump in κ observed during the increase in SPS temperature from 1600 °C to 1700 °C (Fig. 11.8b) [42]. The initial limited increase is ascribed to the interruption of amorphous matrix with crystallites of $\beta - SiC$ and free carbon phase (that is segregated). However, the subsequent dramatic jump in κ with increase in SPS temperature from 1600 °C to 1700 °C is attributed to the significant increase in size of $\beta - SiC$ crystallites and carbon clusters. Similar increase in κ with increase in processing temperature is also observed for SiOC elsewhere, and is explained by the formation of SiC via carbothermal reduction (Fig. 11.8c) [25].

An interesting aspect with regard to all these studies is the enhanced contribution of crystalline phases (especially β–SiC), obtained via phase separation from the amorphous matrix for the aforementioned PDC systems. In contrast, the contribution of free carbon phase to improving κ of these ceramics appears to be quite limited, even with significant increase in processing temperature [25, 42]. This may be

Table 11.4 Thermal conductivity of PDC systems

PDC system	Thermal conductivity, κ, $W \times m^{-1} \times K^{-1}$	Measurement temperature	Refs.
SiOC	0.027–2.7 (pure) 1.7–5.6 (upon addition of fillers or reinforcements)	20–1300 °C	[25, 40, 42, 92, 94, 100, 101, 103]
SiCN	2.1–5.5 (from pure to addition of fillers and reinforcements)	100–900 °C	[102]
SiBCN	0.647–3.5 (pure) 0.5–2.0 (with sago as filler)	20–1300 °C	[91, 104]

Table 11.5 Seebeck coefficient values

PDC system	S, μV/K	Range of measurement temperature, °C and K	Refs.
SiCN	−10	Room temperature (20 °C, 293 K)	[63]
SiOC	−10 to −12	30–150 °C (303–453 K)	[25]
SiOC/h − BN	−33		
SiBCN	0–2	−223 to 27 °C (50–300 K)	[63, 70]

possibly due to the lack of a fully continuous percolating structure of free carbon phase for these PDC systems.

With regard to measurement temperature, PDCs show a contrasting trend for κ across existing literature (Fig. 11.8). For instance, studies show invariant κ at low measurement temperatures ($\leq$1000 °C) for SiOC [25, 100, 101], as well as for SiBCN [91], but also highlight that SiBCN shows variation in κ at higher measurement temperatures (>1000 °C). The latter trend is partly explained by the role of higher crystallinity for SiBCN sample, since crystalline materials tend to exhibit reduction in κ via enhanced phonon scattering with increase in measurement temperature. However, this explanation is complemented with speculation regarding the exact role of SiO_2- a phase observed in SiOC but found absent in non-oxide PDCs-that remains to be fully analyzed and understood [25]. Finally, PDCs also exhibit significant increase in κ upon incorporation of conductive fillers, such as CNT in SiCN [102] and Ba in SiOC [40], especially with increase in measurement temperature. While CNT forms a strong percolating network in the ceramic that enables electrons to participate in conduction, and, thereby, enhance κ of SiCN, presence of Ba enables the formation of percolating network of free carbon and its association with β−SiC crystallites to improve κ of SiOC [40, 102].

Table 11.4 shows thermal conductivity of different PDC systems studied till date.

11.3.3 Seebeck Coefficient (S)

Seebeck coefficient (S) of PDC systems has been reported or analyzed only in three studies, among which two report values for SiCN/SiBCN systems [63, 70], while the third has reported S for SiOC [25]; these values are provided in Table 11.5. Additionally, the third study [25] also reports S of SiOC filled with hexagonal boron nitride ($h-BN$), and these values are also given in Table 11.5.

As can be seen, while SiOC and SiCN exhibit S in the range from -10 to -12 μV/K, SiBCN shows much lower S (up to 2 μV/K). The lower value of S for SiBCN (compared to SiCN) is ascribed to the respective behavior of boron (B) and nitrogen (N) as p- and n-type dopants, in the SiC network [105]. Addition of B is explained in this study to create holes that compensate for additional electrons introduced in SiC network via incorporation of N (n-type dopant). With regard to SiOC, its S shows marginal increase in absolute value from -10 to -12 μV/K with increase in temperature from 303 K to 423 K. This behavior is understood as the outcome of interactions between electrons (hopping across free carbon phase) and phonons (transferring in the amorphous phase) [25]. Interestingly, addition of $h-BN$ is observed to enhance S of SiOC at all temperatures in the same study [25] while also causing difference in Seebeck behavior of $h-BN$-filled SiOC vis-à-vis pure SiOC. This difference, whereby $h-BN$-filled SiOC shows initial decrease in S followed by its subsequent increase with rise in measurement temperature, is attributed to the wide bandgap of $h-BN$ [25, 106]. However, the same study points out the need for more research to assess the contribution of amorphous matrix to S values of PDCs.

Interestingly, the last study [25] also highlights simultaneous increase in both S and σ of SiOC with increase in measurement temperature. This is attributed to the occurrence of electrical conduction in PDCs via hopping of charge carriers [53, 63]. Also, anomaly is observed in the nature of charge carriers as indicated by S and Hall effect [25, 51], which is explained by localization of charge carriers due to hopping conduction [107]. Given the importance of hopping as a mechanism for electrical conduction in PDCs (as discussed earlier), similar findings may be reported in future for other PDC systems.

11.4 Summary and Future Work

A vast majority of initiatives directed toward the development of high ZT materials have focused solely on engineering the crystal structure of highly crystalline materials while completely ignoring amorphous materials due to inherent poor electrical conductivity. In contrast, despite the amorphous nature, PDCs exhibit electrical conductivity in the semiconducting regime along with intrinsically poor thermal conductivity (<2.5 W $\times$ m^{-1} $\times$ K^{-1}). This behavior offers a strong platform for considering PDCs as a potential thermoelectric material that exhibits "phonon-glass electron-crystal" behavior. This potential is significantly enhanced by the observation of simultaneous increase in electrical conductivity and Seebeck coefficient of these PDCs with increase in measurement temperature. However, given the low values for both parameters, this potential for PDCs needs to be further explored on a sustained basis, especially with regard to Seebeck effect that continues to remain much of an enigma till date, particularly at higher temperature regimes. This is critical in order to develop mechanisms that can help enhance both Seebeck coefficient and ZT values of PDCs. Further research works also needs to be undertaken on studying thermoelectric properties of any PDC that is processed under the same conditions for developing better understanding of process-property relationship, especially, regarding the impact of variation in prominent parameters, such as polymeric precursor, chemical composition of ceramic, processing temperature, and free carbon content, on thermoelectric properties of various PDC systems. Such research remains pivotal to harnessing the true potential of PDCs for commercial thermoelectric applications.

Acknowledgments The authors would like to acknowledge the financial support by Robert Patrick Jenkins Professorship and Dean's Faculty Fellow Professorship. A.B.K would like to acknowledge the financial support provided through the scholarship fund of Southern Automotive Women's Forum.

References

1. K. Biswas, J. He, I.D. Blum, C.-I. Wu, T.P. Hogan, D.N. Seidman, et al., High-performance bulk thermoelectrics with all-scale hierarchical architectures. Nature **489**, 414–418 (2012). https://doi.org/10.1038/nature11439
2. C.B. Vining, An inconvenient truth about thermoelectrics. Nat. Mater. **8**, 83–85 (2009). https://doi.org/10.1038/nmat2361
3. T.M. Tritt, M.A. Subramanian, Thermoelectric materials , phenomena, and applications: a bird's eye view. MRS Bull. **31**, 188–198 (2006). https://doi.org/10.1557/mrs2006.44
4. J.P. Heremans, M.S. Dresselhaus, L.E. Bell, D.T. Morelli, When thermoelectrics reached the nanoscale. Nat. Nanotechnol. **8**, 471–473 (2013). https://doi.org/10.1038/nnano.2013.129
5. L.D. Hicks, M.S. Dresselhaus, Effect of quantum-well structures on the thermoelectric figure of merit. Phys. Rev. B **47**, 12727–12731 (1993). https://doi.org/10.1103/PhysRevB.47.12727
6. L.D. Hicks, M.S. Dresselhaus, Thermoelectric figure of merit of a one-dimensional conductor. Phys. Rev. B **47**, 16631–16634 (1993). https://doi.org/10.1103/PhysRevB.47.16631
7. A. Mehdizadeh Dehkordi, M. Zebarjadi, J. He, T.M. Tritt, Thermoelectric power factor: enhancement mechanisms and strategies for higher performance thermoelectric materials. Mater. Sci. Eng. R Rep. **97**, 1–22 (2015). https://doi.org/10.1016/j.mser.2015.08.001
8. M. Zebarjadi, K. Esfarjani, M.S. Dresselhaus, Z.F. Ren, G. Chen, Perspectives on thermoelectrics: from fundamentals to device applications. Energ. Environ. Sci. **5**, 5147–5162 (2012). https://doi.org/10.1039/C1EE02497C
9. S.I. Kim, K.H. Lee, H.A. Mun, H.S. Kim, S.W. Hwang, J.W. Roh, et al., Dense dislocation arrays embedded in grain boundaries for high-performance bulk thermoelectrics. Science **348**, 109–114 (2015). https://doi.org/10.1126/science.aaa4166
10. S. Wang, H. Li, R. Lu, G. Zheng, X. Tang, Metal nanoparticle decorated n-type Bi_2Te_3 -based materials with enhanced thermoelectric performances. Nanotechnology **24**, 285702 (2013). https://doi.org/10.1088/0957-4484/24/28/285702
11. K.F. Hsu, S. Loo, F. Guo, W. Chen, J.S. Dyck, C. Uher, et al., Cubic AgPbmSbTe2+m: bulk thermoelectric materials with high figure of merit. Science **303**, 818–821 (2004). https://doi.org/10.1126/science.1092963

12. P.F.P. Poudeu, J. D'Angelo, A.D. Downey, J.L. Short, T.P. Hogan, M.G. Kanatzidis, High thermoelectric figure of merit and nanostructuring in bulk p-type Na1−xPbmSbyTem+2. Angew. Chem. Int. Ed. **45**, 3835–3839 (2006). https://doi.org/10.1002/anie.200600865

13. Y. Pei, J. Lensch-Falk, E.S. Toberer, D.L. Medlin, G.J. Snyder, High thermoelectric performance in PbTe due to large nanoscale Ag2Te precipitates and La doping. Adv. Funct. Mater. **21**, 241–249 (2011). https://doi.org/10.1002/adfm.201000878

14. A. Saramat, G. Svensson, A.E.C. Palmqvist, C. Stiewe, E. Mueller, D. Platzek, et al., Large thermoelectric figure of merit at high temperature in Czochralski-grown clathrate Ba[sub 8]Ga [sub 16]Ge[sub 30]. J. Appl. Phys. **99**, 023708 (2006). https://doi.org/10.1063/1.2163979

15. T. Caillat, J.-P. Fleurial, A. Borshchevsky, Preparation and thermoelectric properties of semiconducting Zn 4 Sb 3. J. Phys. Chem. Solid **58**, 1119–1125 (1997)

16. X. Shi, J.J. Yang, J.R. Salvador, M. Chi, J.Y. Cho, H. Wang, et al., Multiple-filled skutterudites: high thermoelectric figure of merit through separately optimizing electrical and thermal transports. J. Am. Chem. Soc. **133**, 7837–7846 (2011). https://doi.org/10.1021/ja111199y

17. L.-D. Zhao, S.-H. Lo, Y. Zhang, H. Sun, G. Tan, C. Uher, et al., Ultralow thermal conductivity and high thermoelectric figure of merit in SnSe crystals. Nature **508**, 373–377 (2014). https://doi.org/10.1038/nature13184

18. W. Liu, X. Tan, K. Yin, H. Liu, X. Tang, J. Shi, et al., Convergence of conduction bands as a means of enhancing thermoelectric performance of n-type Mg2Si(1-x)Sn(x) solid solutions. Phys. Rev. Lett. **108**, 166601 (2012). https://doi.org/10.1103/PhysRevLett.108.166601

19. X. Yan, W. Liu, H. Wang, S. Chen, J. Shiomi, K. Esfarjani, et al., Stronger phonon scattering by larger differences in atomic mass and size in p-type half-Heuslers Hf1−xTixCoSb0.8Sn0.2. Energ. Environ. Sci. **5**, 7543 (2012). https://doi.org/10.1039/c2ee21554c

20. X.W. Wang, H. Lee, Y.C. Lan, G.H. Zhu, G. Joshi, D.Z. Wang, et al., Enhanced thermoelectric figure of merit in nanostructured n-type silicon germanium bulk alloy. Appl. Phys. Lett. **93**, 193121 (2008). https://doi.org/10.1063/1.3027060

21. H. Liu, X. Shi, F. Xu, L. Zhang, W. Zhang, L. Chen, et al., Copper ion liquid-like thermoelectrics. Nat. Mater. **11**, 422–425 (2012). https://doi.org/10.1038/nmat3273

22. E.S. Toberer, C.A. Cox, S.R. Brown, T. Ikeda, A.F. May, S.M. Kauzlarich, et al., Traversing the metal-insulator transition in a Zintl phase: rational enhancement of thermoelectric efficiency in $Yb_{14}Mn_{1-x}Al_xSb_{11}$. Adv. Funct. Mater. **18**, 2795–2800 (2008). https://doi.org/10.1002/adfm.200800298

23. J.H. Sui, J.C.F. Li, J.Q. He, Y.-L.L. Pei, D. Berardan, H. Wu, et al., Texturation boosts the thermoelectric performance of BiCuSeO oxyselenides. Energ. Environ. Sci. **6**, 2916 (2013). https://doi.org/10.1039/c3ee41859f

24. NHTSA, CAFE - Fuel Economy. (National Highway Traffic Safety Administration, 2013), http://www.nhtsa.gov/fuel-economy/. Accessed 24 June 2016

25. A.B. Kousaalya, X. Zeng, M. Karakaya, T. Tritt, S. Pilla, A.M. Rao, Polymer-derived silicon oxycarbide ceramics as promising next-generation sustainable thermoelectrics. ACS Appl. Mater. Interfaces **10**, 2236–2241 (2018). https://doi.org/10.1021/acsami.7b17394

26. P. Colombo, G. Mera, R. Riedel, G. Domenicosorarù, Polymer-derived ceramics: 40 years of research and innovation in advanced ceramics. J. Am. Ceram. Soc. **93**, 1805–1837 (2010). https://doi.org/10.1111/j.1551-2916.2010.03876.x

27. F.W. Ainger, J.M. Herbert, The preparation of phosphorus-nitrogen compounds as non-porous solids, in *Spec. Ceram*, ed. by P. Popper, 1st edn., (Academic Press, New York, 1960), pp. 168–182

28. P.G. Chantrell, P. Popper, Inorganic polymers and ceramics, in *Spec. Ceram*, ed. by P. Popper, 1st edn., (Academic Press, New York, 1965), pp. 87–103

29. W. Verbeek, Production of shaped articles of homogeneous mixtures of silicon carbide and nitride. US3853567A, 1973

30. R. Riedel, G. Mera, R. Hauser, A. Klonczynski, Silicon-based polymer-derived ceramics: synthesis properties and applications-a review. J. Ceram. Soc. Jpn. **114**, 425–444 (2006). https://doi.org/10.2109/jcersj.114.425

31. E. Ionescu, G. Mera, R. Riedel, L.A. An, R. Riedel, C. Konetschny, et al., Polymer-derived ceramics (PDCs): materials design towards applications at ultrahigh-temperatures and in extreme environments. MAX Phases Ultra-High Temp. Ceram. Extrem. Environ. **81**, 203–245. https://doi.org/10.4018/978-1-4666-4066-5.ch007

32. R. Sujith, A.B. Kousaalya, R. Kumar, Coarsening induced phase transformation of hafnia in polymer-derived Si-Hf-C-N-O ceramics. J. Am. Ceram. Soc. **94**, 2788–2791 (2011). https://doi.org/10.1111/j.1551-2916.2011.04719.x

33. E. Ionescu, C. Linck, C. Fasel, M. Müller, H. Kleebe, R. Riedel, Polymer-derived SiOC/ZrO$_2$ ceramic nanocomposites with excellent high-temperature stability. J. Am. Ceram. Soc. **93**, 241–250 (2010). https://doi.org/10.1111/j.1551-2916.2009.03395.x

34. R. Kumar, F. Phillipp, F. Aldinger, Oxidation induced effects on the creep properties of nano-crystalline porous Si–B–C–N ceramics. Mater. Sci. Eng. A **445–446**, 251–258 (2007). https://doi.org/10.1016/J.MSEA.2006.09.024

35. N.V. Ravi Kumar, S. Prinz, Y. Cai, A. Zimmermann, F. Aldinger, F. Berger, et al., Crystal-lization and creep behavior of Si–B–C–N ceramics. Acta Mater. **53**, 4567–4578 (2005). https://doi.org/10.1016/J.ACTAMAT.2005.06.011

36. R. Riedel, A. Kienzle, W. Dressler, L. Ruwisch, J. Bill, F. Aldinger, A silicoboron carbonitride ceramic stable to 2,000°C. Nature **382**, 796–798 (1996). https://doi.org/10.1038/382796a0

37. J. Cordelair, P. Greil, Electrical characterization of polymethylsiloxane/MoSi2-derived com-posite ceramics. J. Am. Ceram. Soc. **84**, 2256–2259 (2004). https://doi.org/10.1111/j.1151-2916.2001.tb00998.x

38. G.A. Slack, New materials and performance limits for thermoelectric cooling, in *CRC Handbook of Thermoelectrics*, ed. by D. M. Rowe, 1st edn., (CRC Press, Boca Raton, FL, 1995), pp. 399–432

39. K.J. Kim, J.-H.H. Eom, Y.-W.W. Kim, W.-S.S. Seo, Electrical conductivity of dense, bulk silicon-oxycarbide ceramics. J. Eur. Ceram. Soc. **35**, 1355–1360 (2015). https://doi.org/10.1016/j.jeurceramsoc.2014.12.007

40. J.-H. Eom, Y.-W. Kim, K.J. Kim, W.-S. Seo, Improved electrical and thermal conductivities of polysiloxane-derived silicon oxycarbide ceramics by barium addition. J. Eur. Ceram. Soc. **38**, 487–493 (2018). https://doi.org/10.1016/J.JEURCERAMSOC.2017.09.045

41. M.A. Mazo, A. Nistal, A.C. Caballero, F. Rubio, J. Rubio, J.L. Oteo, Influence of processing conditions in TEOS/PDMS derived silicon oxycarbide materials. Part 1: microstructure and properties. J. Eur. Ceram. Soc. **33**, 1195–1205 (2013). https://doi.org/10.1016/J.JEURCERAMSOC.2012.11.022

42. M.A. Mazo, A. Tamayo, A.C. Caballero, J. Rubio, Electrical and thermal response of silicon oxycarbide materials obtained by spark plasma sintering. J. Eur. Ceram. Soc. **37**, 2011–2020 (2017). https://doi.org/10.1016/J.JEURCERAMSOC.2017.01.003

43. F. Dalcanale, J. Grossenbacher, G. Blugan, M.R. Gullo, A. Lauria, J. Brugger, et al., Influence of carbon enrichment on electrical conductivity and processing of polycarbosilane derived ceramic for MEMS applications. J. Eur. Ceram. Soc. **34**, 3559–3570 (2014). https://doi.org/10.1016/J.JEURCERAMSOC.2014.06.002

44. P. Moni, M. Wilhelm, K. Rezwan, The influence of carbon nanotubes and graphene oxide sheets on the morphology, porosity, surface characteristics and thermal and electrical proper-ties of polysiloxane derived ceramics. RSC Adv. **7**, 37559–37567 (2017). https://doi.org/10.1039/C7RA01937H

45. M. Shibuya, M. Sakurai, T. Takahashi, Preparation and characteristics of a vapor-grown carbon fiber/ceramic composite using a methylsilicone precursor. Compos. Sci. Technol. **67**, 3338–3344 (2007). https://doi.org/10.1016/J.COMPSCITECH.2007.03.023

46. S. Martínez-Crespiera, E. Ionescu, H.-J. Kleebe, R. Riedel, Pressureless synthesis of fully dense and crack-free SiOC bulk ceramics via photo-crosslinking and pyrolysis of a

polysiloxane. J. Eur. Ceram. Soc. **31**, 913–919 (2011). https://doi.org/10.1016/j.jeurceramsoc.2010.11.019

47. G.M. Renlund, S. Prochazka, R.H. Doremus, Silicon oxycarbide glasses: part II. Structure and properties. J. Mater. Res. **6**, 2723–2734 (1991). https://doi.org/10.1557/JMR.1991.2723

48. A. Tamayo, M.A. Mazo, F. Rubio, J. Rubio, Structure properties relationship in silicon oxycarbide glasses obtained by spark plasma sintering. Ceram. Int. **40**, 11351–11358 (2014). https://doi.org/10.1016/J.CERAMINT.2014.03.111

49. W. Duan, X. Yin, Q. Li, X. Liu, L. Cheng, L. Zhang, Synthesis and microwave absorption properties of SiC nanowires reinforced SiOC ceramic. J. Eur. Ceram. Soc. **34**, 257–266 (2014). https://doi.org/10.1016/J.JEURCERAMSOC.2013.08.029

50. K. Lu, D. Erb, M. Liu, Thermal stability and electrical conductivity of carbon-enriched silicon oxycarbide. J. Mater. Chem. C **4**, 1829–1837 (2016). https://doi.org/10.1039/C6TC00069J

51. K.J. Kim, J.H. Eom, T.Y. Koh, Y.W. Kim, W.S. Seo, Effects of carbon addition on the electrical properties of bulk silicon-oxycarbide ceramics. J. Eur. Ceram. Soc. **36**, 2705–2711 (2016). https://doi.org/10.1016/j.jeurceramsoc.2016.04.034

52. C. Shen, E. Barrios, L. Zhai, Bulk polymer-derived ceramic composites of graphene oxide. ACS Omega **3**, 4006–4016 (2018). https://doi.org/10.1021/acsomega.8b00492

53. Y. Wang, T. Jiang, L. Zhang, L. An, Electron transport in polymer-derived amorphous silicon oxycarbonitride ceramics. J. Am. Ceram. Soc. **92**, 1603–1606 (2009). https://doi.org/10.1111/j.1551-2916.2009.03044.x

54. H.-Y. Ryu, Q. Wang, R. Raj, Ultrahigh-temperature semiconductors made from polymer-derived ceramics. J. Am. Ceram. Soc. **93**, 1668–1676 (2010). https://doi.org/10.1111/j.1551-2916.2010.03623.x

55. V.L. Nguyen, C. Zanella, P. Bettotti, G.D. Sorarù, Electrical conductivity of SiOCN ceramics by the powder-solution-composite technique. J. Am. Ceram. Soc. **97**, 2525–2530 (2014). https://doi.org/10.1111/jace.12963

56. J. Cordelair, P. Greil, Electrical conductivity measurements as a microprobe for structure transitions in polysiloxane derived Si–O–C ceramics. J. Eur. Ceram. Soc. **20**, 1947–1957 (2000). https://doi.org/10.1016/S0955-2219(00)00068-6

57. P. Colombo, T. Gambaryan-Roisman, M. Scheffler, P. Buhler, P. Greil, Conductive ceramic foams from preceramic polymers. J. Am. Ceram. Soc. **84**, 2265–2268 (2004). https://doi.org/10.1111/j.1151-2916.2001.tb01000.x

58. J. Kaspar, M. Graczyk-Zajac, S. Choudhury, R. Riedel, Impact of the electrical conductivity on the lithium capacity of polymer-derived silicon oxycarbide (SiOC) ceramics. Electrochim. Acta **216**, 196–202 (2016). https://doi.org/10.1016/J.ELECTACTA.2016.08.121

59. A. Klonczynski, G. Schneider, R. Riedel, R. Theissmann, Influence of boron on the microstructure of polymer derived SiCO ceramics. Adv. Eng. Mater. **6**, 64–68 (2004). https://doi.org/10.1002/adem.200300525

60. Y. Chen, F. Yang, L. An, On electric conduction of amorphous silicon carbonitride derived from a polymeric precursor. Appl. Phys. Lett. **102**, 2319021–2319024 (2013). https://doi.org/10.1063/1.4809825

61. G. Shao, W. Peng, C. Ma, W. Zhao, J. Guo, Y. Feng, et al., Enhanced electric conductivity of polymer-derived SiCN ceramics by microwave post-treatment. J. Am. Ceram. Soc. **100**, 842–847 (2017). https://doi.org/10.1111/jace.14590

62. L.-H. Hu, R. Raj, Semiconductive behavior of polymer-derived SiCN ceramics for hydrogen sensing. J. Am. Ceram. Soc. **98**, 1052–1055 (2015). https://doi.org/10.1111/jace.13520

63. C. Haluschka, C. Engel, R. Riedel, Silicon carbonitride ceramics derived from polysilazanes part II. Investigation of electrical properties. J. Eur. Ceram. Soc. **20**, 1365–1374 (2000). https://doi.org/10.1016/S0955-2219(00)00009-1

64. C. Shen, J.E. Calderon, E. Barrios, M. Soliman, A. Khater, A. Jeyaranjan, et al., Anisotropic electrical conductivity in polymer derived ceramics induced by graphene aerogels. J. Mater. Chem. C **5**, 11708–11716 (2017). https://doi.org/10.1039/C7TC03846A

65. H. Mei, Y. Xu, Y. Sun, Q. Bai, L. Cheng, Carbon nanotube buckypaper-reinforced SiCN ceramic matrix composites of superior electrical conductivity. J. Eur. Ceram. Soc. **36**, 1893–1898 (2016). https://doi.org/10.1016/J.JEURCERAMSOC.2016.02.045

66. V.S. Pradeep, D.G. Ayana, M. Graczyk-Zajac, G.D. Soraru, R. Riedel, High rate capability of SiOC ceramic aerogels with tailored porosity as anode materials for li-ion batteries. Electrochim. Acta **157**, 41–45 (2015). https://doi.org/10.1016/j.electacta.2015.01.088

67. E. Ionescu, A. Francis, R. Riedel, Dispersion assessment and studies on AC percolative conductivity in polymer-derived Si–C–N/CNT ceramic nanocomposites. J. Mater. Sci. **44**, 2055–2062 (2009). https://doi.org/10.1007/s10853-009-3304-3

68. Y. Chen, X. Yang, Y. Cao, L. An, Effect of pyrolysis temperature on the electric conductivity of polymer-derived silicoboron carbonitride. J. Eur. Ceram. Soc. **34**, 2163–2167 (2014). https://doi.org/10.1016/J.JEURCERAMSOC.2014.03.012

69. P.A. Ramakrishnan, Y.T. Wang, D. Balzar, L. An, C. Haluschka, R. Riedel, et al., Silicoboroncarbonitride ceramics: a class of high-temperature, dopable electronic materials. Appl. Phys. Lett. **78**, 3076–3078 (2001). https://doi.org/10.1063/1.1370540

70. A.M. Hermann, Y. Wang, P.A. Ramakrishnan, C. Haluschka, R. Riedel, Structure and electronic transport properties of Si-(B)-C-N ceramics. J. Am. Ceram. Soc. **84**, 2260–2264 (2001). https://doi.org/10.1111/j.1151-2916.2001.tb00999.x

71. Y. Wang, L. Zhang, W. Xu, T. Jiang, Y. Fan, D. Jiang, et al., Effect of thermal initiator concentration on the electrical behavior of polymer-derived amorphous silicon carbonitrides. J. Am. Ceram. Soc. **91**, 3971–3975 (2008). https://doi.org/10.1111/j.1551-2916.2008.02782.x

72. Y. Yu, F. Xia, Q. Huang, J. Fang, L. An, Electrical conductivity of silicon carbonitride-reduced graphene oxide composites. J. Am. Ceram. Soc. **100**, 5113–5119 (2017). https://doi.org/10.1111/jace.15025

73. B. Ma, Y. Wang, Fabrication of dense polymer-derived silicon carbonitride ceramic bulks by precursor infiltration and pyrolysis processes without losing piezoresistivity. J. Am. Ceram. Soc. **101**, 2752–2759 (2018). https://doi.org/10.1111/jace.15442

74. B. Ma, Y. Wang, K. Wang, X. Li, J. Liu, L. An, Frequency-dependent conductive behavior of polymer-derived amorphous silicon carbonitride. Acta Mater. **89**, 215–224 (2015). https://doi.org/10.1016/J.ACTAMAT.2015.02.020

75. D. Shopova-Gospodinova, Z. Burghard, T. Dufaux, M. Burghard, J. Bill, Mechanical and electrical properties of polymer-derived Si–C–N ceramics reinforced by octadecylamine – modified single-wall carbon nanotubes. Compos. Sci. Technol. **71**, 931–937 (2011). https://doi.org/10.1016/J.COMPSCITECH.2011.02.013

76. F. Ye, L. Zhang, X. Yin, Y. Zhang, L. Kong, Y. Liu, et al., Dielectric and microwave-absorption properties of SiC nanoparticle/SiBCN composite ceramics. J. Eur. Ceram. Soc. **34**, 205–215 (2014). https://doi.org/10.1016/J.JEURCERAMSOC.2013.08.005

77. F. Ye, L. Zhang, X. Yin, Y. Zhang, L. Kong, Q. Li, et al., Dielectric and EMW absorbing properties of PDCs-SiBCN annealed at different temperatures. J. Eur. Ceram. Soc. **33**, 1469–1477 (2013). https://doi.org/10.1016/J.JEURCERAMSOC.2013.01.006

78. Y. Cao, Y. Gao, R. Zhao, L. An, Coupling effect of temperature and stress on the electronic behavior of amorphous SiAlCO. J. Am. Ceram. Soc. **99**, 1881–1884 (2016). https://doi.org/10.1111/jace.14260

79. Y. Cao, X. Yang, L. An, Electric conductivity and microstructure evolution of polymer-derived SiAlCO ceramics. Ceram. Int. **42**, 4033–4038 (2016). https://doi.org/10.1016/J.CERAMINT.2015.11.073

80. G.D. Soraru, G. Kacha, R. Campostrini, A. Ponzoni, M. Donarelli, A. Kumar, et al., The effect of B-doping on the electrical conductivity of polymer-derived Si(B)OC ceramics. J. Am. Ceram. Soc. **100**, 4611–4621 (2017). https://doi.org/10.1111/jace.14986

81. S. Trassl, M. Puchinger, E. Rössler, G. Ziegler, Electrical properties of amorphous SiCxNyHz-ceramics derived from polyvinylsilazane. J. Eur. Ceram. Soc. **23**, 781–789 (2003). https://doi.org/10.1016/S0955-2219(02)00155-3

82. N.F. Mott, Conduction in non-crystalline materials. Philos. Mag. **19**, 835–852 (1969). https://doi.org/10.1080/14786436908216338

83. X. Li, K. Wang, B. Ma, H. Hong, M. Zhang, J. Liu, et al., Effect of acrylic acid additive on electric conductivity of polymer-derived amorphous silicon carbonitride. Ceram. Int. **41**, 7971–7976 (2015). https://doi.org/10.1016/j.ceramint.2015.02.139

84. C. Haluschka, H.-J. Kleebe, R. Franke, R. Riedel, Silicon carbonitride ceramics derived from polysilazanes Part I. Investigation of compositional and structural properties. J. Eur. Ceram. Soc. **20**, 1355–1364 (2000). https://doi.org/10.1016/S0955-2219(00)00010-8
85. P.W. Anderson, Model for the electronic structure of amorphous semiconductors. Phys. Rev. Lett. **34**, 953–955 (1975). https://doi.org/10.1103/PhysRevLett.34.953
86. M.H. Cohen, Review of the theory of amorphous semiconductors. J. Non Cryst. Solids **4**, 391–409 (1970). https://doi.org/10.1016/0022-3093(70)90068-2
87. D.S. Grierson, A.V. Sumant, A.R. Konicek, T.A. Friedmann, J.P. Sullivan, R.W. Carpick, Thermal stability and rehybridization of carbon bonding in tetrahedral amorphous carbon. J. Appl. Phys. **107**, 033523 (2010). https://doi.org/10.1063/1.3284087
88. A.C. Ferrari, J. Robertson, Interpretation of Raman spectra of disordered and amorphous carbon. Phys. Rev. B **61**, 14095–14107 (2000). https://doi.org/10.1103/PhysRevB.61.14095
89. A.C. Ferrari, J. Robertson, Resonant Raman spectroscopy of disordered, amorphous, and diamondlike carbon. Phys. Rev. B **64**, 075414 (2001). https://doi.org/10.1103/PhysRevB.64.075414
90. A.B. Kousaalya, R. Kumar, S. Packirisamy, Characterization of free carbon in the as-thermolyzed Si-B-C-N ceramic from a polyorganoborosilazane precursor. J. Adv. Ceram. **2**, 325–332 (2013). https://doi.org/10.1007/s40145-013-0079-4
91. A.B. Kousaalya, R. Kumar, B.T.N. Sridhar, Thermal conductivity of precursor derived Si–B–C–N ceramic foams using *Metroxylon sagu* as sacrificial template. Ceram. Int. **41**, 1163–1170 (2015). https://doi.org/10.1016/j.ceramint.2014.09.044
92. J. Feng, Y. Xiao, Y. Jiang, J. Feng, Synthesis, structure, and properties of silicon oxycarbide aerogels derived from tetraethylortosilicate /polydimethylsiloxane. Ceram. Int. **41**, 5281–5286 (2015). https://doi.org/10.1016/J.CERAMINT.2014.11.111
93. J.L. Oteo, M.A. Mazo, C. Palencia, F. Rubio, J. Rubio, Synthesis and characterization of silicon oxycarbide derived nanocomposites obtained through ceramic processing of TEOS/PDMS preceramic materials. J. Nano Res. **14**, 27–38 (2011). https://doi.org/10.4028/www.scientific.net/JNanoR.14.27
94. M.A. Mazo, C. Palencia, A. Nistal, F. Rubio, J. Rubio, J.L. Oteo, Dense bulk silicon oxycarbide glasses obtained by spark plasma sintering. J. Eur. Ceram. Soc. **32**, 3369–3378 (2012). https://doi.org/10.1016/J.JEURCERAMSOC.2012.03.033
95. C. Godet, Physics of bandtail hopping in disordered carbons. Diamond Relat. Mater. **12**, 159–165 (2003). https://doi.org/10.1016/S0925-9635(03)00017-7
96. C. Godet, Hopping model for charge transport in amorphous carbon. Philos. Mag. B **81**, 205–222 (2001). https://doi.org/10.1080/13642810108216536
97. C. Godet, Variable range hopping revisited: the case of an exponential distribution of localized states. J. Non Cryst. Solids **299–302**, 333–338 (2002). https://doi.org/10.1016/S0022-3093(01)01008-0
98. W.D. KINGERY, M.C. McQUARRIE, Thermal conductivity: I, concepts of measurement and factors affecting thermal conductivity of ceramic materials. J. Am. Ceram. Soc. **37**, 67–72 (1954). https://doi.org/10.1111/j.1551-2916.1954.tb20100.x
99. D.T. Morelli, G.A. Slack, *High Lattice Thermal Conductivity Solids. High Thermal Conductivity Materials* (Springer-Verlag, New York, 2006), pp. 37–68. https://doi.org/10.1007/0-387-25100-6_2
100. C. Stabler, A. Reitz, P. Stein, B. Albert, R. Riedel, E. Ionescu, Thermal properties of SiOC glasses and glass ceramics at elevated temperatures. Materials **11**, 279 (2018). https://doi.org/10.3390/ma11020279
101. A. Gurlo, E. Ionescu, R. Riedel, D.R. Clarke, The thermal conductivity of polymer-derived amorphous Si-O-c compounds and nano-composites. J. Am. Ceram. Soc. **99**, 281–285 (2016). https://doi.org/10.1111/jace.13947
102. J. Yang, J. Sprengard, L. Ju, A. Hao, M. Saei, R. Liang, et al., Three-dimensional-linked carbon fiber-carbon nanotube hybrid structure for enhancing thermal conductivity of silicon carbonitride matrix composites. Carbon **108**, 38–46 (2016). https://doi.org/10.1016/J.CARBON.2016.07.002

103. L. Qiu, Y.M. Li, X.H. Zheng, J. Zhu, D.W. Tang, J.Q. Wu, et al., Thermal-conductivity studies of macro-porous polymer-derived SiOC ceramics. Int. J. Thermophys. **35**, 76–89 (2014). https://doi.org/10.1007/s10765-013-1542-8

104. O. Majoulet, F. Sandra, M.C. Bechelany, G. Bonnefont, G. Fantozzi, L. Joly-Pottuz, et al., Silicon–boron–carbon–nitrogen monoliths with high, interconnected and hierarchical porosity. J. Mater. Chem. A **1**, 10991 (2013). https://doi.org/10.1039/c3ta12119d

105. A.M. Hermann, Y.-T. Wang, P.A. Ramakrishnan, C. Haluschka, R. Riedel, D. Balzar, et al., Structure and electronic transport properties of Si-(B)-C-N ceramics. J. Am. Ceram. Soc. **84**, 2260–2264 (2004). https://doi.org/10.1111/j.1151-2916.2001.tb00999.x

106. G. Cassabois, P. Valvin, B. Gil, Hexagonal boron nitride is an indirect bandgap semiconductor. Nat. Photonics **10** (2016)

107. P. Nagels, M. Rotti, R. Gevers, Thermoelectric power due to variable-range hopping. J. Non Cryst. Solids **59–60**, 65–68 (1983). https://doi.org/10.1016/0022-3093(83)90526-4

Part III
Performance Evaluation and Measurement Techniques

Chapter 12
Grain Boundary Engineering for Thermal Conductivity Reduction in Bulk Nanostructured Thermoelectric Materials

Adam A. Wilson, Patrick J. Taylor, Daniel S. Choi, and Shashi P. Karna

Abstract Bulk thermoelectric (TE) materials have recently seen significant enhancement in the measured dimensionless figure of merit ZT by nanostructuring the constituent materials. This is usually attributed to phonon scattering at grain boundaries, with increased grain boundary density leading to significant suppression of phonon propagation from one grain to the next while maintaining electron transport. However, to date, the reduction in thermal conductivity has been observed solely at the bulk scale. Controlling and understanding morphology and size distribution of the nanostructured grains remain a challenge. There is general lack of experimental validation of local effects of grain boundary scattering at micro- and nanoscale. This chapter discusses two strategies by which we may tune the grain size and quality of the local domains of Bi_2Te_3-based materials: shockwave consolidation and AC electric field-assisted sintering technology (FAST) via Gleeble system. These two strategies give a wide range of mean grain boundary size, from less than 100 nm to more than 500 nm. We use a multi-scale approach to measure the thermal conductivity of these samples on macroscopic/bulk scale and mesoscopic/deep submicron scale. To determine thermal conductivity over this wide dimensional scale, we leverage the ultrahigh-resolution capabilities offered by scanning thermal microscopy, the microscale capabilities of frequency-domain thermoreflectance, and the bulk-scale one-dimensional (1D) steady-state method. Despite local variations, the values on average agree well with one another, and added local thermal resolution may offer insight to future efforts to better tune materials for optimal TE performance.

A. A. Wilson (✉) · P. J. Taylor · S. P. Karna
US Army Research Laboratory, Adelphi, MD, USA
e-mail: adam.a.wilson6.civ@mail.mil

D. S. Choi
Oak Ridge Associated Universities, Oak Ridge, TN, USA

S. Skipidarov, M. Nikitin (eds.), *Novel Thermoelectric Materials and Device Design Concepts*, https://doi.org/10.1007/978-3-030-12057-3_12

12.1 Introduction: Grain Boundary Scattering in Bulk Nanostructured Thermoelectric Materials

Recent advancements in the ability to create and manipulate nanostructures have suggested that materials with grain size on the order of nanometers lead to higher density of interfaces between grains, which could yield materials with increased TE figure of merit ZT [1, 2]. TE figure of merit is a unitless benchmark of efficiency and is expressed by:

$$ZT = \frac{\sigma S^2}{\kappa_L + \kappa_e} T, \tag{12.1}$$

where σ is electrical conductivity, S is Seebeck coefficient, T is temperature, and κ is thermal conductivity (separated into lattice (phonon) κ_L and electronic κ_e parts, respectively). Enhanced ZT is achieved either by optimizing the power factor $PF = \sigma S^2$ while maintaining relatively low κ or by minimizing the contribution of κ_L to total κ while maintaining relatively large PF. This chapter discusses strategies to minimize κ_L while leaving PF unaffected. Specifically, we show that κ may be reduced by increasing grain boundary density in bulk nanostructured materials. Two different methods of densification are demonstrated: shockwave consolidation and alternating current (AC) field-assisted sintering technology (FAST) by Gleeble system. The methods yield bulk nanostructured materials with individual grain sizes spanning from nanoscale to microscale. Using a multi-scale approach spanning from nanoscale to bulk for thermal characterization, we show that κ is significantly reduced at grain boundaries.

Several recent studies have leveraged a bulk nanostructure approach to facilitate a reduction in κ_L component of thermal conductivity and observed enhancement in ZT [2–7]. Poudel et al. demonstrated a reduction in κ of bulk $Bi_{0.5}Sb_{1.5}Te$, by ball-milling and hot-pressing the material to form nanostructured bulk material [7]. Thermal conductivity was reduced by about 30% compared with commercially available state of the art, which led to enhancement in ZT of up to 40%. Mehta et al. found that by processing $Bi_{0.5}Sb_{1.5}Te$ materials in a microwave, nanocrystalline bulk composites could be formed that yield κ_L of 0.29 W $\times$ m^{-1} $\times$ K^{-1}, leading to increase in ZT of about 30%, and that, by this strategy, several other Bi_2Te_3-based material systems can be dramatically improved in thermoelectric efficiency compared with conventionally prepared counterparts of these systems [2].

In these studies, reduction in κ relied on controlling the grain size so that the grain boundary density is high. This leads to significant phonon scattering at grain boundaries, while electrons remain unscattered [8, 9]. The common effect of each effort leveraged to obtain enhancement in ZT is to fabricate materials so that the consolidation occurs quickly, under pressure, in a controlled fashion. These recent successes motivate continued investigation using this strategy for enhanced efficiency. However, to continue these investigations, better local, high-resolution thermal characterization is necessary. While these studies demonstrate experimentally the reduction in κ, analysis is performed only at the bulk scale, which does not directly address the effect of grain boundary density and its effect on the local κ at micro- or nanoscale. By locally resolving thermal property variation with preparation conditions and directly observing the effects of grain boundary scattering, a much deeper level of insight is enabled, which is presented in this work.

12.2 Tuning Grain Boundary Density: Shockwave Consolidation and AC FAST

This chapter emphasizes recent studies conducted by the authors that highlight the capability to create grain boundaries of varying size and density. Highlighted here are two strategies: shockwave consolidation and AC FAST. Each relies on rapid formation of grain boundaries, which enables the formation of such boundaries at the nanoscale.

12.2.1 Shockwave Consolidation

The goal of nanostructuring TE materials is to create local grain boundaries which induce added thermal resistance in the material by suppressing phonon propagation at the interface between grains. The shockwave consolidation approach offers a means to rapidly solidify the source nanopowder material (which is derived by atomizing pre-alloyed materials) into a fully dense solid material. The advantage of using shockwave consolidation is that it suppresses the kinetics of grain growth for small grains and requires minimal time at temperature to prevent recrystallization and of grain growth [10, 11]. The approach leverages a controlled explosion, which imparts an impulse into the material, nearly instantly compacting it to its fully dense state, permanently locking in the nanostructured texturing [10]. However, since the material is cylindrically constrained, the impulse tends to lead to an intense concentration of compressive stress, and heat, in the center of the material. This facilitates unwanted grain growth and crystallization in that region compared with the outer region of the material [11]. Such a gradient in grain size allows for unique, impactful approach to investigate the effect of nanostructuring on TE properties of the material. In this case we highlight the results of a recent study which demonstrates grain growth in $Bi_{0.5}Sb_{1.5}Te$ system prepared by shockwave consolidation [11]. Figure 12.1 depicts highly inhomogeneous distribution of crystal grain size, highlighting the differences in grain boundary density at two different locations within the material.

Fig. 12.1 Grain boundary distribution demonstrating significant variation in average grain size, depending on distance from central axis; average grain size near the center is equal to 553 nm (left inset), while it is equal to 193 nm (right inset) near the edge (adapted from [11] with permission from SPIE, copyright 2015)

At both locations, the grain size follows a distribution. Near the center, the range was found to be from ~100 nm to ~5 μm, with average value of about 553 nm. Near the edge, the average value was found to be about 197 nm, with a substantial portion of grain boundaries found to be less than 100 nm. When phonons reach grain boundaries, some are reflected, leading to scattering of heat, so it is expected that κ of the sample with the smaller grains, and thus greater density of grain boundaries, will be much lower. Typical values for interface thermal conductance in phonon-dominated transport regimes range from 10–100 MW $\times$ m^{-2} $\times$ K^{-1} [12, 13], which is the equivalent thermal resistance of including ~10–100 nm of SiO$_2$ in the conduction path. Adding this amount of thermal resistance over a significantly smaller length can significantly reduce κ_L contribution to overall κ. More detail on the grain boundary density determination, and its effect on thermal properties, will be discussed in the "quantitative scanning hot probe" section.

12.2.2   AC FAST

Like the more ubiquitous direct current (DC) spark plasma sintering (SPS), AC field-assisted sintering technique (FAST) provided by Gleeble system is used in this experiment for sintering hydrothermally synthesized Bi_2Te_3 nanopowder samples into highly dense pellet without compromising nanostructure/grain size within the material. Recently, there has been much interest in utilizing AC instead of DC to drive the heating for densification of materials. The exact relationship between sintering quality (e.g., grain size, densification) and the type of current that is applied to drive the heating needs to be established; still, the following hypothesis may elucidate the advantages of AC current used in Gleeble system over DC current used in SPS method in producing higher-quality samples.

During the sintering process, two main factors are believed to affect the quality of the sintered product: thermal and current (field) effects. Thermal effect relates to Joule heating, as well as Peltier effect, while current effect relates to electromigration or current-induced mass transport. In DC sintering, because current is being driven in one direction, significant electromigration and Peltier effect produce relatively large temperature and electric field gradients, which lead to uneven sintering and change in grain size. Conversely, in Gleeble system, because of the polarity switching in AC, electromigration and Peltier effect are mitigated and minimized during sintering process, leading to a more even distribution of heat, as well as preservation of the grain size compared to DC SPS. As of this writing, more studies are being performed at the US Army Research Laboratory (ARL) to further shed light on the effects of AC vs. DC in sintering processes.

We start from hexagonal nanoplatelet configuration of Bi_2Te_3: $h - Bi_2Te_3$, which was prepared via solvothermal (hydrothermal) synthesis. One millimole of bismuth (III) chloride $BiCl_3$, 1.5 mmol of tellurium powder Te, 0.4 g of sodium hydroxide NaOH, and 0.5 g of polyvinylpyrrolidone PVP were added to 18 mL of ethylene glycol and mixed for few hours until all solids were dissolved. The resulting solution was then transferred to a stainless steel hydrothermal vessel (autoclave) with Teflon liner. The vessel was heated to 180 °C for 40 h in a box oven and then cooled to room temperature. The resulting black powder was collected via centrifugation at 12,000 RCF for 15 min and then washed with DI water, acetone, and isopropyl alcohol. The washed precipitants were dried overnight and collected.

Powdered Bi_2Te_3 was sintered and pressed into a disc via Gleeble thermal-mechanical physical simulator system. Then, 5.0 g of Bi_2Te_3 powder was placed in 1-in. graphite die sandwiched between two nickel plates. The system was heated to 390 °C (35 °C/s) at 1.38×10^{-3} torr and with zero-second hold time at the target temperature. The resulting disc is shown in Fig. 12.2a. The disc was then coated with wax and cut into uniform, rectangular pellets for thermal and thermoelectric measurements. The dimension of the pellets is outlined in Fig. 12.2b, c.

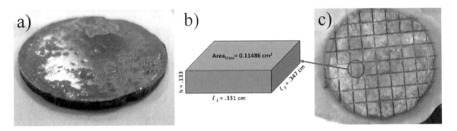

Fig. 12.2 (a) Bi_2Te_3 disc prepared by Gleeble method, (b) schematic depicting the area to be diced out for measurement, and (c) diced disc after cutting for appropriate form factor

Fig. 12.3 TEM and SAED images of hexagonal Bi_2Te_3 nanoplates

Figure 12.3 shows transmission electron microscopy (TEM) and selected area electron diffraction (SAED) images of $h - Bi_2Te_3$ nanoplates. The images show individual hexagonal Bi_2Te_3 crystals with sizes ranging from less than 50 nm to about 300 nm. Corresponding SAED patterns show the fastest growth of the hexagonal crystals to be in $\langle 11\overline{2}0 \rangle$ direction, while the slowest growth occurs at the top and bottom faces $\langle 0001 \rangle$.

Figure 12.4 shows X-ray diffraction (XRD) spectra of synthesized $h - Bi_2Te_3$ nanoplates. The identified peaks are indexed as rhombohedral Bi_2Te_3 crystals with space group $(R - 3\,m)$ indicating single-phase Bi_2Te_3 crystals.

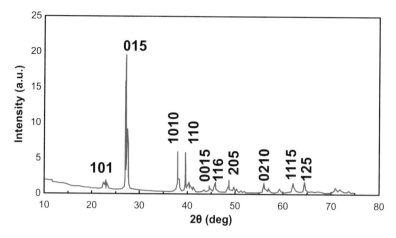

Fig. 12.4 XRD spectra of synthesized Bi_2Te_3

12.3   Quantitative Scanning Hot Probe

The scanning hot probe (SHP) technique leverages a heated atomic force microscopy (AFM) probe, which is brought into contact with the sample, transferring some heat into the sample and cooling off the probe. The temperature is monitored, and the change in temperature when the probe is in contact with the sample correlates with the sample's thermal properties. SHP technique boasts extremely high-resolution thermal mapping capabilities (on the order of 10s nm [14], compared with, at best, ~500 nm for laser-based measurement techniques) [15]. We use a similar technique as described by Zhang et al. [16, 17] and Wilson et al. [18–21]. Since the full derivation for the thermal model has been described in detail in previous publications [16, 18, 19, 22], the details essential for understanding are reported here.

The probe is heated via Joule heating and may be modeled as a line source of volumetric heat generation. In the case of the probe used in this work, palladium Pd filament is deposited on SiN_x cantilever tip. Pd filament is connected to gold trace pads, which are much wider than the filament. Therefore, the gold pads are assumed to be a perfect heat sink and are assumed to be fixed at room temperature (i.e., $T(x = 0) = T_\infty$), which is consistent with previous studies [23–27]. In the case of DC heating, the temperature profile of the probe may be described as [16]:

$$\frac{d^2 T^*}{dx^2} - \left(\frac{\text{pr} \times h_{\text{eff}}}{\kappa_p \times A_{c,\text{SiN}}} - \frac{I^2 \times \rho_0 \times \text{TCR}}{\kappa_p \times A_{c,\text{SIN}} \times A_{c,\text{Pd}}} \right) T^* + \frac{I^2 \times \rho_0}{\kappa_p \times A_{c,\text{SIN}} \times A_{c,\text{Pd}}} = 0,$$

$$(12.2)$$

where $T^* = T(x) - T_\infty$, pr, and $A_{c,\,\text{Pd}}$ are perimeter and cross-sectional area of Pd heating element, respectively; κ_p is thermal conductivity of SiN_x (note that thermal conductivity of Pd is neglected due to the thinness of the film, consistent with

Puyoo's model [27]); h_{eff} is effective heat transfer coefficient, accounting for all heat loss to ambient; I is current across the probe leg of the circuit; ρ_0 is room temperature electrical resistivity of heating element; $A_{c, SiN}$ is cross-sectional area of SiN_x part of the probe; and TCR is temperature coefficient of resistance of Pd.

The probe interacts with the sample via its contact point but also through the air gap to the sample (if operating in ambient conditions) [25] and potentially through a water meniscus (if measuring a hydrophilic sample in ambient conditions with relative humidity above ~20%) [28, 29]. For simplicity, we can define a distance, b, such that the probe's temperature is held constant in that region and that all the heat reaching the sample leaves the probe through that region. In this case, we can solve for energy balance in the tip region of half-probe, which yields the following boundary conditions:

$$-\kappa_p \frac{dT^*}{dx}\Big|_{(L-b)<x<L} + I^2\rho_0(1 + TCR \times T^*|_{x=L-b})\frac{b}{A_{c, SiN}} = \frac{Q_S}{2}, \quad (12.3)$$

where Q_S is total heat transfer rate to the sample and is divided by 2 here, since we are solving the temperature distribution in one leg only and invoking symmetry for the other leg.

Ultimately, the probe's temperature distribution is described as:

$$T^*(x) = C_1 e^{\lambda x} - C_2 e^{-\lambda x} + \frac{\Gamma}{\lambda^2}, \quad (12.4)$$

where $\lambda = \frac{I^2 \times \rho_0}{\kappa_p \times A_{c, SiN} \times A_{c, Pd}}$, $\Gamma = \frac{pr \times h_{eff}}{\kappa_p \times A_{c, SiN}} - \frac{I^2 \times \rho_0 \times TCR}{\kappa_p \times A_{c, SiN} \times A_{c, Pd}}$, and constants C_1 and C_2 may be solved for by applying the boundary conditions (Eq. (12.3) and $T(0) = 0$). From this, we may find the average probe temperature (which we can compare to calculated experimental values). When the probe is in Wheatstone bridge configuration with reference resistors and potentiometer, the probe temperature is related to the measured probe resistance R by:

$$\Delta T = \frac{R(I) - (R_{nom} + R_{contacts})}{R_{nom} \times TCR}, \quad (12.5)$$

and $R(I)$ can be calculated from how much the voltage changes with 1 Ohm change in resistance (as provided by the manufacturer) [30]:

$$V = K \times G \times \left(\frac{R_P + 1 \text{ Ohm}}{R_L + R_P + 1 \text{ Ohm}} - \frac{R_P}{R_L + R_P}\right) \times V_0, \quad (12.6)$$

where K is constant accounting for the gain of the electronics, G is applied gain to the system, V_0 is heating voltage applied to the entire circuit, R_L is resistance of two current limiting resistors in the bridge cable, and R_P is probe's electrical resistance.

To use the probe to measure thermal properties of the sample, the probe geometry must be known or estimated, and ambient thermal exchange coefficient h_{eff}, thermal exchange radius to the sample b, and contact thermal resistance between probe and sample R_C^{th} must be calibrated. To calibrate the probe for these values, several strategies have been proposed. However, two have emerged as having promise for robust and widespread usage due to repeatability and simplicity: the intersection method and the implicit method. The implicit method relies on empirical curve fitting to back calculate what unknown sample's thermal conductivity value is. This is the most widely used probe calibration strategy [31, 32], which offers simple, straightforward approach to measuring sample thermal conductivity with a probe. However, if care is not taken to account for differences in the probe-sample interaction from one measurement to the next [33] or of samples of different types [34], this technique can lead to misleading results. To calibrate using this method, at least five samples with known values of thermal conductivity are measured, and differential probe voltage response (i.e., the difference between the voltage drop across the probe leg of the Wheatstone bridge circuit, V_S, and the reference leg, V_r) is recorded. This may be plotted vs. thermal conductivity, and the resulting curve may be fitted to obtain an equation that projects the thermal conductivity for a given measured probe voltage [25]. This calibration is depicted in Fig. 12.5. Note that in this work we use a noncontact method, which has recently been developed and employed in studies [17, 18, 22, 35, 36].

In contrast, the intersection method uses at least two samples with known thermal conductivity and assumes thermal exchange radius and contact thermal resistance are constant values (i.e., not dependent on the sample measured). Though recent works have suggested that thermal exchange parameters may depend on sample properties [19], further studies have demonstrated that the expected experimental error, if the probe is calibrated with low thermal conductivity sample (around or

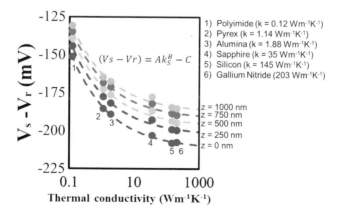

Fig. 12.5 Implicit calibration of SHP using noncontact mode, several reference samples with known thermal conductivity are measured, and the differential probe voltage signal (V_S-V_r) is fit to find the value of thermal conductivity corresponding to a given probe voltage value for a given input current and nominal probe resistance

below 1 W×m^{-1}×K^{-1}) and high thermal conductivity sample (above ~50 W×m^{-1}×K^{-1}), should be less than about 20% [18]. For this reason, Pyrex ($\kappa = 1.18$ W × m^{-1} × K^{-1} [37]) and silicon ($\kappa = 148$ W × m^{-1} × K^{-1} [38]) were chosen as reference samples. Taking the integral average of Eq. (12.4), we can compare the experimentally determined value of average probe temperature with that of the analytical model. By assuming a value of b and setting the thermal resistance in the tip region of the probe to be:

$$R_{tip}^{th} = R_C^{th} + \frac{1}{4\kappa_{ref}b}, \qquad (12.7)$$

where κ_{ref} is thermal conductivity of the reference sample, it is possible to find value of R_C^{th} such that the average probe temperature projected by the integral average of Eq. (12.4) matches that determined experimentally, represented in Eq. (12.5). Thus, for every b there is R_C^{th}, which solves the heat transfer equation. If this is repeated for a second calibration sample, one gets another set of b and R_C^{th}, which also solves the heat transfer equation. Assuming these are constant values, one can take the intersection of two curves and use those values of b and R_C^{th} as calibrated probe thermal exchange parameters. This is depicted in Fig. 12.6.

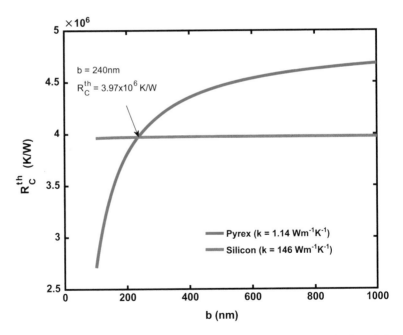

Fig. 12.6 Intersection method for probe calibration allows explicit determination of thermal exchange radius b and contact thermal resistance R_C^{th}

Table 12.1 Tip thermal resistance, calibrated probe-sample thermal exchange parameters, and determined values of sample thermal conductivity

Sample	R_P^{th} ($\times 10^6$ K/W)	b (nm)	R_C^{th} ($\times 10^6$ K/W)	κ_{int} (W $\times$ m^{-1} $\times$ K^{-1})	κ_{imp} (W $\times$ m^{-1} $\times$ K^{-1})
Gleeble-1	8.31	240	3.97	0.24 ± 0.06	0.20
SW-2	4.93	240	3.97	1.08 ± 0.27	1.06
SW-1	4.64	240	3.97	1.55 ± 0.39	2.10

For the sake of this work, several reference samples were measured that were fit to find the implicit curve fit at several noncontact probe-sample clearance values, as depicted in Fig. 12.5. Contact mode calibration was also performed using the intersection method with Pyrex and silicon as calibration samples. These were the chosen calibration samples because of the conclusions of a study that demonstrated a wide range of sample κ may be determined within ~25% experimental uncertainty if κ of the calibration samples cover a wide range [18]. For this reason, the uncertainty in κ found by the intersection method is taken to be 25% for each of three samples. The determined values of bulk κ obtained by the intersection method agree within 26% of that obtained by the implicit method. Table 12.1 summarizes the value of thermal conductivity determined using SHP technique on these samples. The sample sintered using Gleeble system is denoted "Gleeble-1," while the samples prepared by shockwave consolidation are denoted "SW-1" and "SW-2" for the inner and outer sections of the material, respectively. Note that all measurements (unless otherwise noted) were performed at room temperature (295 K), and estimated temperature of the probe was 60–80 °C in the tip region. Therefore, properties are assumed to be near the room temperature value and are reported as such later in this chapter.

Here R_P^{th} is the measured probe-to-sample thermal resistance; κ_{int} is thermal conductivity determined by intersection method; κ_{imp} is thermal conductivity determined by implicit method.

In addition to be a local thermal conductivity measurement tool, SHP technique offers the power of local qualitative thermal analysis. It has been shown that SHP is capable of directly measuring the average size of local grains and demonstrates the effect of grain boundaries on the overall measured thermal resistance of the probe. Figure 12.7 depicts thermal images taken on each of shockwave-consolidated Bi$_2$Te$_3$ samples and Gleeble sample. The grain size distribution was analyzed from these images, and determined grain sizes are compared with the values found solely by optical analysis (depicted in Fig. 12.1). The grain size distribution found via optical means compared with those found via SHP is summarized in Table 12.2.

Fig. 12.7 SHP maps (signal
normalized for each image)
of three samples; SW-1 and
Gleeble-1 have smaller
grain sizes than SW-2

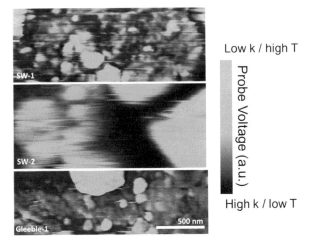

Table 12.2 Grain size (mean, d_{ave}, and mode, d_{mode}) determined optically ($d_{optical}$) and by SHP
(d_{SHP})

Sample	$d_{ave,\ optical}$ (nm)	$d_{ave,\ SHP}$ (nm)	$d_{mode,\ optical}$ (nm)	$d_{mode,\ SHP}$ (nm)
Gleeble-1	50–300	131	50–300	48
SW-2	197	123	98	85
SW-1	553	353	308	422

12.4 Frequency-Domain Thermoreflectance

Frequency-domain thermoreflectance (FDTR) has recently emerged as a versatile
and accurate quantitative measurement technique for films and interfaces [39–
43]. FDTR technique leverages a temperature-dependent change in surface reflectivity to infer sample thermal properties. It requires a blanket metal film to be
deposited with known change in reflectivity with temperature dR/dT. In this case,
80 nm gold transducer was deposited on the sample measured. To determine the
thermal properties, one must calculate the phase response of the sample due to the
heating from the pump, as detected by the probe. These are related by:

$$\Phi = \left(tan \frac{Im(H(\omega))}{Re(H(\omega))} \right)^{-1} + \Phi_0, \tag{12.8}$$

where H is complex frequency response of sample surface temperature to periodic
heat flux with Gaussian intensity distribution, as described by Schmidt et al. [39],
and Φ_0 is the phase signal coming from optics and electronics in the system and is
subtracted to determine true phase response of the sample to periodic heating

a)

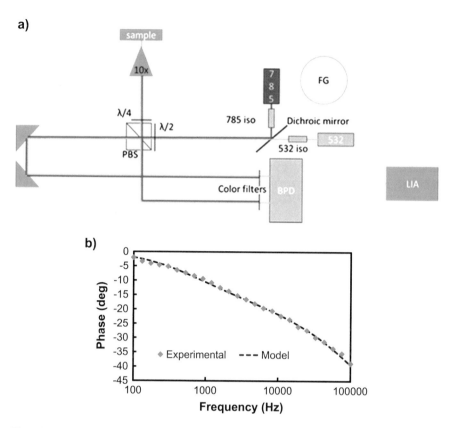

Fig. 12.8 (**a**) Schematic of FDTR system and (**b**) phase vs. frequency, fit via Eq. (12.8) to determine thermal conductivity of nanostructured Bi_2Te_3 sample; κ determined to be 0.23 ± 0.08 $W \times m^{-1} \times K^{-1}$

induced by the pump. In our implementation of FDTR, we use the pump path length as a reference for the phase, have both pump and probe reach the detector collinearly, and use a series of filters mounted to a programmable stage to filter selectively either the pump or probe. Figure 12.8a depicts the schematic of FDTR setup used in this work.

The phase signal is sensitive to substrate thermal conductivity (especially at frequencies below 1 MHz), as well as interface thermal conductance (at frequencies in the range 1–10 MHz). In this case, since we are interested in bulk κ measurements with lateral resolution on the order of microns (since the laser beam radii are 8 and 5.5 μm for pump and probe, respectively, as measured by a beam profiler), low-frequency (100–100 kHz) phase data are sufficient for measuring thermal properties of these samples. The surface of Gleeble-1 sample was polished until the average roughness was less than 100 nm, and gold transducer layer with nominally 80 nm thickness (on top of 5 nm Ti adhesion layer) was deposited by electron beam evaporation on top surface to prepare the sample for measurement by FDTR. To model appropriately the thermal properties of the transducer layer, four-point probe electrical resistivity test was performed on Pyrex reference sample, which was placed in the same deposition run. The electrical resistivity was then correlated to the film's κ via Wiedemann-Franz law [44], and value of κ was found to be 152 W $\times$ m^{-1} $\times$ K^{-1}, consistent with several other studies with evaporated gold thin films with similar thickness [15, 43, 45]. A stylus profiler was used to determine the actual thickness of the transducer layer and was determined to be 83 $\pm$ 5 nm. Several measurements were performed at locations on the sample that gave a strong signal. The resulting phase measurements were then fitted with the model based on Eq. (12.8). The standard deviation in the measurement results was taken to be the experimental uncertainty in this case. Figure 12.8b shows plot representative phase vs. frequency and the best fit to sample κ, yielding 0.23 $\pm$ 0.08 W $\times$ m^{-1} $\times$ K^{-1}.

12.5 Bulk Thermal and Thermoelectric Property Characterization

Determining bulk κ demonstrates the overall observed effect of grain boundary scattering due to nanostructuring of the material. To determine κ, we used steady-state isothermal technique as described in [46]. Since the goal of this effort was to enhance efficiency of TE materials, the remaining constituent properties and ZT of shockwave consolidated samples were also measured as described in previous works [47, 48]. When comparing κ values obtained at room temperature by FDTR or SHP to that obtained by steady-state isothermal technique, the values agree within margin of uncertainty.

To determine the overall value of ZT, values of σ, S, and κ were all measured separately. Value σ was measured by four-point probe technique where sample voltage and applied current curves are used to determine the resistance of the sample. With the knowledge of sample dimensions and assuming contact resistance is much smaller than the resistance of the sample, electrical resistivity was calculated. This is reported in Fig. 12.9 for shockwave consolidated samples. In this case, SW-1 is denoted as the nanostructured material, and SW-2 is denoted as the "diced ensemble." The electrical resistivity was also measured for Gleeble sample, but the value was quite large ($\sim$16 mOhm $\times$ cm).

Value of S is determined by imposing a varying temperature gradient across the material. The temperature at sample's hot-side T_H and cold-side T_C surfaces is measured with embedded thermocouples. The varying open-circuit voltage that results is measured as the value of $T_H - T_C$ is varied. In this way S may be determined as slope of the voltage with respect to temperature difference at mean temperature, i.e., $\frac{T_H + T_C}{2}$. For all these measurements, a correction of +1.9 μV/K must be used because of the contribution from the copper lead wires. S as function of temperature was measured for shockwave consolidated samples and is reported in Fig. 12.10. Because Gleeble sample was unintentionally doped, S reflected ambipolar contributions and cannot be directly compared.

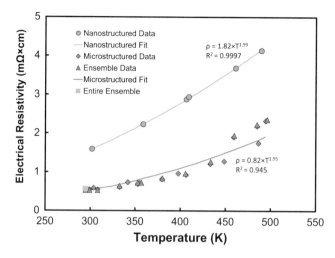

Fig. 12.9 Temperature-dependent electrical resistivity of shockwave consolidated samples. Reprinted from [11] with permission from SPIE publishing

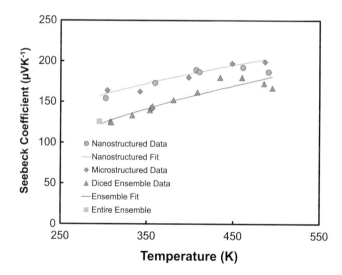

Fig. 12.10 Temperature-dependent Seebeck coefficient of shockwave consolidated samples. Reprinted from [11] with permission from SPIE publishing

Value κ is determined by the steady-state isothermal technique, where Peltier heating/cooling within TE material can be leveraged to cancel the heat flow and induce a homogeneous temperature distribution across the sample. This leads to minimizing heat loss due to radiation, a known source of large error in measurements

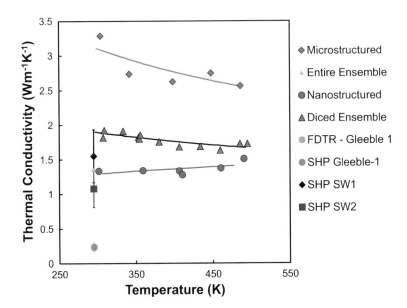

Fig. 12.11 Temperature-dependent thermal conductivity of shockwave consolidated samples and the room temperature values measured by SHP and FDTR (adapted from [11] with permission from SPIE publishing)

Table 12.3 Total thermal conductivity and estimated electronic and phonon components with surroundings at room temperature ($T_S = 300$ K)

Sample	κ (W × m^{-1} × K^{-1})	κ_e (W × m^{-1} × K^{-1})	κ_L (W × m^{-1} × K^{-1})	Grain size (nm)
Gleeble-1	0.23	0.03	0.20	50–300
SW-1	3.29	1.35	1.94	200–2500
SW-2	1.06	0.46	0.60	50–1000
$n - Bi_2Te_3$-1 [2]	1.23	0.64	0.59	N/A
$n - Bi_2Te_3$-2 [2]	0.80	0.42	0.38	N/A
$p - Bi_{0.5}Sb_{1.5}Te$ [2]	0.46	0.17	0.29	10–100

of this variety [49, 50]. Value κ determined in this way demonstrated that there was a very wide range of values between the microstructured sample, and the overall properties of the ensemble structure that contains both nanostructured material and large-grained material. This is plotted in Fig. 12.11, and the data measured by FDTR and SHP at room temperature are also included for reference. Since values reported in Fig. 12.11 are total κ, which is made of both κ_e and κ_L components, and since the values of σ were reported in Fig. 12.9, κ_L contribution to total κ may be estimated for these samples. Table 12.3 summarizes room temperature data for total κ, estimated κ_e (calculated by Wiedemann-Franz law), and estimated κ_L contribution (the difference

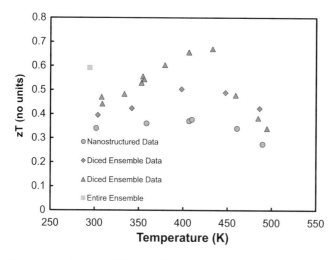

Fig. 12.12 Temperature-dependent ZT of shockwave consolidated samples. Reprinted from [11] with permission from SPIE publishing

between total κ and κ_e part). Lorenz number was assumed to be 1.49×10^{-8} WOhmK^{-2} for Gleeble sample, which corresponds to the non-degenerate limit [51], and 2.44×10^{-8} WOhmK^{-2} for SW samples, which corresponds to the degenerate limit. The data reported in Table 12.3 include shockwave consolidated samples and Gleeble sample together with reference data from the literature [2]. Gleeble sample shows a smaller grain size than shockwave consolidated samples and comparable to those reported by Mehta et al. [2] and is among the lowest values reported in the literature for κ_L of Bi$_2$Te$_3$-based materials.

With each of the constituent properties known, values ZT of shockwave consolidated samples are determined by using the values of κ, σ, and S at each temperature in Eq. (12.1). The values of ZT for shockwave consolidated samples are reported in Fig. 12.12.

12.6   Summary and Outlook

Thermoelectric properties of micro- and nanostructured bulk TE materials vary significantly as a function of grain boundary density. Thermal conductivity is particularly impacted by the grain size distribution. In this chapter, we have demonstrated that by tuning the grain boundary density by either shockwave consolidation or AC FAST by Gleeble system, thermal conductivity may significantly vary, ranging from less than 0.25 W $\times$ m^{-1} $\times$ K^{-1} to over 3 W $\times$ m^{-1} $\times$ K^{-1}, depending on doping level. With a multi-scale approach to thermal analysis, insights which can lead to grain boundary engineering are gleaned, which will ultimately yield optimally tuned high ZT materials. For this multi-scale approach, we leveraged nanoscale-resolution capabilities of scanning hot probe, microscale resolution of frequency-domain thermoreflectance, and bulk scale characterization using 1D steady-state approach. Three measurements in cluster allow for a more holistic picture of the effect of grain boundary density on observed thermal conductivity of the material. It is clear that there is a path forward with the nanostructuring approach for enhancement of ZT beyond what has been realized to date. If we can take the lessons learned from this study and apply those lessons in general to nanostructuring materials to improve TE efficiency, we may realize optimum nanostructuring for maximum efficiency materials. In the case of shockwave consolidated samples, the grain boundaries had a gradient in size that led to a gradient in thermal properties, ranging from less than $1.5 \text{ W} \times \text{m}^{-1} \times \text{K}^{-1}$ to over $3 \text{ W} \times \text{m}^{-1} \times \text{K}^{-1}$ and with estimated lattice contribution of $0.87–1.94 \text{ W} \times \text{m}^{-1} \times \text{K}^{-1}$. In general, however, the material was over-doped, leading to suboptimal TE performance. In the case of Gleeble samples, the sample exhibited anomalously low thermal conductivity (~ 0.2 $\text{W} \times \text{m}^{-1} \times \text{K}^{-1}$); nearly all of which may be attributed to the lattice, which is one of the lowest reported values of lattice thermal conductivity in Bi_2Te_3-based materials. However, the material was underdoped and was electrically insulating. Grain growth mechanisms and optimal grain size distributions remain yet to be fully elucidated. However, the recent promising results showing significant ZT enhancement, combined with novel nanoscale thermal metrology techniques, prime the stage for significant advancement in the field of nanostructured bulk thermoelectrics in the future.

References

1. G.J. Snyder, E.S. Toberer, Complex thermoelectric materials. Nat. Mater. **7**, 105–114 (2008)
2. R.J. Mehta, Y. Zhang, C. Karthik, B. Singh, R.W. Siegel, T. Borca-Tasciuc, et al., A new class of doped nanobulk high-figure-of-merit thermoelectrics by scalable bottom-up assembly. Nat. Mater. **11**, 233–240 (2012)
3. J. Maiz, M. Muñoz Rojo, B. Abad, A.A. Wilson, A. Nogales, D.-A. Borca-Tasciuc, et al., Enhancement of thermoelectric efficiency of doped PCDTBT polymer films. RSC Adv. **5**, 66687–66694 (2015)
4. T. Cardinal, M. Kwan, T. Borca-Tasciuc, G. Ramanath, Multifold electrical conductance enhancements at metal-bismuth telluride interfaces modified using an organosilane monolayer. ACS Appl. Mater. Interfaces **9**, 2001–2005 (2017)
5. E.S. Choi, J.S. Brooks, D.L. Eaton, M.S. Al-Haik, M.Y. Hussaini, H. Garmestani, et al., Enhancement of thermal and electrical properties of carbon nanotube polymer composites by magnetic field processing. J. Appl. Phys. **94**, 6034–6039 (2003)
6. B. Zhang, J. Sun, H.E. Katz, F. Fang, R.L. Opila, Promising thermoelectric properties of commercial PEDOT:PSS materials and their Bi2Te3 powder composites. ACS Appl. Mater. Interfaces **2**, 3170–3178 (2010)
7. B. Poudel, Q. Hao, Y. Ma, Y. Lan, A. Minnich, B. Yu, et al., High-thermoelectric performance of nanostructured bismuth antimony telluride bulk alloys. Science **320**, 634–638 (2008)
8. M.S. Dresselhaus, G. Chen, M.Y. Tang, R.G. Yang, H. Lee, D.Z. Wang, et al., New directions for low-dimensional thermoelectric materials. Adv. Mater. **19**, 1043–1053 (2007)

9. H.S. Kim, W. Liu, G. Chen, C.W. Chu, Z. Ren, Relationship between thermoelectric figure of merit and energy conversion efficiency. Proc. Natl. Acad. Sci. U. S. A. **112**, 8205–8210 (2015)

10. N.S. Prasad, P. Taylor, D. Nemir, Shockwave consolidation of nanostructured thermoelectric materials, in *SPIE Nanophotonics and Macrophotonics for Space Environments VIII* (2014), p. 92260J

11. N. Prasad, D. Nemir, J. Beck, J. Maddux, P. Taylor, Inhomogeneous Thermoelectric materials: improving overall zT by localized property variations, in *Proc. SPIE 9493, Energy Harvesting and Storage: Materials, Devices and Applications VI* (Baltimore, MD, 2015), p. 949305

12. P.E. Hopkins, Thermal transport across solid interfaces with nanoscale imperfections: effects of roughness, disorder, dislocations, and bonding on thermal boundary conductance. ISRN Mech. Eng. **2013**, 1–19 (2013)

13. M.E. DeCoster, K.E. Meyer, B.D. Piercy, J.T. Gaskins, B.F. Donovan, A. Giri, et al., Density and size effects on the thermal conductivity of atomic layer deposited TiO_2 and Al_2O_3 thin films. Thin Solid Films **650**, 71–77 (2018)

14. H. Chae, G. Hwang, O. Kwon, Fabrication of scanning thermal microscope probe with ultra-thin oxide tip and demonstration of its enhanced performance. Ultramicroscopy **171**, 195–203 (2016)

15. J. Yang, C. Maragliano, and A. J. Schmidt, "Thermal property microscopy with frequency domain thermoreflectance," Rev. Sci. Instrum., vol. 84, p. 104904, 2013

16. Y. Zhang, C.L. Hapenciuc, E.E. Castillo, T. Borca-Tasciuc, R.J. Mehta, C. Karthik, et al., A microprobe technique for simultaneously measuring thermal conductivity and Seebeck coefficient of thin films. Appl. Phys. Lett. **96**, 062107 (2010)

17. Y. Zhang, E.E. Castillo, R.J. Mehta, G. Ramanath, T. Borca-Tasciuc, A noncontact thermal microprobe for local thermal conductivity measurement. Rev. Sci. Instrum. **82**, 024902 (2011)

18. A.A. Wilson, T. Borca-Tasciuc, Quantifying non-contact tip-sample thermal exchange parameters for accurate scanning thermal microscopy with heated microprobes. Rev. Sci. Instrum. **88**, 074903 (2017)

19. A.A. Wilson, M. Munoz Rojo, B. Abad, J.A. Perez, J. Maiz, J. Schomacker, et al., Thermal conductivity measurements of high and low thermal conductivity films using a scanning hot probe method in the 3omega mode and novel calibration strategies. Nanoscale **7**, 15404–15412 (2015)

20. A.A. Wilson, T. Borca-Tasciuc, M. Martín-González, O. Caballero-Calero, M. Muñoz Rojo, in *Scanning Hot Probe Technique for Thermoelectric Characterization of Films. Materials Research Society Fall Meeting* (Boston, MA, 2013)

21. A.A. Wilson, M. Muñoz Rojo, B. Abad, M. Martin-Gonzalez, D. Borca-Tasciuc, T. Borca-Tasciuc, Investigating thermal exchange parameters between a heated microprobe and sample, in *International Conference of Thermoelectrics/European Conference of Thermoelectrics* (Dresden, 2015)

22. A.A. Wilson, Analysis of non-contact and contact probe-to-sample thermal exchange for quantitative measurements of thin film and nanostructure thermal conductivity by the scanning hot probe method, Doctor of Philosophy Dissertation, Mechanical, Aerospace and Nuclear Engineering, Rensselaer Polytechnic Institute, Troy, NY, 2017

23. K. Kim, J. Chung, J. Won, O. Kwon, J.S. Lee, S.H. Park, et al., Quantitative scanning thermal microscopy using double scan technique. Appl. Phys. Lett. **93**, 203115 (2008)

24. S. Lefèvre, J.B. Saulnier, C. Fuentes, S. Volz, Probe calibration of the scanning thermal microscope in the AC mode. Superlattice. Microst. **35**, 283–288 (2004)

25. S. Lefèvre, S. Volz, J.-B. Saulnier, C. Fuentes, N. Trannoy, Thermal conductivity calibration for hot wire based dc scanning thermal microscopy. Rev. Sci. Instrum. **74**, 2418–2423 (2003)

26. E. Puyoo, S. Grauby, J.M. Rampnoux, E. Rouviere, S. Dilhaire, Thermal exchange radius measurement: application to nanowire thermal imaging. Rev. Sci. Instrum. **81**, 073701 (2010)

27. E. Puyoo, S. Grauby, J.-M. Rampnoux, E. Rouvière, S. Dilhaire, Scanning thermal microscopy of individual silicon nanowires. J. Appl. Phys. **109**, 024302 (2011)

28. M. Bartosik, L. Kormos, L. Flajsman, R. Kalousek, J. Mach, Z. Liskova, et al., Nanometer-sized water bridge and pull-off force in AFM at different relative humidities: reproducibility measurement and model based on surface tension change. J. Phys. Chem. B **121**, 610–619 (2017)

29. A.A. Wilson, D.J. Sharar, Temperature-dependent adhesion mechanisms of metal and insulator probe-sample contact pairs, Presented at the 17th IEEE Intersociety Conference on Thermal and Thermomechanical Phenomena in Electronic Systems, San Diego, CA, 2018

30. Anasys, *SThM installation and operation manual* (Calibrating the Probe, Santa Clara, CA, 2013)

31. J. Juszczyk, A. Kazmierczak-Balata, P. Firek, J. Bodzenta, Measuring thermal conductivity of thin films by scanning thermal microscopy combined with thermal spreading resistance analysis. Ultramicroscopy **175**, 81–86 (2017)

32. A. A. Wilson, M. Graziano, M. Rivas, D. Baker, and B. Hanrahan, Effective thermal conductivity of iridium oxide nanostructures by a combined non-contact and contact mode scanning hot probe technique, in *Electronic and Advanced Materials Conference* (Orlando, FL, 2018)

33. A. Kaźmierczak-Bałata, J. Bodzenta, M. Krzywiecki, J. Juszczyk, J. Szmidt, P. Firek, Application of scanning microscopy to study correlation between thermal properties and morphology of BaTiO3 thin films. Thin Solid Films **545**, 217–221 (2013)

34. J. Juszczyk, M. Wojtol, J. Bodzenta, DC experiments in quantitative scanning thermal microscopy. Int. J. Thermophys. **34**, 620–628 (2013)

35. B. Abad, D.A. Borca-Tasciuc, M.S. Martin-Gonzalez, Non-contact methods for thermal properties measurement. Renew. Sust. Energ. Rev. **76**, 1348–1370 (2017)

36. A.A. Wilson, T. Borca-Tasciuc, H. Wang, C. Yu, Thermal conductivity of double-wall carbon nanotube-polyaniline composites measured by a non-contact scanning hot probe technique, in *IEEE 16th Intersociety Conference on Thermal and Thermomechanical Phenomena in Electronic Systems* (Orlando, FL, 2017), p. 456

37. I. Williams, R. Shawyer, Certification Report for a Pyrex Glass Reference Material for Thermal Conductivity between-75° C and 195°C, Commission of the European Communities, 1991

38. D.M. Rowe, V.S. Shukla, The effect of phonon-grain boundary scattering on the lattice thermal conductivity and thermoelectric conversion efficiency of heavily doped fine-grained, hot-pressed silicon germanium alloy. J. Appl. Phys. **52**, 7421–7426 (1981)

39. A.J. Schmidt, R. Cheaito, M. Chiesa, A frequency-domain thermoreflectance method for the characterization of thermal properties. Rev. Sci. Instrum. **80**, 094901 (2009)

40. J.P. Feser, J. Liu, D.G. Cahill, Pump-probe measurements of the thermal conductivity tensor for materials lacking in-plane symmetry. Rev. Sci. Instrum. **85**, 104903 (2014)

41. Z. Ge, D.G. Cahill, P.V. Braun, Thermal conductance of hydrophilic and hydrophobic interfaces. Phys. Rev. Lett. **96**, 186101 (2006)

42. P.E. Hopkins, J.R. Serrano, L.M. Phinney, Comparison of thermal conductivity and thermal boundary conductance sensitivities in continuous-wave and ultrashort-pulsed thermoreflectance analyses. Int. J. Thermophys. **31**, 2380–2393 (2010)

43. J. Yang, E. Ziade, A.J. Schmidt, Uncertainty analysis of thermoreflectance measurements. Rev. Sci. Instrum. **87**, 014901 (2016)

44. R. Franz, G. Wiedemann, Ueber die Wärme-Leitungsfähigkeit der Metalle. Ann. Phys. **165**, 497–531 (1853)

45. G. Chen, P. Hui, Thermal conductivities of evaporated gold films on silicon and glass. Appl. Phys. Lett. **74**, 2942–2944 (1999)

46. P.J. Taylor, J.R. Maddux, P.N. Uppal, Measurement of thermal conductivity using steady-state isothermal conditions and validation by comparison with thermoelectric device performance. J. Electron. Mater. **41**, 2307–2312 (2012)

47. P.J. Taylor, A. Wilson, J.R. Maddux, T. Borca-Tasciuc, S.P. Moran, E. Castillo, et al., Novel measurement methods for thermoelectric power generator materials and devices, in *Thermoelectrics for Power Generation - A Look at Trends in the Technology*, ed. by S. Skipidarov, M. Nikitin, (InTech, Rijeka, 2016), pp. 389–434

48. P. Taylor, J. Maddux, A.A. Wilson, Evaluation of thermoelectric devices by the slope-efficiency method, ARL Technical Reports, September 2016
49. C. Miers, A. Marconnet, Uncertainty quantification for a high temperature Z-meter characterization system. Presented at the IEEE 17th intersociety conference on thermal and thermomechanical phenomena in electronic systems, San Diego, CA, 2018
50. Z. Ouyang, D. Li, Modelling of segmented high-performance thermoelectric generators with effects of thermal radiation, electrical and thermal contact resistances. Sci. Rep. **6**, 24123 (2016)
51. M. Thesberg, H. Kosina, N. Neophytou, On the Lorenz number of multiband materials. Phys. Rev. B **95**, 125206 (2017)

Chapter 13
Novel Measurements and Analysis for Thermoelectric Devices

Patrick J. Taylor, Adam A. Wilson, Terry Hendricks, Fivos Drymiotis, Obed Villalpando, and Jean-Pierre Fleurial

Abstract Thermoelectric power generation is the premiere solid-state energy conversion technology for two very interesting niche applications: (1) low-temperature, $\Delta T < \sim 700$ K, conversion of heat energy into electrical energy and (2) extremely small geometrical form factor cooling and refrigeration. However, evaluation, interpretation, and analysis of thermoelectric (TE) devices are not straightforward. In this work, we introduce two new methods of analyses that provide new simple experimental methods to obtain validation of TE device performance. For cooling devices, we introduce a new test where small ΔT divergence from room temperature ($\Delta T <$ 20 K) is determined in the range where ΔT changes linearly with respect to current. The test compares ΔT and the transient cooling/heating with theoretical predictions based on measured properties and thus can confirm materials' measurements with very high accuracy. The second new analysis and empirical test allow for directly determining and, hence, validating TE figure of merit ZT_{max}. The significance of this new method is that it provides fast experimental method to confirm the validity of basic materials' measurements. The measured conversion efficiency is used to extract TE device ZT_{max} which can be compared to individual material ZT_{max} measurement claims. Therefore, this approach minimizes systemic error. We demonstrate the efficacy of this method by three cases of thermoelectric power generation modules (TEGs) fabricated from different materials: low-cost module produced by the former Alphabet Energy that demonstrates $ZT = 0.4$, commercial module from Marlow that shows $ZT = 0.7$, and specialty module made by NASA JPL that has $ZT = 0.95$.

P. J. Taylor (✉)
US Army Research Laboratory, Adelphi, MD, USA
e-mail: patrick.j.taylor36.civ@mail.mil

A. A. Wilson
National Academy of Sciences/National Research Council, Washington, DC, USA

T. Hendricks · F. Drymiotis · O. Villalpando · J.-P. Fleurial
NASA Jet Propulsion Laboratory, Pasadena, CA, USA

© This is a U.S. Government work and not under copyright protection in the U.S.; foreign copyright protection may apply 2019
S. Skipidarov, M. Nikitin (eds.), *Novel Thermoelectric Materials and Device Design Concepts*, https://doi.org/10.1007/978-3-030-12057-3_13

13.1   Introduction

The field of TE materials research is replete with materials measurement error. This problem stems from the lack of device analyses and easy means of testing/confirmation. One difficulty with testing is that the material properties are all temperature dependent. So, using values measured at, say, room temperature will not be sufficient to predict device performance over a range of temperatures above (power generation) or below (cooling) this temperature point.

In this work, we introduce two new analyses that can help fill the void of validation and testing methodologies. One analysis involves comparing measured temperature changes due to Peltier heating and Peltier cooling to theoretical predictions based on measurement of individual properties. Because the devices are tested at small electrical current (I) and, therefore, small ΔT, the equations are dominated by the terms that are linear with respect to electrical current (in this case, Joule heating is small). By using Peltier device to pump heat into or out of a known thermal mass and monitoring the time dependence of the temperature change, very accurate theoretical prediction of heating and cooling can be compared to experimental data.

A second analysis will concern large ΔT testing that is more representative of TEG performance. In performance testing, the efficiency is usually found to be a linear function of ΔT across the device. We derive some simplified relationships to explain that linearity and use that to develop new useful metric that yields device ZT value that can be compared to individual material ZT values.

It is believed that these two new analyses can help reduce the error and provide a physical reality check on reported material ZT values.

13.2   Device Testing in the Linear Range: Small ΔT

The first new analysis leverages the characteristics of two experimental conditions within the test: (1) small steady-state temperature difference, ΔT_{SS}, and measurements as a function of current (Eq. 13.2) and (2) nonsteady-state temperature difference, $\Delta T(t)$, and Peltier heating and Peltier cooling (Eq. 13.1). Because this analysis can be performed close to the ambient temperature of the environment, T_e, it lends itself to testing of TE cooling materials. The component of the first analysis that is most sensitive and, therefore, the focus of this work is seen in Eq. (13.1), the nonsteady-state case, where $\Delta T(t)$.

It should be noted that this simple thermodynamic analysis is specifically derived for the case of single TE device consisting of one n-type and one p-type leg interconnected by a metal junction. So, TE properties being confirmed are those of the individual legs measured separately. Specifically, the performance of this device is governed by Seebeck coefficients (S_n and S_p), electrical resistivity values (ρ_n and ρ_p), and thermal conductivity values (κ_n and κ_p) of n-type and p-type materials, respectively. The geometry of two legs is defined by the length-to-cross-sectional-area (l/A) ratio of the legs.

The temporal behavior results from passing the appropriate polarity of electrical current I through the single device that causes either pumping heat out of (cooling) or into (heating) of the metal junction. The rate of heat pumping is, therefore, dependent on the molar amount of metal material, m (moles), at the metal junction, and that metal's specific heat C_p.

A metric that enables the comparison between the calculated temperature and the measured temperature of the cold junction can be obtained by determining R^2, the coefficient of determination. Knowing R^2 for the comparison between calculated ΔT_{SS} and observed ΔT_{SS} provides a quantitative "goodness" factor that allows for the confirmation of validity or invalidity of materials' measurements.

The time-dependent temperature difference $\Delta T(t)$ for either cooling or heating from the ambient environment temperature, T_e, depends on ΔT_{SS}, according to [1]:

$$\Delta T(t) = \Delta T_{SS} \left[1 - \exp \frac{-\left(\frac{A}{l}\right)(\kappa_n + \kappa_p) - (S_n - S_p)}{mC_p} \right] \tag{13.1}$$

where ΔT_{SS} is given as:

$$\Delta T_{SS} = I \left[\frac{I\left(R_C + \frac{l}{2A}(\rho_n + \rho_p)\right) - (S_n - S_p)T_e}{(\kappa_n + \kappa_p)\left(\frac{A}{l}\right) - (S_n - S_p)I} \right], \tag{13.2}$$

One device was constructed from n-type $(Bi_2Te_3)_{0.90}(Sb_2Te_3)_{0.05}(Sb_2Se_3)_{0.05}$ and p-type $(Sb_2Te_3)_{0.72}(Bi_2Te_3)_{0.25}(Sb_2Se_3)_{0.03}$ materials having the following independently measured properties shown in Table 13.1.

Contact resistance R_C is $<10^{-5}$ Ohm/cm^2, and, so, it is negligible for this device, so these properties can be substituted in theoretical expressions for $\Delta T(t)$ and ΔT_{SS} along with the specific information about the metal cold junction: $m = 0.211$ moles and specific heat $C_p = 24.46$ J/(mol $\times$ K).

The device was mounted in a vacuum test stand, so that convective heat flow is negligible, and the time-dependent temperature data was collected. Time starts at zero for this test when DC electrical current is first switched on and supplied to the device, and ΔT_{SS} is approached after 1600 s. Two values of DC current were used: 0.6 A (Fig. 13.1 (left)) and 1.1 A (Fig. 13.1 (right)).

Because the materials' properties are temperature dependent, test results presented in Fig. 13.1 will specifically focus on two moderate ΔT ranges: (1) ± 8 °C with respect to the constant temperature of the heat sink maintained at

Table 13.1 Measured thermoelectric properties of device components	TE material	ρ, Ohm $\times$ cm	S, μV/K	κ, W $\times$ m^{-1} $\times$ K^{-1}	l/A, 1/cm
	n-type	0.00104	−210	1.40	0.94/0.40
	p-type	0.00143	+243	1.32	0.96/0.37

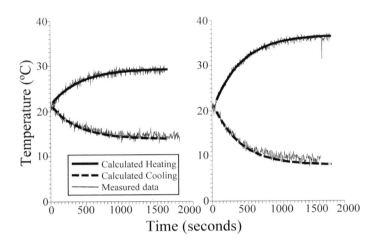

Fig. 13.1 Evaluation of thermoelectric devices in the linear region

T_e, when 0.6 A current is applied, and (2) ±16 °C with respect to T_e when 1.1 A current is applied. The maximum ΔT for the device was ~70 °C.

For the plots on Fig. 13.1 (left), there are two separate experimental results: the first using direct polarity current (heating) and the second using reversed polarity current (cooling). One polarity heats the junction, and the opposite polarity cools it, and T_e is kept constant at 21 °C. The total $\Delta T(t) = \pm8$ °C for both polarities, and the excellent agreement between the experimental device measurements and theoretical predictions based on materials' measurements is indicated by obtained R^2 value of 0.99.

This allows to conclude that S, κ, and ρ measurements are validated and confirmed to be correct.

For the plots on Fig. 13.1 (right), there are again two separate experimental results: the first using direct polarity current (heating) and the second using reversed polarity current (cooling). T_e is kept constant at 21 °C for both runs. For ±16 °C, acceptable agreement is still obtained, but for the case of cooling, the prediction is no longer accurate to R^2 value of 0.99; it is slightly less. For Peltier heating case, the agreement remains excellent, because the materials' properties are determined very close to 30 °C.

Thus, the materials' measurements are once again confirmed for Peltier heating case. The agreement for Peltier cooling case is 20 °C away from where the materials were characterized. Therefore, Peltier cooling case is evaluated at 20 °C away from where the materials were characterized, so the total temperature window of something close to ±16 °C appears to be about the acceptable limit for the range where the statistical deviation between theory and experiment begins to diverge from $R^2 = 0.99$.

The key point is that this data analysis technique can be useful to confirm TE property measurements within $\Delta T = \pm10$ °C, because the calculated and measured temperature performance are in agreement with $R^2 = 0.99$.

In the next section, we describe a different analysis for validating measurements outside of this range.

13.3   Device Testing in Nonlinear Range: Large ΔT

The efficiency with which TEG can convert heat energy to electricity can be predicted, in part, by making individual measurements of basic TE properties of the constituent materials: S, κ, and ρ. However, just like the previous example of cooling devices, experimental measurement of those materials' properties can result in significant error $> \pm 20\%$ [2]. That error leads to significant overestimates of actual device performance.

For example, efforts to quantify the error of one property, κ, showed it can be as large as $\pm 15\%$. There have been initiatives to characterize these properties with significantly improved accuracy [3–19].

What is lacking in the field of TEGs is a straightforward method of confirming measurements and validating ZT claims at elevated temperature device operation. Therefore, one goal of this section is to demonstrate a simple method of confirming measurements and validating ZT claims. A secondary goal is the description of a straightforward technique for accurate measuring output power and efficiency of TEG modules.

13.3.1   Slope-Efficiency Method: Rapid Measurement of Device ZT_{max}

The maximum electrical power output P_{max} of any TEG module is obtained when it is operated in impedance-matched condition. The impedance-matched condition occurs when internal device electrical resistance R_{int} equals exactly to electrical resistance of external load R_{load}. When $R_{int} = R_{load}$, then total system resistance is equal to $2R_{int}$, and open-circuit voltage V_{OC} drops by half leading to:

$$P_{max} = \frac{V_{OC}^2}{4R_{int}}, \tag{13.3}$$

For TEG consisting of some number i of individual thermocouples connected in series and each having n-type thermoelement and p-type thermoelement, Seebeck effect relates V_{OC} to ΔT induced by the heat source as described by:

$$V_{OC} = \sum_{1}^{i} \left(S_n(T) + S_p(T) \right)_i \Delta T, \tag{13.4}$$

where $S_n(T)$ and $S_p(T)$ are absolute values of Seebeck coefficients of n-type and p-type materials of each individual thermoelement, which are temperature dependent. Thus, the sum of Seebeck coefficients of i thermocouples is the ensemble-average proportionality between V_{OC} and ΔT. Likewise, R_{int} is the sum of resistances of

i thermocouples, and it is ensemble-average ρ of n-type (ρ_n) and p-type (ρ_p) thermoelements times those respective length-to-area $\frac{l}{A}$ values:

$$R_{int} = \sum_1^i \left(\rho_n \frac{l}{A} + \rho_p \frac{l}{A} \right)_i , \qquad (13.5)$$

P_{max} can be expressed in terms of Seebeck coefficients:

$$P_{max} = \frac{\left(\sum_1^i (S_n + S_p)_i \right)^2 \Delta T^2}{4 R_{int}}, \qquad (13.6)$$

This expression highlights the first important point: P_{max} increases as ΔT^2. So, for the maximum ΔT across the device, the maximum electrical power output is obtained.

The efficiency Φ with which TEG can convert heat flow Q to electrical power is also important because the most electrical power possible from a given amount of heat flow is desirable. A new expression for the efficiency of TEG can be obtained starting with expression for P_{max}. The efficiency is the ratio of electrical power generated per amount of input heat flow:

$$\Phi = \frac{P_{max}}{Q}, \qquad (13.7)$$

Equation (13.7) can be rewritten, assuming for simplicity a solitary $p - n$ couple ($i = 1$), as:

$$\Phi = \frac{(S_n + S_p)^2 \Delta T^2}{4 R_{int} Q}, \qquad (13.8)$$

The net flow of heat through the device Q has many components including Peltier heat, Thomson heat, Joule heat, and Fourier heat. Including all of those terms causes intractable mathematical issues that do not add to the overall perspective of the analysis. Fortunately, pure rigorous treatment that incorporates all those terms isn't required. There are simplifications that can capture the physics without the unnecessary intractable mathematics. Because the derivative of the equation is the final goal, the small terms that we ignore will become even more negligible. So, we consider only the major contribution to Q as the primary simplification and describe further simplifications and justification after Eq. (13.14).

The primary simplification is that the heat flow is dominated by κ of the materials from which TEG is constructed. This simplification is justified by the relatively small contribution from Peltier, Thomson, and Joule heat terms. The evidence which allows this simplification is that the overall efficiency for TEG conversion efficiency is limited to $\leq 10\%$ such that $\geq 90\%$ of the heat is passively lost through conduction.

This makes sense as common TE materials have total κ that is dominated by κ_L contribution [20–24]. Because Peltier, Thomson, and Joule heat account for $\leq 10\%$ of the total, this simplification yields expression:

$$\Phi = \frac{\left(S_n + S_p\right)^2 \Delta T^2}{4 R_{int}\left(\left(\kappa_n + \kappa_p\right)\frac{A}{l}\Delta T\right)}, \tag{13.9}$$

and then expresses R_{int} as described earlier:

$$\Phi = \frac{\left(S_n + S_p\right)^2 \Delta T^2}{4\left(\rho_n \frac{l}{A} + \rho_p \frac{l}{A}\right)\left(\left(\kappa_n + \kappa_p\right)\frac{A}{l}\Delta T\right)}, \tag{13.10}$$

For planar TEG devices, the values l of both n-type and p-type thermoelements are equal; however, cross-sectional areas A of n-type and p-type may be quite different. Identifying cross-sectional area of n-type as A_n and that of p-type as A_p allows a simplification, yielding Φ in terms of measurable materials' properties and ΔT:

$$\Phi = \frac{1}{4}\left(\frac{\left(S_n + S_p\right)^2}{\left(\frac{\rho_n}{A_n} + \frac{\rho_p}{A_p}\right)\left(\kappa_n A_n + \kappa_p A_p\right)}\right)\Delta T, \tag{13.11}$$

The proportionality between Φ and ΔT will be termed Z_{device}:

$$Z_{device} = \left(\frac{\left(S_n + S_p\right)^2}{\left(\frac{\rho_n}{A_n} + \frac{\rho_p}{A_p}\right)\left(\kappa_n A_n + \kappa_p A_p\right)}\right), \tag{13.12}$$

Note that when area-to-length ratios are optimized for maximum efficiency, this relationship reduces to the common, well-known expression for device Z_{max}:

$$Z_{max} = \left(\frac{S_n + S_p}{\sqrt{\kappa_n \rho_n} + \sqrt{\kappa_p \rho_p}}\right)^2, \tag{13.13}$$

TEG efficiency can be measured as function of ΔT, and the slope of that data should equal to:

$$\frac{\partial \Phi}{\partial \Delta T} = \frac{1}{4}\left(\frac{\left(S_n + S_p\right)^2}{\left(\frac{\rho_n}{A_n} + \frac{\rho_p}{A_p}\right)\left(\kappa_n A_n + \kappa_p A_p\right)}\right) = \frac{Z_{device}}{4}, \tag{13.14}$$

This expression highlights a second important point that Φ should linearly increase as a function of ΔT according to the slope indicated by $\frac{1}{4}$ of the quantity in parentheses. This makes sense, because Q function increases linearly with ΔT and P_{max} increases as ΔT^2. Taking the ratio yields a simple linear dependence on ΔT.

It is important to note that TE materials' properties are all temperature dependent, so taking the derivative would necessarily yield higher-order terms. However, we make use of the following simplifications: (1) the temperature dependence of κ_e depends on mobility of charge carriers and ρ depends on the inverse of that mobility, so these dependencies roughly cancel. (2) S does have a relatively small but finite temperature dependence; however, derivative of S should yield a temperature dependence of T^{-1} which approximately cancels with T^{-1} temperature dependence of κ_L in the denominator. The remaining terms are essentially constant over the operational temperature range of interest such that the derivative of Eq. (13.14) would have a solution that is approximately constant. Thus, it is not unreasonable to expect a basically linear slope. The linearity of that slope is in fact observed experimentally over a range of temperature, as will be shown.

A new index to determine $Z_{device}T_{max}$ of any TEG device can be obtained by measuring the slope of the efficiency as a function of ΔT. To calculate $Z_{device}T_{max}$, the slope of efficiency multiplied the product of four times the maximum temperature T_{max}, whose definition follows:

$$Z_{device}T_{max} = 4\left(\frac{(S_n + S_p)^2}{\left(\frac{\rho_n}{A_n} + \frac{\rho_p}{A_p}\right)(\kappa_n A_n + \kappa_p A_p)}\right)T_{max}, \qquad (13.15)$$

T_{max} is material dependent and is the upper temperature limit where transport of charge carriers is dominated by extrinsic carriers set by the intentional doping density. Above T_{max}, intrinsic charge carriers, that are thermally excited across the band gap, degrade the thermoelectric transport by increasing recombination. Increased recombination degrades TEG efficiency with respect to ΔT because charge carriers recombine before energy conversion. When TEG is operated outside this linear regime, the basic materials' properties can no longer be described by these functions, and the slope converges to zero; the efficiency rolls over and becomes flat. Therefore, the slope-efficiency method yields a measure of maximum $Z_{device}T_{max}$ as per Eq. (13.15).

For example, in the case of (Bi, Sb)$_2$(Te, Se)$_3$ materials which energy gap is 0.15 eV, concentration of intrinsic charge carriers at room temperature is $<10^{18}$ cm^{-3}. To avoid the deleterious effects of recombination from intrinsic charge carriers, the materials are extrinsically doped to mid-10^{19} cm^{-3}. However, as the temperature of operation is raised, then concentration of intrinsic charge carriers increases exponentially. At 500 K, concentration of intrinsic charge carriers rises to $\sim 8 \times 10^{18}$ cm^{-3} and at 600 K- to $\sim 10^{19}$ cm^{-3} which is known T_{max} upper limit to the linear operation range of (Bi, Sb)$_2$(Te, Se)$_3$ devices.

T_{max} is the temperature where the ratio of concentrations of intrinsic and extrinsic charge carriers is ~10% or less.

The overall significance of this analysis is it allows unique means to rapidly obtain $Z_{device}T_{max}$ and confirm properties and individual measurements. Measurements can be confirmed by measuring slope of efficiency as function of ΔT and ZT can be obtained and compared to theoretical ZT as calculated by individual measurements.

13.3.2 Experimental Configuration

The temperature difference across thermoelectric module, as defined by hot-side temperature T_{hot} and cold-side temperature T_{cold}, causes electrical power output from the resulting heat flow through the device. The magnitude of that power is directly measured by collecting V_{OC} and current-voltage data $V(I)$ ranging from V_{OC} condition until the short circuit current condition. The slope of the current-voltage curve yields R_{int}. The experimental V_{OC} and R_{int} data are used with Eq. (13.3) to exactly determine the electrical power output.

To characterize thermodynamic conversion efficiency, the magnitude of heat flow through thermoelectric module must be known. The specific strategy employed in this work for determining the heat flow through the device is schematically shown in Fig. 13.2. In this specific strategy, heat flow through the device is given as the sum of heat flow that passes through a heat flow meter after it leaves the module

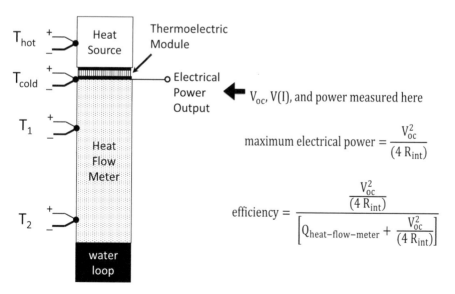

$$\text{maximum electrical power} = \frac{V_{oc}^2}{(4\,R_{int})}$$

$$\text{efficiency} = \frac{\dfrac{V_{oc}^2}{(4\,R_{int})}}{\left[Q_{\text{heat-flow-meter}} + \dfrac{V_{oc}^2}{(4\,R_{int})}\right]}$$

Fig. 13.2 Schematic experimental apparatus for measuring current-voltage data, maximum electrical power, and efficiency of TEG

($Q_{\text{heat-flow-meter}}$) added to the magnitude of generated electrical power output, Eq. (13.3). All other heat flows are negligible because the test is performed in vacuum, and the small temperature difference between the cold side and the environment eliminates radiative heat loss at the cold side.

Having eliminated heat flow errors at the cold side, Fourier heat flow law can characterize that heat by steady-state temperature difference ($T_1 - T_2$) along the length of the heat flow meter made of material of known thermal conductivity, in this case oxygen-free high-purity copper. The efficiency is defined then as the power output divided by the sum of Fourier heat flow in Q meter plus the power output.

By increasing in heat input from heat source in Fig. 13.2, a range of temperature differences across thermoelectric device can be investigated. The temperature difference data can be self-consistent if cold-side temperature T_{cold} is maintained constant. In the present strategy, constant T_{cold} is easily preserved by cooling water loop at the base of the heat flow meter.

In the individual cases that follow, we measure the power and efficiency of some currently available thermoelectric modules. Each module was measured using the experimental strategy as described above. The obtained data was then used to demonstrate the efficacy of the new slope-efficiency methodology to analyze and quantify the performance of the modules in terms of ZT.

13.3.3 Case I: Analysis of Module from Alphabet Energy

A module assembled by TE materials produced by the former Alphabet Energy was subjected to the present analysis. The materials were designed to be extremely low-cost, not high performance. According to Alphabet Energy, the module technology has a hot-side temperature limit of 873 K beyond which slope linearity breaks down, so hot-side temperature of 573 K is sufficient to fully capture the linear slope of interest. In this test cold side was maintained at constant 343 K (70 °C).

Figure 13.3 shows current-voltage and current-power data sets that were collected with $T_{\text{cold}} = 343$ K with maximum temperature difference of 229 K to capture the linear slope. On the left panel of Fig. 13.3, the slopes indicated by the best-fit equations of current-voltage data are R_{int} values as a function of T_{hot}. The y-intercepts are V_{OC}. The right-side panel shows the collected power output from the module ranging from open circuit, passing through the parabolic power at maximum efficiency, to stopping at short circuit. There are three ΔT values of 135 K, 179 K, and 229 K.

The efficiency was determined and is plotted in Fig. 13.4. In this range, the efficiency is linear, and the slope is $1.45 \times 10^{-4}\,\text{K}^{-1}$. Therefore, for this module, the determined $Z_{\text{device}}T_{\text{max}} = 0.5$.

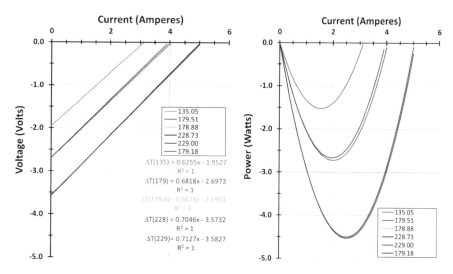

Fig. 13.3 Current-voltage load lines (left) and current-power curves (right) obtained with $T_{\text{cold}} = 70\,^{\circ}\text{C}$. The highest $T_{\text{hot}} = 573$ K for which $\Delta T = 229$ K

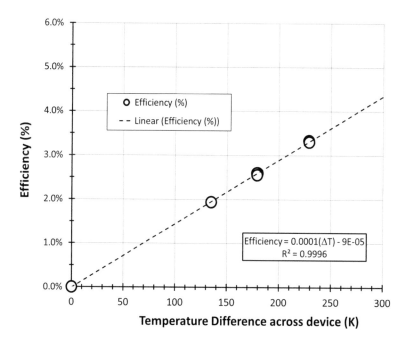

Fig. 13.4 Slope-efficiency data to determine $Z_{\text{device}}T_{\max}$. $T_{\text{cold}} = 343$ K and T_{hot} limit $= 873$ K. $ZT_{\max} = 0.5$

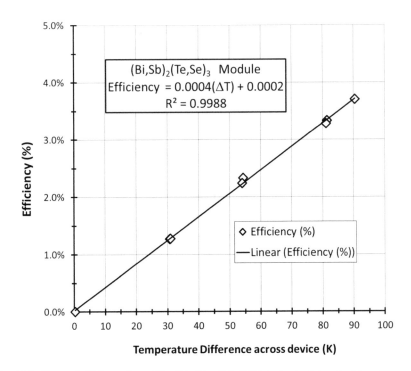

Fig. 13.5 Slope of efficiency from (Bi, Sb)$_2$(Te, Se)$_3$ TEG device to determine $Z_{device}T_{max}$

13.3.4 Case II: Analysis of Commercial (Bi, Sb)$_2$(Te, Se)$_3$ Module

The efficiency measured on commercial (Bi, Sb)$_2$(Te, Se)$_3$ TEG device from Marlow Industries is presented in Fig. 13.5. This device is designed for high performance for applications requiring especially large value of thermal impedance, such as energy harvesting. The result of high thermal impedance is that the maximum hot-side temperature is a relatively low value causing a relatively narrow optimum performance range: roughly from room temperature to 425 K.

The current-voltage data, current-power data, and efficiency were collected as a function of ΔT. The slope-efficiency of 4.26×10^{-4} K^{-1} is shown in inset in Fig. 13.5. As expected, the slope is highly linear function of ΔT until deviation from nonlinearity begins at 405 K. $Z_{device}T_{max}$ can be obtained by the simple relationship and observed maximum temperature of roughly 405 K:

$$Z_{device}T_{max} = 4\left(\frac{\partial \Phi}{\partial \Delta T}\right)T_{max}, \tag{13.16}$$

Obtained value of $Z_{device}T_{max} = 0.7$ is consistent with established values for such commercial devices designed for high performance.

13.3.5 Case III: Analysis of NASA JPL Skutterudite Module

The efficiency of TEG module fabricated from skutterudite materials was measured up to 873 K. The geometrical form factor of the module is not the conventional square shape because of the intended application. Instead the module is rectangular to accommodate integration with a special heat exchanger. Because of the module's unconventional shape, a new heat flow meter having the same rectangular shape was designed and fabricated.

The heat flow meter was designed to ensure a planar heat flow along the length. MATLAB was used to model the relative shape and dimensions in order to ensure that the heat flow was planar, as evidenced by horizontal, isothermal lines. Using the thermal properties of oxygen-free copper and assuming the heat rejection kept cold side of the module at roughly 280 K, a suitable design was found. A planar heat flow is indicated by the horizontal color-coded isotherms shown on left side of Fig. 13.6. The design was implemented and milled from a monolithic block of oxygen-free copper as shown on the right side of Fig. 13.6.

The rectangular skutterudite module was mounted onto new Cu heat flow meter which was wrapped with open-cell foam to thermally isolate that from external heat even though all tests were performed in vacuum. Commercial Grafoil sheet was inserted at each interface, and a compressive force of 200 PSI was applied to reduce thermal interface resistance values and facilitate easy heat flow into and out of the module. TEG module interfaced to hot and cold side heat exchangers is shown in Fig. 13.7. For this measurement, cold-side temperature was constant maintained at $T_{cold} = 280$ K using a water loop, and hot-side temperature was increased to $T_{hot} > 800$ K.

Thus, ΔT across TEG device was as high as $\Delta T \sim 550$ K. For ΔT values <300 K, the slope is small and is nonlinear with respect to ΔT. A range of $\Delta T < 300$ K and below is outside the normal ΔT operation range for skutterudite materials.

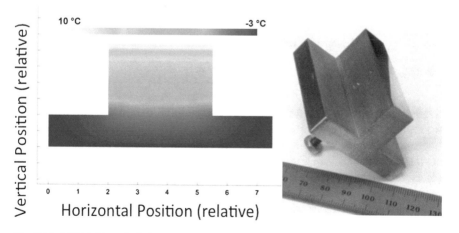

Fig. 13.6 MATLAB analysis for planar heat flow and heat flow meter

Fig. 13.7 Measurement
configuration for NASA JPL
TEG module

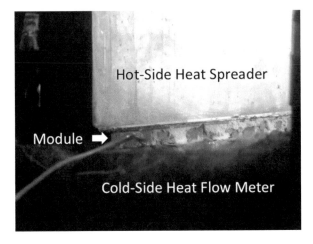

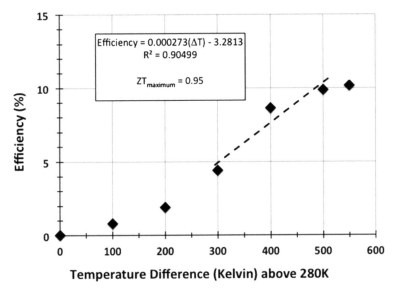

Fig. 13.8 Slope efficiency from skutterudite TEG module with $Z_{device}T_{max} = 0.95$

Skutterudites operate well in the range ΔT from ~300 K to 600 K. Thus, measure of
the representative slope-efficiency is determined where efficiency is expected to be
linear with respect to ΔT, in the range of $\Delta T \sim 400$ K. Note that there is some
curvature to the data in Fig. 13.8. This curvature results from several sources:
(1) experimental error caused by the use of new Q meter; (2) the efficiency of this
module is $\geq 10\%$ which means that Peltier, Thomson, and Joule heat terms are
increasing in significance with respect to thermally conducted heat – the material
quality is very good; and (3) the significantly higher operation temperature for TEG
device.

Therefore, representative value of $Z_{device}T_{max}$ can be obtained by using Eq. (13.14) and observed T_{max} for linear device behavior, which for TEG device being measured is equal to 873 K. The obtained slope is equal to 2.73×10^{-4} K^{-1} resulting in value $Z_{device}T_{max} = 0.95$, which is consistent with established maximum ZT values independently determined from the constituent n-type and p-type skutterudite materials in isolation.

For individual skutterudite materials of the current study, the maximum ZT for n-type skutterudite leg is ZT = 1.18 at 775 K, and the maximum ZT for p-type - ZT = 0.90 at temperature of 850 K. The materials and device fabricated from skutterudite peak at nearly identical temperatures. The slope-efficiency value for the device is intermediate to both of those individual material values yet remarkably close to both. This excellent agreement is a testimony to the accuracy of this analysis, to the high performance of new skutterudite module design, and to the robust perfection with which the modules are assembled.

13.4 Summary and Outlook

We have introduced two new analyses that can enable validation and testing methodologies to confirm thermoelectric measurements. One analysis leverages the linearity obtained for small electrical currents to drive Peltier heating and Peltier cooling magnitudes. Small currents yield both small but significant Joule heating, so that the major heat flows are those based on Peltier heat and thermally conducted heat. The obtained small ΔT values can be compared to theoretical predictions based on measurement of individual properties. Because the devices are tested at small ΔT, the parasitic heat flows are small, linear equations hold, and very accurate device predictions can be obtained and compared to experiment.

A second analysis is a new method for large ΔT testing that is more representative of performance in service. In large ΔT performance testing, the efficiency is usually found to be a linear function of ΔT across the device within certain temperature ranges. We derive some simple relationships to explain that linearity and use that to develop new useful metric that yields a device ZT value that can be compared to individual material ZT values. The obtained TEG device ZT should have a value which magnitude is intermediate to the individual p-type and n-type material ZT values.

These two new analyses can reduce the error and provide a physical reality check on material ZT values. It is extremely useful in comparing advertised and quoted performance levels between different TE module technologies using different design techniques and TE material classes.

References

1. P.J. Taylor, W.A. Jesser, F.D. Rosi, Z. Derzko, A model for the non-steady-state temperature behaviour of thermoelectric cooling semiconductor devices. Semicond Sci. Technol. **12**, 443 (1997)
2. H. Wang, W.D. Porter, H. Böttner, J. König, L. Chen, S. Bai, et al., Transport Properties of bulk thermoelectrics—an international round-robin study, part I: seebeck coefficient and electrical resistivity. J. Electron. Mater. **42**, 654–664 (2013)
3. E.E. Castillo, C.L. Hapenciuc, T. Borca-Tasciuc, Thermoelectric characterization by transient Harman method under nonideal contact and boundary conditions. Rev. Sci. Instrum. **81**, 044902 (2010)
4. P.J. Taylor, A. Wilson, J.R. Maddux, T. Borca-Tasciuc, S.P. Moran, E. Castillo, et al., Novel measurement methods for thermoelectric power generator materials and devices, in *Thermoelectrics for Power Generation - A Look at Trends in the Technology*, ed. by S. Skipidarov, M. Nikitin, (InTech, Rijeka, Croatia, 2016), pp. 389–434
5. A.A. Wilson, Analysis of non-contact and contact probe-to-sample thermal exchange for quantitative measurements of thin film and nanostructure thermal conductivity by the scanning hot probe method, Doctor of Philosophy Dissertation, Mechanical, Aerospace and Nuclear Engineering, Rensselaer Polytechnic Institute, Troy, NY (2017)
6. A.A. Wilson, T. Borca-Tasciuc, Quantifying non-contact tip-sample thermal exchange parameters for accurate scanning thermal microscopy with heated microprobes. Rev. Sci. Instrum. **88**, 074903 (2017)
7. A.A. Wilson, T. Borca-Tasciuc, M. Martín-González, O. Caballero-Calero, and M. Muñoz Rojo, Scanning hot probe technique for thermoelectric characterization of films, in *Materials Research Society Fall Meeting*, Boston, MA, 2013
8. A.A. Wilson, M. Muñoz Rojo, B. Abad, M. Martin-Gonzalez, D. Borca-Tasciuc, and T. Borca-Tasciuc, Investigating thermal exchange parameters between a heated microprobe and sample, in *International Conference of Thermoelectrics/European Conference of Thermoelectrics* (Dresden, 2015)
9. A.A. Wilson, M. Munoz Rojo, B. Abad, J.A. Perez, J. Maiz, J. Schomacker, et al., Thermal conductivity measurements of high and low thermal conductivity films using a scanning hot probe method in the 3omega mode and novel calibration strategies. Nanoscale **7**, 15404–15412 (2015)
10. P.J. Taylor, J.R. Maddux, P.N. Uppal, Measurement of thermal conductivity using steady-state isothermal conditions and validation by comparison with thermoelectric device performance. J. Electron. Mater. **41**, 2307–2312 (2012)
11. R.J. Buist, A new method for testing themoelectric materials and devices, in *11th International Conference on Thermoelectrics* (Arlington, TX, 1992)
12. J. Juszczyk, A. Kazmierczak-Balata, P. Firek, J. Bodzenta, Measuring thermal conductivity of thin films by scanning thermal microscopy combined with thermal spreading resistance analysis. Ultramicroscopy **175**, 81–86 (2017)
13. T. Borca-Tasciuc, A.R. Kumar, G. Chen, Data reduction in 3ω method for thin-film thermal conductivity determination. Rev. Sci. Instrum. **72**, 2139–2147 (2001)
14. G. Chen, T. Borca-Tasciuc, Experimental techniques for thin-film thermal conductivity characterization, in *Thermal Conductivity*, ed. by T. M. Tritt, (Kluwer Academic/Plenum Publishers, New York, NY, 2004), pp. 205–2327
15. A.J. Schmidt, R. Cheaito, M. Chiesa, A frequency-domain thermoreflectance method for the characterization of thermal properties. Rev. Sci. Instrum. **80**, 094901 (2009)
16. Y. Zhang, C.L. Hapenciuc, E.E. Castillo, T. Borca-Tasciuc, R.J. Mehta, C. Karthik, et al., A microprobe technique for simultaneously measuring thermal conductivity and seebeck coefficient of thin films. Appl. Phys. Lett. **96**, 062107 (2010)
17. C. Dames, Measuring the thermal conductivity of thin films: 3 omega and related electrothermal methods. Ann. Rev. Heat Transfer **16**, 7–49 (2013)

18. M.L. Bauer, P.M. Norris, General bidirectional thermal characterization via the 3omega technique. Rev. Sci. Instrum. **85**, 064903 (2014)
19. D. Zhao, X. Qian, X. Gu, S.A. Jajja, R. Yang, Measurement techniques for thermal conductivity and interfacial thermal conductance of bulk and thin film materials. J. Electr. Pack. **138**, 040802 (2016)
20. H.J. Goldsmid, The Thermal Conductivity of Bismuth Telluride. Proc. Phys. Soc. B **69**, 203–209 (1956)
21. Z.H. Dughaish, Lead telluride as a thermoelectric material for thermoelectric power generation. Physica B **322**, 205–223 (2002)
22. X. Shi, H. Kong, C.P. Li, C. Uher, J. Yang, J.R. Salvador, et al., Low thermal conductivity and high thermoelectric figure of merit in n-type BaxYbyCo4Sb12 double-filled skutterudites. Appl. Phys. Lett. **92**, 182101 (2008)
23. C.M. Bhandari, D.M. Rowe, Boundary scattering of phonons. J. Phys. C Solid State Phys. **11**, 1787–1794 (1978)
24. D.M. Rowe, V.S. Shukla, The effect of phonon-grain boundary scattering on the lattice thermal conductivity and thermoelectric conversion efficiency of heavily doped fine-grained, hot-pressed silicon germanium alloy. J. Appl. Phys. **52**, 7421–7426 (1981)

Part IV
Device Design, Modeling and Simulation

Chapter 14
Modeling and Optimization of Thermoelectric Modules for Radiant Heat Recovery

Je-Hyeong Bahk and Kazuaki Yazawa

Abstract In this chapter, we present a detailed methodology for modeling and optimization of multielement thermoelectric (TE) power generation modules for radiant heat recovery application. Radiative heat sources such as concentrated solar irradiation and radiation from hot steel casting in steel production processes are considered as examples in this application. Large temperature difference across TE elements is typically created by the strong radiation heat transfer due to high-temperature source. Therefore, temperature-dependent TE material properties are considered for accurate simulation. Iterative simulation method based on one-dimensional finite element method is employed to obtain the precise temperature profile along heat flow direction in each of TE elements. Optical parameters such as emissivity of gray surfaces and shape factors are considered to quantify the accurate heat input to estimate the conversion efficiency under various conditions of the source and TE module. Using fractional area coverage and thickness of element as key design parameters, the power output of the system is optimized with significantly reduced material mass in use, ultimately for enhanced power-per-cost.

Symbols Nomenclature

A_i – Surface area of object i ($i = 1$ for heat source, $i = 2$ for heat absorber) (m^2)
A_n – Cross-sectional area of n-type TE element (m^2)
A_p – Cross-sectional area of p-type TE element (m^2)
A_{TE} – Cross-sectional area of TE element (m^2)
A_{total} – Total area of TE module or total area of heat absorber (m^2)
a – Area ratio of heat absorber area to heat source area
C – Solar concentration (suns)

J.-H. Bahk (✉)
Department of Mechanical and Materials Engineering, University of Cincinnati, Cincinnati, OH, USA
e-mail: bahkjg@uc.edu

K. Yazawa
Birck Nanotechnology Center, Purdue University, West Lafayette, IN, USA

© Springer Nature Switzerland AG 2019
S. Skipidarov, M. Nikitin (eds.), *Novel Thermoelectric Materials and Device Design Concepts*, https://doi.org/10.1007/978-3-030-12057-3_14

C_i – Radiation coefficient for blackbody emissive power from object i

E_{bi} – Blackbody emissive power of object i (W/m^2)

$E_{b\lambda}(\lambda)$ – Monochromatic blackbody emissive power as a function of wavelength (W/m^3)

F – Fill factor of TE module

F_{ij} – Radiation shape factor from object i to object j

G – Solar irradiation (W/m^2)

$G_\lambda(\lambda)$ – Monochromatic irradiation as a function of wavelength (W/m^3)

h_{conv} – Convection heat transfer coefficient (W $\times$ m^{-2} $\times$ K)

I – Electrical current (A)

J_i – Radiosity of object i (W/m^2)

K_i – Thermal conductance of ith segment (W/K)

κ_{filler} – Thermal conductivity of TE module filler (W $\times$ m^{-1} $\times$ K^{-1})

L – Length or distance (m)

L_{slab} – Distance between hot steel slab and TE module (m)

L_{TE} – Thickness of TE element (m)

N – Number of segments in TE element

$N_{TE,pair}$ – Number of TE element pairs in TE module

P_{out} – Power output of TE module (W)

$Q_{conv,out}$ – Heat loss by convection from hot side of TE module (W)

$Q_{in,n}$ – Heat input to n-type element (W)

$Q_{in,p}$ – Heat input to p-type element (W)

$Q_{in,TE}$ – Total heat input to all TE elements (W)

$Q_{in,system}$ – Total system heat input before heat losses (W)

Q_{J_i} – Joule heat at ith node (W)

Q_{K_i} – Incoming conduction heat to ith node (W)

$Q_{lateral}$ – Lateral heat exchange in hot-side plate between a pair of TE elements (W)

Q_{P_i} – Peltier heat at ith node (W)

$Q_{rad,in}$ – Radiant heat input to TE module (W)

$Q_{rad,out}$ – Heat loss by radiation from hot side of TE module (W)

R_c – Contact resistance at junction between electrode and TE element (Ohm)

R_{el} – Electrical resistance of electrode (Ohm)

R_i – Electrical resistance of ith segment (Ohm)

R_{ij} – Radiation space resistance from object i to object j (m^{-2})

R_{int} – Internal resistance of TE module (Ohm)

R_L – Load resistance (Ohm)

R_{si} – Radiation surface resistance of object i (m^{-2})

S_i – Average Seebeck coefficient in ith segment (V/K)

T – Temperature (K)

T_{amb} – Ambient temperature (K)

T_{bot} – Bottom-side temperature of TE module (K)

T_C – Cold-side temperature of TE element (K)

T_H – Hot-side temperature of TE element (K)

T_i – Temperature of ith node or ith object (K)

T_{top} – Topside temperature of TE module or temperature of heat absorber (K)
V_{OC} – Total open-circuit voltage from TE module (V)
V_{OC_i} – Open-circuit voltage from ith segment (V)
V_{out} – Total output voltage to the load (V)
w – Width (m)
w_{slab} – Width of hot steel slab
ZT – Thermoelectric figure-of-merit of TE material
α – Absorptivity of heat absorber
$\alpha_\lambda(\lambda)$ – Monochromatic absorptivity as a function of wavelength
$\varepsilon_\lambda(\lambda)$ – Monochromatic emissivity as a function of wavelength
ε_i – Emissivity of object i ($i = 1$ for heat source, $i = 2$ for heat absorber)
η – Efficiency of TE system
η_{TE} – Efficiency of TE module
κ_i – Average thermal conductivity in ith segment (W $\times$ m^{-1} $\times$ K^{-1})
λ – Wavelength of radiation (m)
ρ_i – Reflectivity of object i ($i = 1$ for heat source, $i = 2$ for heat absorber)
σ_{SB} – Stefan-Boltzmann constant ($=5.670 \times 10^{-8}$ W $\times$ m^{-2} $\times$ K^{-4})
σ_i – Average electrical conductivity in ith segment (S/m)
τ – Transmissivity of solar concentrator lens
ψ_h – Conductive heat transfer coefficient of hot-side plate (Wm^{-2}K)
ψ_c – Conductive heat transfer coefficient of cold-side plate (Wm^{-2}K)

14.1 Introduction

Since the dawn of modern thermoelectrics in the 1950s, numerous applications have emerged and flourished for energy conversion with thermoelectric devices [1, 2]. Examples include radioisotope thermoelectric generators (RITEGs) for deep space missions [3], vehicle exhaust waste heat recovery with TEGs [4], compact refrigeration [5], portable energy harvesters, [6] and self-powered sensor nodes [7]. Recently, wearable energy harvesting from human body heat has been gaining great attention as a new application for thermoelectrics in this smart wearable electronics era [8]. Most of these applications, however, rely on the physical attachment of TE module onto the heat source surface in order to promote efficient heat transfer from the source to TE module by conduction. Unfortunately, radiant heat conversion by thermoelectrics has not been investigated as much thus far.

Radiant heat is transferred by electromagnetic radiation in a free space with no medium involved. Radiation heat transfer is everywhere so long as there is a temperature difference between the objects. Although it is usually small at low- to medium-temperature ranges below ~600 K, radiative emissive power increases rapidly with increasing temperature, following the fourth power of absolute temperature ($\sim T^4$), thus potentially useful as a strong heat source for thermoelectric power generation in high-temperature industrial applications. Recent advances in

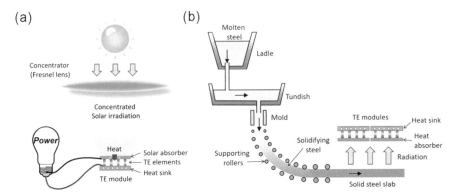

Fig. 14.1 Schematics of thermoelectric radiant heat recovery system (**a**) in concentrating solar irradiation and (**b**) in industrial hot steel casting processes

development of low-cost, high-efficiency optical concentrators such as parabolic mirrors or Fresnel lenses enable high concentration of radiation to magnify the heat flux to TE module for enhanced conversion efficiency and power generation with minimal cost added [9, 10]. Furthermore, radiant heat is a noncontact heat source that can be harvested by TEG from a distance. TE module would impose minimal impacts on the performance of the source system due to its noncontact nature. Additional contact resistances and interface damages/cracks due to thermal stresses that are commonly observed in physically attached TE modules [11] would not be an issue in radiant heat recovery TE system.

One of the most important and abundant radiant heat is solar irradiation. Solar energy is sustainable and easily accessible for energy harvesting. When solar energy is absorbed by solar absorber and converted to heat, this heat can then be converted to electricity by TEG [12, 13]. This technology is known as so-called solar thermoelectric technology. Recently, good reviews on solar thermoelectric technology have been published [14, 15]. A schematic of concentrated solar thermoelectric system is shown in Fig. 14.1a. Solar radiation is typically concentrated into a smaller area covered by TEG module as shown in the schematic in order to increase the operating temperature and thus the conversion efficiency. A heat sink might be necessary to keep sufficiently high the temperature difference between hot and cold sides of TEG module, which, however, consumes power for its operation and thus reduces in system efficiency. Recently, peak system efficiency of 7.4% has been demonstrated for concentrating solar TEG made of segmented TE elements in high-vacuum operation [16]. This result suggests a great potential of solar thermoelectric systems to become a promising alternative solar energy technology.

Another important industrial application with strong radiant heat available for harvesting is found in hot steel casting process in steel manufacturing [17, 18]. Steel manufacturing processes consume a huge amount of electric power. The autonomous operation of the processing lines, e.g., hot rolling mills or continuous casting lines, requires hundreds of sensors such as hot metal detectors, light barrier switches, and proximity switches to ensure reliability and robustness of production [19]. TEG module capable of generating power out of the heat radiated from hot steel slabs, as schematically shown in Fig. 14.1b, can in turn power many such sensors without resorting to the grid power. This will ensure lower electrical power requirement, higher utilization of input power, and more environment-friendly operation.

Since very large temperature difference ~200 K or higher is typically applied to TEG module in radiant heat recovery applications, variations of material properties with temperature over the wide temperature range are critical in module performance. Unfortunately, the effects of temperature-dependent material properties have not been much studied nor accurately accounted for in module simulations in literature so far. A good review of thermoelectric simulation has been published in [20]. Also, temperature boundary conditions are typically used for module simulations [21, 22], e.g., constant topside temperature, but these assumptions are no longer valid in radiant heat recovery applications because the boundary temperatures are not fixed but rather determined after solving for the entire temperature profile across TE module with accurate quantification of heat inputs and losses. For precise performance prediction and module optimization, simulation tool is essential that is capable of accounting for all these effects and unique features. We have published our simulation tool for solar TE systems on nanoHUB.org for public use [23].

In this chapter, we present our simulation methodology and module optimization for TE systems for radiant heat recovery. We employ iterative, one-dimensional finite element methods to accurately determine distributions of temperature along individual TE elements with accounting of temperature-dependent material properties. Heat inputs from the radiative heat exchanges between the objects involved and heat losses due to convection and back radiation, as well as TE filler effects, are quantified for accurate simulation. Our detailed module simulation methodology will be discussed in Sect. 14.2. In Sects. 14.3 and 14.4, we will present the simulation results for concentrating solar thermoelectric systems and hot steel casting application, respectively. We will identify optimal module designs and impacts of various design parameters on power output and efficiency, for these two applications. Like any other renewable energy technologies, presented simulations could be helpful for estimation of power-per-cost value, which can only be achieved by careful design of TE module and whole system [24]. We will provide guidelines to achieve high power-per-cost in these applications.

14.2 Theory and Simulation Methodology

Figure 14.2 shows a schematic of multielement TE module for radiant heat recovery. The top surface of TE module is made of heat absorber, i.e., with high emissivity surface, which efficiently absorbs $Q_{rad,in}$ from heat source located with a certain distance. Concentration of heat is possible using optical concentrators as mentioned earlier before the heat reaches the top surface of TE module to increase $Q_{rad,in}$ by factor of concentration C. Some portion of $Q_{rad,in}$ is dissipated to ambient at the top

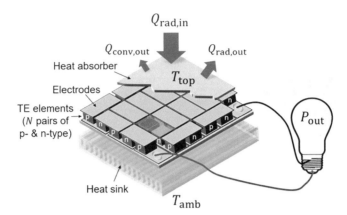

Fig. 14.2 Schematic of multielement TE module for radiant heat recovery. Radiant heat input $Q_{rad,\,in}$ and heat losses by radiation $Q_{rad,\,out}$ and by air convection $Q_{conv,\,out}$ at the topside (heat absorber side) of TE module are displayed. No vacuum enclosure is assumed

surface via back radiation $Q_{rad,out}$ and air convection $Q_{conv,out}$. The whole TE module can be surrounded by high-vacuum enclosure to minimize $Q_{conv,out}$ loss. In this work, however, we assume ambient air surroundings to eliminate the additional installation and maintenance costs for vacuum enclosure. The convection loss can still be much smaller than other parasitic heat losses, so air surrounding may be acceptable without much performance reduction. The remaining heat will pass through TE elements, with some losses by conduction through the air or filler material in the spacing between TE elements, and be partially converted by TE effects to electric power P_{out}, which is delivered to the load. The rest of the heat is then dissipated to ambient by the heat sink at the bottom of the system. In the next subsection, we will discuss the heat transfer model to determine the heat inputs and losses at the topside.

14.2.1 Heat Transfer Model

In solar TE system, in which the heat source is the sun, $Q_{rad,in}$ is given by:

$$Q_{rad,in} = GC\tau\alpha A_{total}, \tag{14.1}$$

where G is solar irradiation power density in W/m^2; C is solar concentration defined as the ratio of solar concentrator area to TE absorber area in the unit of suns; τ is the transmissivity of solar concentrator lens or the reflectivity if concentrator is a mirror type; α is the absorptivity of solar absorber equipped at the top surface of TE module; and A_{total} is the total area of TE absorber plate. The absorptivity α is a weighted average of monochromatic absorptivity $\alpha_\lambda(\lambda)$ of the absorber with monochromatic solar irradiation $G_\lambda(\lambda)$ as the weighting factor over the entire wavelength range given by [25]:

$$\alpha = \frac{\displaystyle\int_0^\infty \alpha_\lambda(\lambda)G_\lambda(\lambda)d\lambda}{\displaystyle\int_0^\infty G_\lambda(\lambda)d\lambda}, \tag{14.2}$$

The heat loss by radiation from TE absorber plate occurs because there is radiation exchange between the plate surface and the surroundings, which is given by:

$$Q_{\text{rad, out}} = \varepsilon\sigma_{\text{SB}}A_{\text{total}}\left(T_{\text{top}}^4 - T_{\text{amb}}^4\right), \tag{14.3}$$

where ε is the emissivity of TE absorber and σ_{SB} is Stefan-Boltzmann constant. Similarly, as the absorptivity, the emissivity is a weighted average of monochromatic emissivity $\varepsilon_\lambda(\lambda)$ of the surface with monochromatic blackbody emissive power $E_{b\lambda}(\lambda)$ as the weighting factor over the entire wavelength range at a given temperature T given by:

$$\alpha = \frac{\displaystyle\int_0^\infty \varepsilon_\lambda(\lambda)E_{b\lambda}(\lambda)d\lambda}{\displaystyle\int_0^\infty E_{b\lambda}(\lambda)d\lambda}, \tag{14.4}$$

In principle, $\alpha_\lambda(\lambda) = \varepsilon_\lambda(\lambda)$ for the same surface at a given wavelength [25]. However, the averaged values of those parameters are not necessarily equal to each other, i.e., $\alpha \neq \varepsilon$, because $G_\lambda(\lambda)$ coming from the sun at very high surface temperature ~5800 K is very different from $E_{b\lambda}(\lambda)$ coming from TE plate surface at much lower temperature below 1000 K. It is desirable for TE absorber to have high α for high $Q_{\text{rad,in}}$ and low ε for low heat loss by $Q_{\text{rad,out}}$.

In general case of radiant heat recovery system, heat source and TEG power system are placed within a certain distance, usually facing each other for high heat transfer between the two. Unlike the case in solar TE systems, $Q_{\text{rad,in}}$ is not constant but determined by the geometry of the system and optical/thermal characteristics of the objects involved. We model both the heat source and TE heat absorber as gray surfaces with emissivity ε_1 and ε_2, respectively. Note that the emissivity is a function of temperature. For simplicity, emissivity is assumed to be constant in the operating temperature range in this chapter.

Unless one object is completely enclosed by the other, the surroundings will also participate in the radiation exchange as the third object. We assume the surroundings to be blackbody at ambient temperature T_{amb}. Then, one can determine the radiation

Table 14.1 Radiation shape factors for several common geometries [26]

Geometry	Shape factor F_{12}
Infinitely long concentric parallel plates	$F_{12} = \sqrt{\left(\frac{L}{w_1}\right)^2 + \left(\frac{w_1 + w_2}{2w_1}\right)^2} - \sqrt{\left(\frac{L}{w_1}\right)^2 + \left(\frac{w_1 - w_2}{2w_1}\right)^2}$
Aligned parallel rectangles	$A = a/L, \; B = b/L$ $F_{12} = \frac{2}{\pi AB} \left\{ ln\sqrt{\frac{(1+A^2)(1+B^2)}{1+A^2+B^2}} + \right.$ $+ A\sqrt{1+B^2}\left(tan\frac{A}{\sqrt{1+B^2}}\right)^{-1} + B\sqrt{1+A^2}\left(tan\frac{B}{\sqrt{1+A^2}}\right)^{-1} -$ $\left. - \frac{A}{tanA} - \frac{B}{tanB}\right\}$
Coaxial parallel disks	$S_{dd} = 1 + \frac{L^2 + r_2^2}{r_1^2}$ $F_{12} = \frac{1}{2}\left\{ S_{dd} - \sqrt{S_{dd}{}^2 - 4\left(\frac{r_2}{r_1}\right)^2}\right\}$
Infinitely long perpendicular plates with common edge	$F_{12} = \frac{1 + \left(\frac{b}{a}\right) - \sqrt{1 + \left(\frac{b}{a}\right)^2}}{2}$

shape factor F_{ij} between each combination of two objects. Table 14.1 summarizes predetermined shape factors for several important geometries [26].

Once the shape factors are determined, radiation circuit can be constructed between three objects as shown in Fig. 14.3. Here, Objects 1, 2, and 3 in the circuit correspond to heat source, top of TE module and surroundings, respectively. Three nodes in the middle triangular circuit are associated with the radiosity J_i ($i = 1, 2, 3$) of each surface, and then each radiosity node is connected to each of the other two radiosity nodes with the appropriate *space resistance* R_{ij} determined by the corresponding shape factor and the surface area as displayed in Fig. 14.3. Each radiosity node is also connected to blackbody emissive power $E_{bi} = \sigma_{SB}T_i^4$ ($i = 1, 2, 3$) associated with temperature of that surface T_i using a *surface resistance* $R_{s,i}$. We used the relation $\rho_i = 1 - \varepsilon_i$, where ρ_i is the reflectivity of ith surface with assumption of no transmission through the surface. Note that the

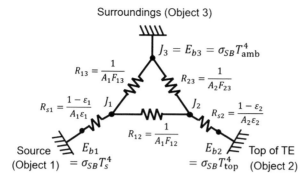

Fig. 14.3 Analogous circuit model for radiation transfer between heat source, TE module, and surroundings. Object 1 is the heat source with emissivity ε_1. Object 2 is the top of TE module with emissivity ε_2. Object 3 is the surroundings assumed black ($\varepsilon_1 = 1$). R_{ij} refers to radiative resistance between two objects. F_{ij} is radiation shape factor from object i to j. E_{bi} is the emissive power at ith object, and J_i is the radiosity. A_1 and A_2 ($=A_{\text{total}}$) are the area of the heat source and TE module, respectively

radiosity of the surroundings is equal to its blackbody emissive power, i.e., $J_3 = E_{b3}$, as it is assumed to be black. So, there is no surface resistance for that node.

From the radiation network shown in Fig. 14.3, the net rate of heat transfer to TE surface can be found as a function of the emissive powers of three nodes by:

$$Q_{\text{rad,in}} - Q_{\text{rad,out}} = A_{\text{total}}(C_1 E_{b1} + C_2 E_{b2} + C_3 E_{b3}), \tag{14.5}$$

where coefficients C_i are found by applying Kirchhoff's circuit law to the radiation resistance network in Fig. 14.3 as:

$$C_1 = \frac{F_{12}\varepsilon_1\varepsilon_2}{a - F_{12}^2(1 - \varepsilon_1)(1 - \varepsilon_2)}, \tag{14.6}$$

$$C_2 = \frac{a\varepsilon_2\left[1 - F_{12}^2(1 - \varepsilon_1)\right]}{a - F_{12}^2(1 - \varepsilon_1)(1 - \varepsilon_2)}, \tag{14.7}$$

$$C_3 = \frac{\varepsilon_2\left[a - F_{12}\varepsilon_1 - F_{12}^2(1 - \varepsilon_1)\right]}{a - F_{12}^2(1 - \varepsilon_1)(1 - \varepsilon_2)}, \tag{14.8}$$

where $a = A_2/A_1$ is area ratio of TE module area A_2 to source area A_1. We assumed no self-radiation, i.e., $F_{ii} = 0$, and used the reciprocity theorem, i.e., $A_i F_{ij} = A_j F_{ji}$, to leave only one shape factor F_{12}, in the equations.

In addition, we model the convective heat loss at heat absorber surface with constant heat transfer coefficient h_{conv} as:

$$Q_{\text{conv,out}} = h_{\text{conv}} A_{\text{total}}\left(T_{\text{top}} - T_{\text{amb}}\right), . \tag{14.9}$$

Finally, the total heat input entering TE elements is found using Eqs. (14.5) and (14.9) as:

$$Q_{in,total} = Q_{rad,in} - Q_{rad,out} - Q_{conv,out}, \qquad (14.10)$$

14.2.2 Modeling of Individual TE Elements

For multielement TE module, we use two main design parameters as independent variables to optimize the performance. The first parameter is the fill factor F which is defined as the fractional area coverage of TE elements over the entire module area given by:

$$F = \frac{N_{TE,\,pair}\left(A_n + A_p\right)}{A_{total}}, \qquad (14.11)$$

where $N_{TE,pair}$ is the number of n-type and p-type TE element pairs and A_n and A_p are cross-sectional area of individual n-type and p-type TE elements. Absorbed heat at the topside by the absorber is concentrated into smaller TE area by factor of $1/F$, which is greater than unity. This factor is called the *thermal concentration factor* [27]. In this work, we fix cross-sectional areas of individual elements and vary the spacing between TE elements to change fill factor. $N_{TE,pair}$ is changed accordingly for different fill factors by Eq. (14.11). The second design parameter is the thickness of TE elements L_{TE}. We use the same thickness for both n-type and p-type TE elements. By adjusting two parameters L_{TE} and F independently, both total thermal resistance and electrical resistance can be tuned independently and broadly to find out the optimal performances.

Our numerical model aims to solve simultaneously coupled thermal and electrical circuit equations for individual TE elements. The heat balance equations determine the temperature profile along TE element, whereas electrical circuit analysis with a load resistance determines electrical current, which again affects the heat balance equation. In order to obtain a detailed numerical analysis, temperature-dependent material properties are taken into account.

In order to obtain the accurate temperature profile and heat transfer rate in TE elements with temperature-dependent material properties taken into account, each TE element is divided into N segments with S_i, σ_i, and κ_i being Seebeck coefficient, electrical conductivity, and thermal conductivity, respectively, inside ith segment as shown in Fig. 14.4a. N must be sufficiently large, or the thickness of individual segments must be sufficiently small, so that temperature change within each segment can be assumed negligibly small. In this case, material properties S_i, σ_i, and κ_i in each segment can be assumed to be constant and obtained at average temperature of the segment, i.e., $(T_{i-1} + T_i)/2$.

Since the temperature in each segment is not known a priori, we make a guess for the initial temperature values at each node between segments and then assign the

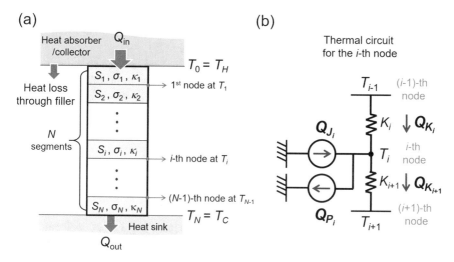

Fig. 14.4 (**a**) One-dimensional finite element model for single TE element divided into N segments for simulation with temperature-dependent material properties. (**b**) Thermal circuit at each node between segments, i.e., ith node, including three components of heat: conduction Q_{K_i}, Peltier/ Thomson Q_{P_i}, and Joule terms Q_{J_i}

material properties in each segment corresponding to the temperature. By solving all the heat balance equations formulated at all the nodes simultaneously, we obtain updated temperature profile across TE element. Then a new set of material properties is allotted at each node based on updated temperatures for the next iteration. This process is repeated until the temperature profile converges within acceptable tolerance. A good initial guess for the temperature profile can be linear temperature distribution inside TE elements, assuming the conduction term is dominant over Joule and Peltier heat terms.

At one point of iteration, Q_{J_i} and Q_{P_i} at ith node and Q_{K_i} across ith segment are given by:

$$Q_{J_i} = \frac{1}{2}I^2 R_i + \frac{1}{2}I^2 R_{i+1}, \tag{14.12}$$

$$Q_{P_i} = (S_{i+1} - S_i)T_i I, \tag{14.13}$$

$$Q_{K_i} = (T_{i-1} - T_i)K_i, \tag{14.14}$$

where I is electric current; $R_i = L_{\text{TE}}/(\sigma_i A_{\text{TE}} N)$ and $K_i = (\kappa_i A_{\text{TE}} N)/L_{\text{TE}}$ are, respectively, electrical resistance and thermal conductance of ith segment; and A_{TE} is cross-sectional area of individual TE elements, i.e., $A_{\text{TE}} = A_n$ for n-type and $A_{\text{TE}} = A_p$ for p-type elements. Figure 14.4b shows thermal resistance network model ith node. Hence, heat balance equation at ith node is:

$$Q_{J_i} - Q_{P_i} + Q_{K_i} - Q_{K_{i+1}} = 0, \tag{14.15}$$

where index i varies from 1 to $(N-1)$ creating total of $(N-1)$ equations to solve. It is noted that Peltier term between segments is non-zero if Seebeck coefficient varies with temperature. This is also called *Thomson effect* [28].

At 0th node or at the hot side of TE element, the heat balance equation has the same form as in Eq. (14.15) but with $i = 0$. At this node, Joule and Peltier terms have only one-side component because there is no Joule or Peltier effects from the top hot plate with no current flow. Also, Seebeck coefficient of electrode is assumed to be negligibly small. Thus:

$$Q_{J_0} = \frac{1}{2}I^2 R_1,\tag{14.16}$$

$$Q_{P_0} = S_1 T_0 I,\tag{14.17}$$

where $T_0 = T_{\mathrm{H}}$. The outgoing conduction term Q_{K_1} has the same form as in Eq. (14.14) but with $i = 1$, and the incoming conduction term Q_{K_0} from the top plate at 0th node is determined by heat transfer coefficient of the top plate ψ_h, and also the heat loss through the gap filler must be taken into account, such that, respectively:

$$Q_{K_1} = (T_0 - T_1)K_1,\tag{14.18}$$

$$Q_{K_0} = \frac{1}{F}A_{\mathrm{TE}}\psi_h\left(T_{\mathrm{top}} - T_0\right) - \frac{(1-F)}{F}\frac{A_{\mathrm{TE}}}{L_{\mathrm{TE}}}\kappa_{\mathrm{filler}}(T_0 - T_N),\tag{14.19}$$

where κ_{filler} is thermal conductivity of the gap filler in TE module, i.e., air. In Eq. (14.19), the first term on the right-hand side of the equation describes concentrated heat input to TE element by the concentration factor $1/F$. The second term is the heat lost through the filler by conduction, which is subtracted from the first term to get the net heat input into TE element.

Similarly, we can formulate the heat balance equation at Nth node or at the cold side of individual TE elements using Eq. (14.15). Here, Joule and Peltier terms have only one-side component from Nth segment of TE element because there is no Joule or Peltier effects from the bottom cold plate with no current flow. Thus:

$$Q_{J_N} = \frac{1}{2}I^2 R_N,\tag{14.20}$$

$$Q_{P_N} = S_N T_N I,\tag{14.21}$$

where $T_N = T_C$. The incoming conduction term to Nth node Q_{K_N} has the same form as in Eq. (14.14) but with $i = N$, and the outgoing conduction term $Q_{K_{N+1}}$ from Nth node toward the cold plate is determined by the heat transfer coefficient of the cold plate ψ_c, such that, respectively:

$$Q_{K_N} = (T_{N-1} - T_N)K_N,\tag{14.22}$$

$$Q_{K_{N+1}} = A_{TE}\psi_c(T_N - T_{bot}) \times \frac{1}{F}, \qquad (14.23)$$

where T_{bot} is assumed to be equal to the ambient temperature, i.e., $T_{bot} = T_{amb}$.

At the topside of module, another heat balance equation holds that the total incoming radiation heat transfer rate $Q_{in,total}$ obtained from Eq. (14.5) flows in the hot plate by conduction, such that:

$$Q_{in,total} = A_{total}\psi_h(T_{top} - T_h), \qquad (14.24)$$

So far, we have constructed $N + 1$ heat balance equations given by Eq. (14.15) at $N + 1$ nodes with index i varying from 0 to N for TE element and another equation given by Eq. (14.24) for heat conduction through the hot plate, making the total number of equations $N + 2$. Since, there are the same number of unknown temperatures, i.e., T_i ($i = 0 \sim N$) and T_{top}, unique solution for temperature profile can be found that satisfies all the equations for a given electric current I. Electric current I will be determined by electric circuit model for the whole module with a load resistance connected, which will be discussed in the next subsection.

As discussed earlier, the found temperature values are reentered to update the material properties in each segment, and this process is iterated until the temperature profile converges.

14.2.3 Iterative Module Simulation

In the previous subsection, the heat balance equations for individual TE elements have been formulated. With these equations, temperature profiles for n-type and p-type elements are independently obtained. Since, n-type and p-type TE materials used have typically different material properties, hence temperature profiles will be different from each other. Thus, the topside temperature T_{top} on top of n-type TE element can be different from that of p-type one. This may not be physically reasonable because the hot plate is made of highly thermally conducting material for efficient heat transfer to TE elements, so temperature flattens quickly at the heat-absorbing surface. In reality, there will be lateral heat flow in the top plate between the regions of n-type and p-type elements to compensate nonuniformity in vertical heat flow between n-type and p-type elements [18]. To account for this effect in our one-dimensional finite element model, we add another heat exchange between two adjacent n-type and p-type TE elements in lateral direction inside the top plate, so that the heat input in each TE element is adjusted with that lateral heat exchange until the top temperature T_{top} becomes uniform.

Therefore, the heat input to n-type TE element previously given by Eq. (14.19) is corrected with additional lateral heat term to become:

$$Q_{\text{in},n} = Q_{K_0,n} = \frac{1}{F} A_n \psi_h \left(T_{\text{top}} - T_0 \right) - \frac{1-F}{F} \frac{A_n}{L_n} \kappa_{\text{filler}} (T_0 - T_N) - Q_{\text{lateral}},$$

(14.25)

By energy conservation, the same amount of heat must be added to the heat input to p-type TE element such that:

$$Q_{\text{in},p} = Q_{K_0,p} = \frac{1}{F} A_p \psi_h \left(T_{\text{top}} - T_0 \right) - \frac{1-F}{F} \frac{A_p}{L_p} \kappa_{\text{filler}} (T_0 - T_N) + Q_{\text{lateral}},$$

(14.26)

The lateral heat flow Q_{lateral} can be either positive or negative depending on direction of heat flows. Eqs. (14.25) and (14.26) are now used instead of Eq. (14.19) for simulation of n-type and p-type elements, respectively, to get final temperature profiles in those. Q_{lateral} must also be found by another iterative process to cause the topside temperature of n-type TE elements $T_{\text{top},n}$ and that of p-type ones $T_{\text{top},p}$ to converge with each other.

Lastly, electric current I flowing through TE module is determined from the analysis of electrical circuit with load resistance R_L. Since, all the elements, electrodes, and load resistances are connected electrically in series, adding all resistances from those along with contact resistances results in the total internal electrical resistance in TE module given by:

$$R_{\text{int}} = \sum \left(\sum_i R_i + 2R_c + 2R_{\text{el}} \right),$$

(14.27)

where $\sum_i R_i$ is internal resistance of each TE element obtained by adding the resistances from all the segments of TE element. R_c is contact resistance per one side of TE element, and R_{el} is electrode resistance per one side of TE element. Symbol Σ indicates the summation over all TE elements. In TE module, open-circuit voltage V_{OC} is created by Seebeck effects inside TE elements. In each segment of TE element, Seebeck voltage is obtained as:

$$V_{\text{OC}_i} = S_i (T_i - T_{i-1}),$$

(14.28)

Then, total V_{OC} is obtained by adding all Seebeck voltages over all segments of all TE elements, thus by two-stage summation as:

$$V_{\text{OC}} = \sum \sum_i V_{\text{OC}_i},$$

(14.29)

Finally, electric current I is obtained by Ohm's law for electric circuit model as:

$$I = \frac{V_{\text{OC}}}{R_{\text{int}} + R_L},$$

(14.30)

In short, the total number of unknown variables that need to be solved simultaneously for simulation of TE element becomes $N + 3$, i.e., T_i ($i = 0$ to N), T_{top}, and I. These unknown variables can be determined by solving simultaneously of $N + 3$ equations (15) for ($i = 0 \sim N$) and (24) and (30). After solving $N + 3$ multiple equations together, obtained temperature profile is plugged back in the same equations to update the material properties and then solved again iteratively until the temperature profile converges. Iterative process is performed independently for each n-type and p-type TE element, and then the second-level iteration is performed to make $T_{top,n}$ and $T_{top,p}$ converge as discussed above to complete TE module simulation. The flow chart for two-level iterative module simulation is displayed in Fig. 14.5.

Finally, TE module performances are calculated from obtained temperature profiles, total heat flow, and electric current. The voltage output, power output,

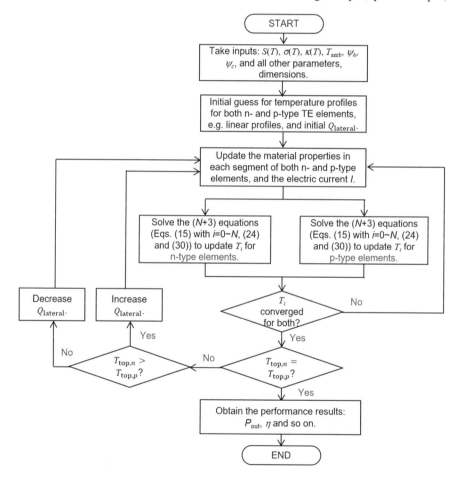

Fig. 14.5 Flow chart of two-level iteration algorithm to solve for temperature profiles and heat transfer in n-type and p-type TE elements and the performance of the entire TE module

total heat input to TE, TE efficiency, and the system efficiency are obtained, respectively, by:

$$V_{\text{out}} = IR_{\text{L}}, \tag{14.31}$$

$$P_{\text{out}} = IV_{\text{out}}, \tag{14.32}$$

$$Q_{\text{in,TE}} = N_{\text{TE,pair}}\left(Q_{\text{in},n} + Q_{\text{in},p}\right), \tag{14.33}$$

$$\eta_{\text{TE}} = \frac{P_{\text{out}}}{Q_{\text{in,TE}}}, \tag{14.34}$$

$$\eta = \frac{P_{\text{out}}}{Q_{\text{in,system}}}, \tag{14.35}$$

where $Q_{\text{in,system}}$ is the total heat input transferred from the source to the system before heat losses occur, i.e., $Q_{\text{in,system}} = \text{GCA}_{\text{total}}$ for solar TE system and $Q_{\text{in,system}} = \varepsilon_1 F_{12}\sigma_{\text{SB}}T_{\text{source}}^4 A_1$ for general radiant heat recovery system with the heat source at T_{source}. The system efficiency η takes into account all optical, radiative, and convection heat losses, so that it is always lower than TE efficiency. The load resistance R_{L} is independent parameter, but for maximum power delivery, it can be set to be equal to the internal resistance, i.e., $R_{\text{L}} = R_{\text{int}}$, which is known as the *electrical load matching* condition [14.28].

14.3 Solar Thermoelectric Generation

In this section, we present detailed simulation results obtained using simulation methods described in the previous section for solar thermoelectric systems. As shown in Fig. 14.1a earlier, our solar TE system consists of multielement TE module, solar absorber attached to the topside of the module, heat sink for efficient heat removal at the bottom side, and solar concentrator, e.g., Fresnel lens, for high-density heat input. We don't assume vacuum enclosure for the system as it can add substantial installation and maintenance costs. Convection heat loss is thus included in the simulation. We will compare the performances in vacuum and in air. Also, we use the same area of absorber as that of TE module without significant thermal concentration by the area shrinkage in effort to minimize the nonuniform tempera-ture distribution over the absorber area. Yet, fill factor less than unity will cause similar effects of thermal concentration in TE module. Multielement module is used to produce not only high efficiency or power output but also high-voltage output required for powering electronics above 1–3 V as the voltage increases in proportion to the number of TE elements. Air gap is used as the filler between TE elements. The fill factor and thickness of TE elements are varied to optimize the performance. Table 14.2 summarizes the system design parameters used in the simulation.

We use the state-of-the-art nanostructured bismuth telluride alloy materials, $Bi_2Te_{2.7}Se_{0.3}$ for n-type and $Bi_{0.5}Sb_{1.5}Te_3$ for p-type TE elements in this study. Figure 14.6 shows the figure-of-merit ZT of TE materials as a function of temperature. The complete material properties are given in [29] and [30], respectively. These materials have peak value ZT = 1.4 for p-type and ZT ~ 1.0 for n-type, both at around 400 K. Beyond peak temperature, ZT value decreases significantly with increasing in temperature. For temperature ranges where the experimental material properties are not available, we calculated the material properties based on linearized Boltzmann transport model based on fitting of the experimental data. Details about the material modeling are given in [32].

Values of absorptivity ($\alpha = 0.9$) and emissivity ($\varepsilon = 0.15$) used for solar absorber are adapted from [16], which is experimentally determined for black-painted

Table 14.2 System parameters used in the numerical analysis for solar TE systems

Material for n-type TE elements	Nanostructured $Bi_2Te_{2.7}Se_{0.3}$ [29]
Material for p-type TE elements	Nanostructured $Bi_{0.5}Sb_{1.5}Te_3$ [30]
Solar irradiation, G	700 W/m^2
Mean transmissivity of solar concentrator, τ	0.94 [13]
Mean absorptivity of solar absorber, α	0.9 [16]
Mean emissivity of solar absorber, ε	0.15 [16]
Temperature of TE bottom, $T_{bot}(=T_{amb})$	300 K
Cross-sectional area of TE elements, $A_n = A_p$	2 × 2 mm^2
Total area of TE module, A_{total}	10 × 10 cm^2
Heat transfer coefficient of hot-side plate, ψ_h	5000 W × m^{-2} × K^{-1}
Heat transfer coefficient of cold-side heat sink, ψ_c	3000 W × m^{-2} × K^{-1} [31]
Convection heat transfer coefficient at hot side, h_{conv}	20 W × m^{-2} × K^{-1}
Thermal conductivity of filler (air), κ_{filler}	0.04 W × m^{-1} × K^{-1}

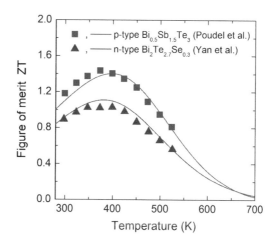

Fig. 14.6 Figure-of-merit ZT as a function of temperature for both n-type [29] and p-type [30] TE materials used in solar TE system simulation. Symbols are experimental data, and lines are theoretical fitting [32]

spectrally selective solar absorber. Detailed information about solar absorber is found in [33]. As discussed earlier in Sect. 14.2.1, these two values are very different from each other because the absorptivity is a mean value given by Eq. (14.2) for the solar irradiation at ~5800 K, while the emissivity is a mean value given by Eq. (14.4) at the absorber temperature around 600–900 K, which is mostly in infrared region.

The hot-side heat transfer coefficient ψ_h is assumed to be 5000 $Wm^{-2} K^{-1}$, assuming highly conductive plate with sufficiently small thickness. At the cold side, high-efficiency water-cooling heat sink is assumed in order to remove such high heat flux. A micro-channel heat sink can achieve net heat transfer coefficient ψ_c as high as 4000 $W \times m^{-2} \times K^{-1}$ or even higher [31]. We choose conservatively value of 3000 $W \times m^{-2} \times K^{-1}$ for ψ_c and later vary two external heat transfer coefficient values to quantify impacts on P_{out}.

14.3.1 Module Optimization

Figure 14.7 shows the module simulation results for 10×10 cm^2 area solar TE module under solar concentration of 60 suns. We varied the thickness of TE elements and the fill factor to optimize the performance. Figure 14.7a shows that hot-side temperature T_H increases and cold-side temperature T_C decreases as the thickness of TE elements increases because thermal resistance of TE elements increases with the thickness, which causes temperature difference applied to TE elements to rise up. Due to the increase in the total thermal resistance in heat flow direction, $Q_{in, TE}$ drops gradually with increasing thickness as shown in Fig. 14.7b. On P_{out}, however, the impact of TE element thickness is mixed. In the beginning, P_{out} increases with increasing in thickness due to increased temperature difference, which causes V_{out} to increase. But, at the same time, $Q_{in, TE}$ decreases due to increased thermal resistance, and, also, I decreases due to increased electrical resistance with increasing thickness. Hence, P_{out} decreases later in high-thickness region. Due to this trade-off, there is optimal thickness that maximizes P_{out} as shown in Fig. 14.7c. The optimal thickness is shifted to larger value as the fill factor F increases because increase in fill factor can neutralize the negative impact of increasing in thickness on $Q_{in, TE}$. As shown in Fig. 14.7b, $Q_{in, TE}$ increases with increasing in fill factor at fixed thickness because increased fill factor reduces the thermal resistance of TE module. Thus, larger thickness is required to maintain the optimal P_{out} for larger fill factor design. Both larger thickness and larger fill factor mean increase in the volume of material used and thus increase in the material cost, which is detrimental.

On the other hand, small thickness with small F can maintain the peak P_{out} high while keeping the material cost lower, thereby reducing the material cost per power. It is important to note that if either thickness or F is too low, the peak P_{out} can be substantially reduced as shown in Fig. 14.7c because the heat input gets too low and, also, the positive effect of increased temperature difference is mitigated by the decreased material figure-of-merit. As discussed above, material ZT drops

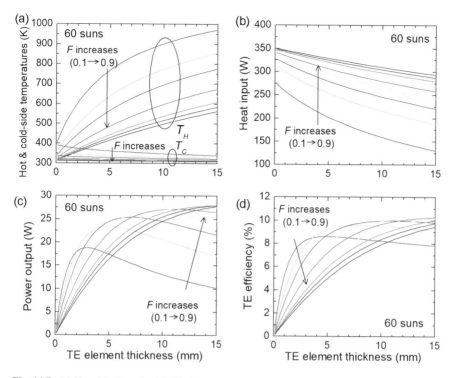

Fig. 14.7 (a) Hot-side T_H and cold-side T_C temperatures of TE elements; (b) total heat input to TE elements $Q_{in,\ TE}$; (c) power output P_{out}; and (d) TE efficiency η_{TE} of $10 \times 10\ cm^2$ solar TE module under solar concentration $C = 60$ suns as a function of TE element thickness with varying fill factor F from 0.1 to 0.9 with step size of 0.1

significantly with temperature above 400 K due to bipolar transport effect. Since hot-side temperature increases beyond 600 K, where material ZT is less than 0.2, the increased temperature difference would not help increase much V_{out} and P_{out} for F lower than 0.2. Reduced η_{TE} for low F confirms this argument (Fig. 14.7d). η_{TE} should have increased with increasing in temperature difference at lower F if average ZT was maintained the same. But, average ZT was reduced due to too high hot-side temperature, which resulted in lower η_{TE} as shown in Fig. 14.7d. Therefore, we may keep F at ~0.2 (green line in Fig. 14.7) and use the thickness slightly less than optimal (<4 mm) to keep P_{out} and η_{TE} high and the material cost-per-power low.

Figure 14.8 shows calculated system efficiency η of the same system. Compared to corresponding η_{TE} shown in Fig. 14.7d, η is much lower due to heat losses at the concentrator and the absorber and through the filler. As F decreases, the reduction in maximum efficiency becomes even more significant, because hot-side temperature increases with decreasing F, which in turn increases in heat losses by back-radiation and convection at the top surface; η above 5% is still possible with F of 0.2 and thickness of 3 mm. Higher η ~ 6.5% is possible if F is larger than 0.5 and thickness greater than 8 mm.

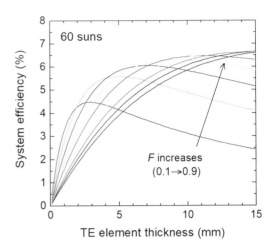

Fig. 14.8 System efficiency η of 10×10 cm^2 solar TE module under solar concentration $C = 60$ suns as a function of TE element thickness with varying fill factor F from 0.1 to 0.9 with step size of 0.1

14.3.2 Performance Variation with Solar Concentration in Air/Vacuum

In the discussion above, we fixed the solar concentration at 60 suns. But the system performance can be further tuned with varying solar concentration. Figure 14.9 shows the simulation results of P_{out} and η of 10×10 cm^2 area solar TE module with varying solar concentration. Here we fixed F at 0.2 and TE element thickness at 3 mm for optimal performance.

As shown in Fig. 14.9, both P_{out} and η rapidly increase with increasing in solar concentration initially. This is due to increased $Q_{in, \, TE}$ and increased temperature difference across TE. However, if solar concentration increases too high, hot-side temperature increases substantially, which increases in heat loss from the top surface, thereby limiting η and η saturates at ~5.5% above 70 suns in the case of usual operation in air without vacuum enclosure. In vacuum, the convection heat loss is removed, so that higher η can be achieved. However, the temperature increases even further in vacuum, which increases in radiation heat loss at the top surface. Increased temperature would also limit the increasing in rate of P_{out} with concentration, because average ZT of TE materials in the module will be decreased in high-temperature range. Therefore, η even slightly decreases beyond 70 suns. Thus, peak $\eta \sim 7\%$ is achieved under moderate solar concentration of ~60 suns. P_{out} is about 30% larger in vacuum than that in air over the entire solar concentration range.

14.3.3 Impact of External Heat Transfer Coefficients

In system performance, external heat transfer coefficients at hot and cold sides of TE module (ψ_h and ψ_c, respectively) are also very important factors. It is known that

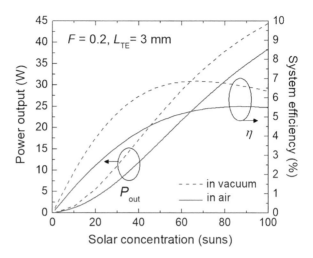

Fig. 14.9 Power output P_{out} (left y- axis) and system efficiency η (right y- axis) as a function of solar concentration for 10×10 cm^2 solar TE module with $F = 0.2$ and TE element thickness $L_{TE} = 3$ mm. Solid curves are properties of the system in air, and dash curves are those in vacuum

maximum P_{out} can be achieved when total external thermal resistance, i.e., $1/(\psi_h A_{total}) + 1/(\psi_c A_{total})$, is equal to total thermal resistance from TE elements assuming that Peltier and Joule heat are much smaller than conduction heat in TE module. This is called the *thermal matching* condition. However, this condition is derived under temperature boundary conditions. In radiant heat recovery applications, however, temperatures at the boundaries are not constant, but rather change a lot with many different system parameters. So, P_{out} may change in more complicated ways with the heat transfer coefficients. Nonetheless, since we are aiming at using thin TE elements to minimize cost-per-power, external thermal resistances need to be sufficiently high.

We have simulated TE module performance with varying both ψ_h and ψ_c and calculated P_{out} is shown as a function of those parameters in Fig. 14.10. Here we used $F = 0.2$ and element thickness of 3 mm under 60 suns. First, P_{out} increases with increasing in ψ_c and becomes almost flattened at high-coefficient region above ~4000 W $\times$ m^{-2} $\times$ K^{-1}. In fact, it is very difficult to achieve high ψ_c beyond this value with active water-cooling heat sink. However, if ψ_c is less than 3000 W m^{-2} K^{-1}, P_{out} is severely suppressed due to the inability in efficient heat removal at the cold side, which increases cold-side temperature of TE elements and limits the temperature difference. Particularly for the materials used in TE module, those high ZT values lie in low temperature range below 500 K. Therefore, it is critical to let those temperature ranges be applied across TE elements with strong heat sink. On the other hand, the effect of ψ_h is not as significant as shown in Fig. 14.10. Since hot-side temperature is already high enough due to high solar concentration and low F, an increase in ψ_h does not help further enhance P_{out} because material ZT is too low at added temperature range induced by increased ψ_h. For example, an increase of ψ_h from 1000 to 10,000 W $\times$ m^{-2} $\times$ K^{-1} can improve P_{out} merely by 5% or less, according to Fig. 14.10.

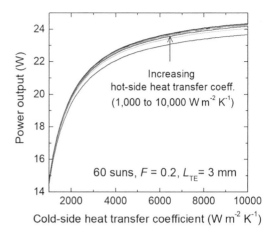

Fig. 14.10 Power output P_{out} as a function of cold-side heat transfer coefficient ψ_c with varying hot-side heat transfer coefficient ψ_h for 10×10 cm^2 solar TE module with $F = 0.2$ and TE element thickness $L_{TE} =$ 3 mm under 60 suns in air

14.4 Radiant Heat Recovery in Hot Steel Casting

In this application, we consider hot steel casting belt carrying steel slabs at temperatures as high as 1200 K or above as schematically shown earlier in Fig. 14.1b. When TE modules are installed above the casting belts, facing down to the slabs in parallel, strong radiation is transferred to TE module. We assume steel slabs continuously sliding along the belts and TE system of the same width as the slabs directly above the slabs. This configuration can be modeled as two infinitely long concentric parallel plates shown as the first geometry in Table 14.1 but with the same width, i.e., $w_1 = w_2$. Thus, the radiation shape factor from the slab to TE system is found from Table 14.1 as:

$$F_{12} = \left(\sqrt{\left(\frac{L_{slab}}{w_{slab}} \right)^2 + 1} \right) - \frac{L_{slab}}{w_{slab}}, \qquad (14.36)$$

where L_{slab} is vertical distance between steel slab and TE module and w_{slab} is width of the slab and TE absorber plate.

Just like in solar TE systems discussed in the previous section, we use multielement TE modules with heat absorber plate at hot side, facing down toward hot steel slabs, and heat sink at cold side for efficient heat removal to ambient. Note that the following simulation results are calculated for 10×10 cm^2 size of TE module. Since width and length of entire TE system are much larger than individual TE module size, multiple modules are installed to cover the entire area. Here, we assume that all TE modules share large heat-absorbing plate at hot side and heat absorber plate is so highly conducting that it can uniformly distribute the heat input to each of the TE modules. Hence, total P_{out} is simply P_{out} of one module multiplied by the number of modules used to cover the entire area. In this case, the total heat input from the slab to each module of area A_{total} is given by:

Table 14.3 System parameters used in simulation for radiant heat recovery system in hot steel casting

Material for n-type TE elements	Nanostructured $Bi_2Te_{2.7}Se_{0.3}$ [29]
Material for p-type TE elements	Nanostructured $Bi_{0.5}Sb_{1.5}Te_3$ [30]
Slab temperature, T_{slab}	1200 K [17]
Slab width, w_{slab}	2 m [17]
Distance between slab and TE system, L_{slab}	2 m
Mean emissivity of hot steel slab, ε_1	0.7 [34]
Mean emissivity of heat absorber, ε_2	0.9 [33]
Mean absorptivity of heat absorber, $\alpha_2 (=\varepsilon_2)$	0.9 [33]
Temperature of TE bottom, $T_{bot} (=T_{amb})$	300 K
Cross-sectional area of TE elements, $A_n = A_p$	2×2 mm^2
Total area of TE module, A_{total}	10×10 cm^2
Heat transfer coefficient of hot-side plate, ψ_h	5000 W $\times$ m^{-2} $\times$ K^{-1}
Heat transfer coefficient of cold-side heat sink, ψ_c	3000 W $\times$ m^{-2} $\times$ K^{-1} [31]
Convection heat transfer coefficient at hot side, h_{conv}	20 W $\times$ m^{-2} $\times$ K^{-1}
Thermal conductivity of filler (air), κ_{filler}	0.04 W $\times$ m^{-1} $\times$ K^{-1}

$$Q_{in,system} = \varepsilon_1 F_{12} \sigma_{SB} T_{slab}^4 A_{total}, \qquad (14.37)$$

where ε_1 and T_{slab} are, respectively, emissivity and temperature of the slab surface. Note that the shape factor F_{12} here was obtained for two parallel plates of the same width, i.e., $A_1 = A_2$. To get the total heat input transferred to smaller module area A_{total} only, area ratio A_{total}/A_2 was multiplied to source area A_1 to simply reduce to A_{total} at the end of Eq. (14.37). This $Q_{in, system}$ is used in Eq. (14.35) to calculate η.

We use the same TE materials used in solar TE for this application. The steel emissivity is as low as 0.2–0.3 below 400 °C, but it increases to 0.7 above 500 °C [34]. We use 0.7 as the emissivity of steel slab that is at 1200 K. We use the same absorber used in solar TE system simulation in the previous section. The absorptivity of absorber is 0.9 in infrared region. The emissivity of absorber must be almost the same (0.9) as its absorptivity because the absorber also emits in infrared region. Other parameters are also assumed to be the same as in the case of solar TE system. Table 14.3 summarizes the system parameters used in this hot steel casting application.

14.4.1 Module Optimization

Figure 14.11 presents calculated module performances with varying TE element thickness and F for 10×10 cm^2 TE module size. As was similarly shown in solar TE system in the previous section, by adjusting thickness and F, one can broadly tune heat input and P_{out} to meet the design goals. There is optimal thickness that maximizes P_{out} for a given F. For example, for $F = 0.2$ (green curves in Fig. 14.11),

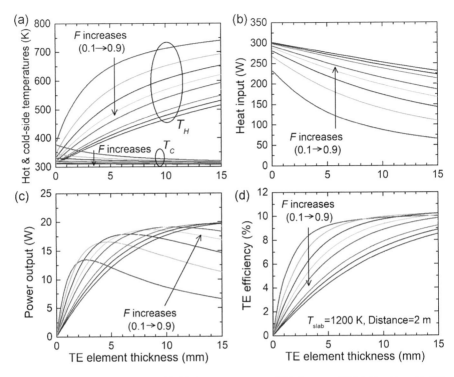

Fig. 14.11 (**a**) Hot-side T_H and cold-side T_C temperatures of TE elements; (**b**) total heat input to TE elements $Q_{in,TE}$; (**c**) power output P_{out}; and (**d**) TE efficiency η_{TE} of 10×10 cm^2 TE module for radiant heat recovery in hot steel casting application. All the properties are displayed as a function of TE element thickness with varying fill factor F from 0.1 to 0.9 with step size of 0.1. The temperature of hot steel slab is 1200 K, and distance between TE module and the slab is 2 m

P_{out} is maximized to reach ~16.5 Watt at TE element thickness ~5 mm. However, P_{out} is quite flat over certain range of thickness near this optimal thickness. This means that by choosing smaller thickness near the optimal value, P_{out} very close to the maximum can be still achieved, while the material cost can be lowered. For example, one can select thickness of 3 mm instead of 5 mm, for which P_{out} is as high as 15.5 Watt, only ~5% lower than the maximum P_{out}.

As F decreases, the temperature difference between hot side and cold side of TE module increases as shown in Fig. 14.11a, which increases in η_{TE} (Fig. 14.11d), except for high-thickness region for $F = 0.1$. In this region, low $F = 0.1$ caused hot-side temperature to be too high, which resulted in lowering average ZT, so that η_{TE} was reduced from that of $F = 0.2$.

The heat input is reduced (Fig. 14.11b) as F decreases, because of increased module thermal resistance. So, P_{out} is reduced accordingly even though η_{TE} was slightly higher. Moreover, heat loss by radiation is increased as hot-side temperature increases, which results in lower η as shown in Fig. 14.12.

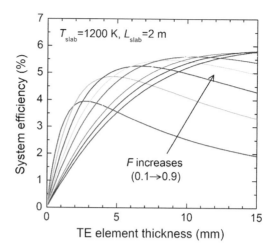

Fig. 14.12 System efficiency η as a function of TE element thickness for $10 \times 10 \text{ cm}^2$ TE module for radiant heat recovery in hot steel casting application. F varied from 0.1 to 0.9 with step size of 0.1. Slab temperature is 1200 K, and distance between TE module and slab is 2 m

As discussed similarly in solar TE systems, we could estimate power-per-cost might be achieved for hot steel casting application.

14.4.2 Performance Variation with Distance Between TE Module and Slab

If distance between TE module and slab is reduced, by placing TE system closer toward the slab, then radiation shape factor increases according to Eq. (14.36), which results in more heat transfer to TE system. As shown in Fig. 14.13a, both heat input and hot-side temperature increase with decreasing distance between TE module and slab. This results in increase in P_{out} (Fig. 14.13b). High P_{out} larger than 25 Watts can be generated for slab distance shorter than 1 m according to the calculation.

Value η also improves with decreasing distance between TE module and slab initially from 4 m until ~1 m and then saturates below 1 m distance. This is due to increased hot-side temperature (Fig. 14.13a) causing more heat losses and thus reducing η. The reduced material figure-of-merit also plays a role in reducing η in this regime. Maximum $\eta \sim 5.3\%$ is obtained here with distance between TE module and the slab from 0.5 m to 1 m.

14.4.3 Performance Variation with Slab Temperature

One can improve P_{out} by increasing in slab temperature too because more radiant heat is transferred to TE system. As shown in Fig. 14.14, P_{out} rapidly increases with slab temperature, almost following $\sim T^4$. At high slab temperature region, η is again saturated at its maximum ~5.3% due to increased heat losses, which in turn lessens

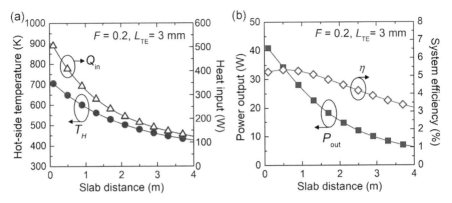

Fig. 14.13 (a) Hot-side temperature T_H (left y- axis) and heat input $Q_{in, TE}$ (right y- axis) and (b) power output P_{out} (left y- axis) and system efficiency η (right y- axis) as a function of distance between TE module and slab for 10×10 cm² TE module for radiant heat recovery in hot steel casting application. Slab temperature is fixed at 1200 K. $F = 0.2$ and TE element thickness $L_{TE} = 3$ mm were used for TE module

Fig. 14.14 Power output P_{out} (left y-axis) and system efficiency η (right y- axis) as a function of slab temperature for 10×10 cm² TE module for radiant heat recovery in hot steel casting. $F = 0.2$ and TE element thickness $L_{TE} = 3$ mm were used. Distance between TE module and slab was fixed at 2 m

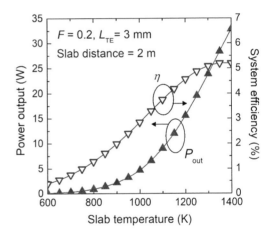

increasing in rate of P_{out} with slab temperature in high-temperature region. At low slab temperature region below 800 K, P_{out} is merely 1 Watt or less for 10×10 cm² TE module, which is much lower than $P_{out} \sim 15$ Watt when slab temperature of 1200 K; η is also lower than 1%, accordingly. This result may imply that TE-based radiant heat recovery may not be efficient solution if temperature of heat source is less than 800–1000 K. Contact-based TE conversion may be considered instead for more efficient energy conversion. Yet, the performance of TE-based radiant heat recovery system can be improved by increasing in shape factor, e.g., placing TE module closer, enclosing the source with TE, and so on, and by improving TE materials, heat absorber, and heat sink.

14.5 Conclusions

In this chapter, we have presented our TE module simulation methodology and theory based on two-level iterative finite element analysis for radiant heat recovery applications. Detailed performance analysis and design optimization have been discussed for two major applications, solar and radiant heat recovery in hot steel casting processes. In both applications, careful analysis on the radiant heat inputs and various heat losses based on optical parameters and geometry of the system was the key to accurate prediction of energy conversion performance. By optimizing TE element thickness and fill factor, thermal and electrical properties of TE module can be broadly tuned to achieve maximum power output, system efficiency, or power-per-cost to make TE radiant heat recovery technology potentially competitive with other leading energy conversion technologies at the market level.

The system efficiency can be further enhanced by improving the heat transfer media such as heat absorber and heat sink used and, also, by improving TE materials' performance. The low performance of nanostructured bismuth telluride alloy materials used in our simulation at high temperatures above 600 K has limited TE efficiency to maximum ~10% in both systems, which in turn limited the system efficiency. Segmented TE elements would be helpful to increase TE efficiency as recently demonstrated [16].

Acknowledgment J.H. Bahk is grateful for the generous faculty start-up supports for this work from the University of Cincinnati.

References

1. L.E. Bell, Cooling, heating, generating power, and recovering waste heat with thermoelectric systems. Science **321**, 1457 (2008)
2. D. Champier, Thermoelectric generators: a review of applications. Energy Convers. Manag. **140**, 167–181 (2017)
3. W. Corliss, D. Harvey, *Radioisotopic Power Generation* (Prentice-Hall, Englewood Cliffs, NJ, 1964)
4. J.C. Bass, N. Elsner, F. Leavitt, Performance of the I kW thermoelectric generator for diesel engines, in *Proceedings of the XIII International Conference on Thermoelectrics* (Kansas City, MO, 1994)
5. M. Zebarjadi, Electronic cooling using thermoelectric devices. Appl. Phys. Lett. **106**, 203506 (2015)
6. R. Vullers, R.v. Schaijk, H.J. Visser, J. Penders, C.V. Hoof, Energy harvesting for autonomous wireless sensor networks. IEEE Solid State Circ. Mag. **2**(2), 29–38 (2010)
7. R. Funahashi, M. Mikami, T. Mihara, S. Urata, N. Ando, A portable thermoelectric-power-generating module composed of oxide devices. J. Appl. Phys. **99**, 066117 (2006)
8. J.-H. Bahk, H. Fang, K. Yazawa, A. Shakouri, Flexible thermoelectric materials and device optimization for wearable energy harvesting. J. Mater. Chem. C **3**, 10362–10374 (2015)
9. Y. Shuai, X.-L. Xia, H.-P. Tan, Radiation performance of dish solar concentrator/cavity receiver systems. Sol. Energy **82**(1), 13–21 (2008)
10. W. Xie, Y. Dai, R. Wang, K. Sumathy, Concentrated solar energy applications using fresnel lenses: a review. Renew. Sust. Energ. Rev. **15**, 2588–2606 (2011)

11. A. Al-Merbati, B. Yilbas, A. Sahin, Thermodynamics and thermal stress analysis of thermo-electric power generator: influence of pin geometry on device performance. Appl. Therm. Eng. **50**(1), 683–692 (2013)
12. D. Narducci, P. Bermel, B. Lorenzi, N. Wang, K. Yazawa, *Hybrid and Fully Thermoelectric Solar Harvesting* (Springer, New York, 2018)
13. M. Telkes, Solar thermoelectric generators. J. Appl. Phys. **25**, 765 (1954)
14. D. Kraemer, B. Poudel, H.-P. Feng, J.C. Caylor, B. Yu, X. Yan, Y. Ma, X. Wang, D. Wang, A. Muto, K. McEnaney, M. Chiesa, Z. Ren, G. Chen, High-performance flat-panel solar thermoelectric generators with high thermal concentration. Nat. Mater. **10**, 532–538 (2011)
15. D. Kraemer, K. Mcenaney, Z. Ren, G. Chen, Concept of solar thermoelectric power conversion, in *Thermoelectrics and its Energy Harvesting*, (CRC Press, New York, 2012), p. 24.1
16. D. Kraemer, Q. Jie, K. McEnaney, F. Cao, W. Liu, L.A. Weinstein, Concentrating solar thermoelectric generators with a peak efficiency of 7.4%. Nat. Energy **1**, 16153 (2016)
17. T. Kuroki, K. Kabeya, K. Makino, T. Kajihara, H. Kaibe, H. Hachiuma, H. Matsuno, A. Fujibayashi, Thermoelectric generation using waste heat in steel works. J. Electron. Mater. **43**(6), 2405–2410 (2014)
18. B. Deo, R. Boom, *Fundamentals of Steelmaking Metallurgy* (Prentice-Hall, New York, NY, 1993)
19. S. Gosh, K. Margatan, A. Chong, J.-H. Bahk, Radiant heat recovery by thermoelectric genera-tors: a theoretical case-study on hot steel casting. Energy Convers. Manag. **175**, 327–336 (2018)
20. S. Lineykin, S. Ben-yaakov, Modeling and analysis of thermoelectric modules. IEEE Trans. Ind. Appl. **43**(2), 505–512 (2007)
21. J.-H. Bahk, A. Abuhamdeh, K. Margatan, Solar thermoelectric module simulator [Online] (2018), https://nanohub.org/tools/solarte
22. Y. Hsiao, W. Chang and S. Chen, "A mathematic model of thermoelectric module with applications on waste heat recovery from automobile engine," Energy, vol. 35, no. 3, pp. 1447-1454, 2010
23. K. Yazawa, A. Shakouri, Cost-efficiency trade-off and the design of thermoelectric power generators. Environ. Sci. Technol. **45**, 7548–7553 (2011)
24. F. Kreith, R.M. Manglik, *Principles of Heat Transfer*, 8th edn. (Cengage, Boston, MA, 2018)
25. T.L. Bergman, A.S. Lavine, F.P. Incropera, D.P. DeWitt, *Fundamentals of Heat and Mass Transfer* (Wiley, Hoboken, NJ, 2017)
26. D. Kraemer, K. McEnaney, M. Chiesa, G. Chen, Modeling and optimization of solar thermo-electric generators for terrestrial applications. Sol. Energy **86**, 1338–1350 (2012)
27. G. Nolas, J. Sharp, J. Goldsmid, *Thermoelectrics Basic Principles and New Materials Devel-opments* (Springer, New York, 2001)
28. C. Goupil (ed.), *Continuum Theory and Modeling of Thermoelectric Elements* (Wiley-VCH, Weinheim, 2016)
29. X. Yan, B. Poudel, Y. Ma, W.S. Liu, G. Joshi, H. Wang, Y. Lan, D. Wang, G. Chen, Z. Ren, Experimental studies on anisotropic thermoelectric properties and structures of n-type Bi2Te2.7Se0.3. Nano Lett. **10**, 3373–3378 (2010)
30. B. Poudel, Q. Hao, Y. Ma, Y. Lan, A. Minnich, B. Yu, X. Yan, D. Wang, A. Muto, D. Vashaee, X. Chen, J. Liu, M.S. Dresselhaus, G. Chen, Z. Ren, High-thermoelectric performance of nanostructured bismuth antimony telluride bulk alloys. Science **320**, 634 (2008)
31. J.-W. Seo, Y.-H. Kim, D. Kim, Y.-D. Choi, K.-J. Lee, Heat transfer and pressure drop characteristics in straight microchannel of printed circuit heat exchangers. Entropy **17**, 3438 (2015)
32. J.-H. Bahk, A. Shakouri, Minority carrier blocking to enhance the thermoelectric figure of merit in narrow-band-gap semiconductors. Phys. Rev. B **93**, 165209 (2016)
33. F. Cao, D. Kraemer, T. Sun, Y. Lan, G. Chen, Z. Ren, Enhanced thermal stability of W-Ni-Al 2 O 3 cermet-based spectrally selective solar absorbers with tungsten infrared reflectors. Adv. Energy Mater. **5**, 1401042 (2015)
34. H. Sadiq, M.B. Wong, J. Tashan, R. Al-Mahaidi, X.L. Zhao, Determination of steel emissivity for the temperature prediction of structural steel members in fire. J. Mater. Civil. Eng. **25**, 167–173 (2013)

Index

A

AC field-assisted sintering technique (FAST), 257, 259, 271
Alloying, 73–76, 204, 207, 224
Alphabet Energy, 286
Angle-resolved photoemission spectroscopy (ARPES), 29, 30, 33, 40, 51
Anharmonicity, 48, 52, 53, 59, 95–97
Anion-centered tetrahedron, 139
Anisotropic interactions, 182
Anisotropy, 19
    electric and thermal conductivity, 9
    electrical conductivity, 15
    preferential orientation, 12
    seebeck coefficient, 16, 20
    temperature dependence, 16
Army Research Laboratory (ARL), 259
Atomic force microscopy (AFM), 152, 261
Average ZT, 88–94, 101

B

Ball milling, 181, 182, 184, 185, 196
Band degeneracy, 64, 77, 84, 86, 89, 145
Band energy offset, 64, 65, 67, 68, 70
Band engineering, 67–72
    charge carriers, 83
    chemical composition, lead chalcogenides, 84
    convergence, 84, 85
    HED, 145
    low deformation potential coefficient, 87, 88
    low effective mass, 86, 87
    pseudocubic approach, 144
    resonant levels, 86

Band range hopping (BRH) behavior, 240
Band structure engineering
    fillers, 182, 183
    substitutions, 183, 184
Bismuth-antimony telluride
    effective thermoelectric material, 4–9
    electron mobility, 4
    properties, 9
    researchers and engineers, 3
    Seebeck coefficient, 10
    sublattice, 10
    tellurium, 4
    tellurium-free thermoelectric materials, 4
    ternary alloys, 10
    thermoelectric efficiency, 11
    thermoelectric generation, 3
    thermoelectric materials
        (*see* Thermoelectric materials)
Boltzmann constant, 5
Boltzmann transport model, 313
BoltzTrap calculation, 51
Bridgman method, 11

C

Chalcogenide, 24, 25
Charge carrier scattering mechanism, 171, 173
Coherent interface, 108, 109
Concentration of charge carriers, 162, 163, 168, 169, 173
Conduction band minimum (CBM), 28, 161
Conduction mechanisms, 235, 236, 241
Crystal structure engineering, 65, 76–77
Current-power curves, 287
Current-voltage data, 285

© Springer Nature Switzerland AG 2019
S. Skipidarov, M. Nikitin (eds.), *Novel Thermoelectric Materials and Device Design Concepts*, https://doi.org/10.1007/978-3-030-12057-3

Current-voltage load lines, 287
Czochralski method, 11

D
Defect engineering, 72–75
Density functional theory (DFT), 68, 152, 161
Density-of-states (DOS), 5, 6, 14, 15, 28, 67,
        68, 86, 117, 141
Differential scanning calorimeter (DSC), 148
Differential thermal analysis (DTA), 194
Doping, 65–67, 71, 75–77

E
Electrical conductivity
        amorphous barrier, 236
        amorphous semiconduction, 236
        direct conduction mechanism, 236
        effect of polymeric precursor, 241
        effect, filler, 242
        effect, processing temperature, 237–241
        free carbon atoms, 237
        free carbon content, 235
        graphitic-like carbon content, 236
        thermoelectric properties, 234
        tunneling percolation, 236
Electrical properties, 216
Electrical transport properties
        SnS, 56, 58
Electron backscattered diffraction (EBSD), 55
Electron paramagnetic resonance (EPR), 146
Electronic structure, 50
Electronic transport properties
        pristine and doped polycrystalline SnSe,
                31–34
        pristine and doped single-crystalline SnSe,
                30, 31
Energy-dispersive X-ray spectrum (EDS), 48, 49
Energy filter effect
        heterojunctions, 221
        mobility of charge carriers, 221, 224
        nanoinclusions, 222, 223
        Seebeck coefficient, 221
        single-phase boundaries, 221
        solid-state reactions, 223
        types of interfaces, 221
Entropy engineering
        HEAs, 147
        spin entropy, 146
Extended X-ray absorption Fi(EXAFS)
        analysis, 184
External heat transfer coefficients, 316, 317

F
Field-assisted sintering technique (FAST), 256,
        257, 259
Figure of merit, 5, 6, 18, 19, 47, 59
Frequency-domain thermoreflectance (FDTR),
        266–268, 270

G
General effective media (GEM), 236
Geometric phase analysis (GPA), 108
Gleeble system, 256, 259, 265, 271
Grain boundary scattering, 268
        nanostructured thermoelectric materials,
                256, 257
Grüneisen parameter, 52

H
Half-Heusler (HH) compounds
        heat-to-power energy conversion
                applications, 204
        phonon transport, 204
        power factor
                band engineering, 216–219
                tuning, scattering mechanisms, 219, 220
        TEGs, 203
Heat transfer model, 302–306
HH matrix and FH nanoinclusion (HH/FH), 222
High-angle annular dark-field scanning
                transmission electron microscopy
                (HAADF-STEM) technique, 169
High-entropy alloys (HEAs), 147
Highly efficient doping (HED), 145
High-pressure torsion (HPT), 181, 183, 185, 196
High-resolution thermal metrology, 256, 261
High-resolution transmission electron
                microscope (HR TEM), 107
Hopping conduction, 240, 245
Hot pressing, 181, 182, 184, 185, 188, 196
Hot steel casting, 300, 301, 318, 319, 321, 322

I
International Centre for Diffraction Data
                (ICDD), 141
Inverse fast Fourier transform (IFFT) images, 108

K
Knudsen effusion mass spectrometry (KEMS),
                194, 195
Korringa-Kohn-Rostoker method, 28

L

Lead chalcogenides
  alloying and doping, 95, 96
  all-scale nanocomposites, 99, 100
  band convergence, 84–86
  band engineering, 83, 84
  charge charges, 93, 94
  composites, 97, 98
  deep defect level engineering, 90–92
  dynamic doping, 92, 93
  high thermoelectric performance, 94
  low deformation potential coefficient, 87, 88
  low effective mass, 86, 87
  microstructural manipulation, 101
  multiphase PbTe materials, 98
  precipitates and strains, 96
  resonant levels, 86
  TE materials, 101, 102
  ZT, 88, 89
Linear range, 278–280
Linear region, 280
Lorenz number, 6
Low-temperature thermoelectric generators, 4

M

Microstructural manipulation, 94–101
Mobility of charge carriers, 217
Monoclinic phase, 140
Multicomponent diamond-like chalcogenides (MDLCs)
  band engineering (*see also* Band engineering)
  conversion efficiency, 137
  device, 150, 154
  eco-friendly and earth-abundant elements, 138
  effective strategies, 137
  electronic structure and lattice dynamics, 141, 143
  energy crisis and environmental pollution, 137
  fabrication and performance test, 150
  in situ displacement reaction, 148
  MDLCs, 138
  mechanical properties, 152, 153
  mosaic nanostructure, 149
  operational stability, 151
  single-crystal SnSe, 138

N

Nano-bulk structuring, 207–209
Nanocomposite, 107

Nanoprecipitation, 105–132
Nanostructuring, 180, 184
Nonlinear range, 281
  Alphabet Energy, 286
  commercial $(Bi,Sb)_2(Te,Se)_3$ module, 288
  experimental configuration, 285, 286
  NASA JPL skutterudite module, 289–291
  slope-efficiency method, 281–285

O

Optimizing concentration of charge carriers, 83

P

PbTe, 24, 64, 65, 98, 105–132
Peltier cooling, 278, 280, 291
Peltier heating, 278, 280, 291
Phonon engineering
  influence of nanoprecipitates, 185, 186, 188
  nanoeffect, 184, 185
Phonon-glass electron-crystal (PGEC), 180
Polycrystalline, 48, 53, 54, 56, 59, 60
Polycrystalline materials, 184
Polymer-derived ceramics (PDCs)
  amorphous matrix and free carbon phase, 234
  cross-linked polymer, 233
  electrical and thermal conductivities, 230
  fossil fuel usage, 230
  microstructure, 233
  phase transformation, 233
  polymeric precursors, 233, 234
  properties, 230
  Seebeck coefficient, 244, 245
  silicon-based polymers, 232
  thermal conductivity, 242–244
  thermoelectric materials, 231, 232
Polyvinylpyrrolidone, 259
Pseudocubic approach, 144
Puyoo's model, 261

Q

Quantum dot, 15
Quantum well, 15
Quantum wire, 15
Quintuple layers (QLs), 8

R

Radiant heat recovery
  applications, 301, 317
  finite element analysis, 323

Radiant heat recovery (*cont.*)
  heat source, 312
  hot steel casting, 318–320, 323
  TE module, 302, 322
  TE system, 300, 301, 322
  TEG power system, 303
Radiation energy harvesting, 299, 300
Radiation shape factor, 303, 304
Radioisotope thermoelectric generators
    (RITEGs), 299
Raman spectroscopy, 146

S
Sb-based skutterudites
  classes of materials, 178
  crystal structure, 178
  crystalline character and electronic
      properties, 180
  development of, 180
  elements, 181
  features, 178
  heat-carrying phonons, 180
  large-scale production, 195, 196
  mechanical properties
    Ashby plot, 191
    behavior, compressive and flexural
        strength, 190
    crystallites, 188
    elastic modulus *vs.* hardness, 190
    equipment, 188
    Hall-Petch relation, 188
    heat treatment, 192
    length change *vs.* temperature, 192
    nanoparticles, 189
    parameters, 188
    static and dynamic hardness, 188
    temperature-dependent electrical
        resistivity, 192
    thermal expansion, 192, 193
    Vickers hardness (HV) values, 188, 189
    young's modulus, 189
  nanostructuring and nanocomposites, 180
  optimal void fillers, 180
  oxidation, 193–195
  pnictogen atoms, 179
  preparation methods, lab scale, 181, 182
  quality of, 178
  rattling vibrations, 179
  sublimation, 193–195
  TE generators, 195, 196
  thermal stability, 193–195
Scanning hot probe (SHP) technique, 261, 263,
    265, 266, 268, 270

Scanning transmission electron microscopy
    high-angle annular dark field
    (STEM-HAADF), 48
Seebeck coefficient, 4–6, 10, 16, 19, 20, 51, 56,
    63, 66–68, 71, 72, 83, 84, 86–89,
    128, 145, 244, 245, 256, 269, 298,
    306, 308
Selected area electron diffraction (SAED),
    208, 260
Semiconductor, 23, 24, 31, 32, 40
Severe plastic deformation (SPD), 185
Shockwave consolidation
    and AC FAST, 259–261
    and alternating current, 256
    controlled explosion, 257
    crystal grain size, 257
    grain boundary distribution, 258
    substantial portion of grain boundaries, 258
    TE materials, 257
    thermal resistance, 258
Single crystal, Sns, 48, 54–60
Single parabolic band (SPB), 34, 217
Skutterudite materials, 14
Skutterudite module, 289–291
Slope-efficiency method, 281–285, 287,
    288, 290
SnTe-based thermoelectrics
    acoustic scattering, 65
    band engineering, 64, 67, 68, 70–72
    band offset, 70
    band structures, 69
    charge carriers, 66, 67
    crystal structure, 76, 77
    defect engineering, 72–75
    dislocations, 77
    Hall concentration, charge carriers, 66, 70
    lattice thermal conductivity, 77
    lowest lattice thermal conductivity, 73, 74
    nontoxic composition, 78
    parameters, 64
    PbTe, 64, 65
    phase diagram, MnTe, 72
    Seebeck coefficient, 63, 70
    synergy, 75, 76
    thermoelectric materials, 63
    ZI, 66
    ZT, 71, 75
Solar photovoltaic panels, 4
Solar thermoelectric system, 300, 301, 312–314
Spark plasma sintering (SPS), 12, 13, 53, 181,
    208, 243
Spin entropy, 146
Synchrotron X-ray diffraction (SXRD)
    measurement, 151

T
TE module, 178, 190, 196
Thermal concentration factor, 306
Thermal conductivity, 47, 48, 52, 53, 56, 58–60
    Boltzmann equation, 205
    Callaway model, 205
    component of, 256
    electronic and phonon components, 270
    experimental error, 263
    Gleeble system, 271
    heat transport, 205
    nano-bulk structuring, 207, 208
    nanoinclusions
        energy filtering effect, 212
        mobility of charge carriers, 212
        oxides, 210
        phonon propagation, 210
        phonon scattering, 212
        second-phase nanoparticles, 212
        semiconductor nanoparticles, 212
        tunable parameters, 210
    perimeter and cross-sectional area, 261
    phase separation, 215, 216
    phonon scattering processes, 205, 207
    point defect scattering, 210
    reference sample, 264
    shockwave consolidated samples, 270
    SHP technique, 265
    temperature-dependent, 207
    Umklapp process, 206
    unknown sample, 263
    value of, 263, 265
    ZrCoBi-based compounds, 207
Thermal properties
    microscopic origin of low lattice, 39–41
    single-crystalline and polycrystalline SnSe,
        36–39
    SnS, 58, 59
    specific heat Cp, 34, 35
Thermodynamic stability, 183
Thermoelectric efficiency, 3, 4, 7, 10–16
Thermoelectric measurements, 277–291
Thermoelectric modeling, 306–309
Thermoelectric power generation
    band bending effect, 131
    band converge and nanostructure, 117
    band convergence, 110
    band structure engineering, 106
    Boltzmann constant, 118
    Boltzmann transport, 130
    Callaway's model, 123
    Debye-Callaway model, 113
    Debye temperature and sound velocity, 115

    effective masses, 119
    electric conductivity, 127, 131
    electrical resistivity, 119
    electron and phonon scattering, 132
    electron concentration, 125
    electron mobility, 129
    electronic and phonon scattering, 123
    energy difference, 111
    experimental and theoretical lattice thermal
        conductivity, 122
    experimental data, 120
    green energy, 106
    Hall concentration, 125, 127
    Hall effect measurements, 117
    Hall mobility, 127
    interface energy barrier, 128
    large-scale production, 106
    Lattice parameter, 120
    lattice thermal conductivity, 106, 120, 122
    low lattice thermal conductivity, 106
    low thermal conductivity, 116
    nanoinclusions, 123, 125, 131, 132
    nanostructuring effect, 106, 121
    PbTe-based materials, 106
    PbTe-PbSe-PbS Quaternary Alloys, 107,
        109, 111
    Pisarenko plot, 113, 120, 127
    quaternary alloy system, 106
    scattering mechanisms, 114
    Seebeck coefficient, 117, 123, 125, 127, 128
    Seebeck coefficients and electric
        conductivity, 112, 113
    semiconductors, 123
    single parabolic band model, 118
    temperature-dependent electric
        conductivity, 129
    temperature-dependent power factor, 113
    temperature-dependent Seebeck coefficient,
        118, 119
    temperature-dependent thermal
        conductivity, 120, 128
    temperature-dependent total thermal
        conductivity, 113
    trade-off relationship, 106
    Wiedemann-Franz law, 120
Thermoelectrics (TE)
    alkaline-earth/divalent rare-earth element
        atoms, 160
    alpha-SnSe, 25
    band structures, 162
    Bi-based Zintl phases, 162
    defect chemistry, 27
    direct bandgap, 161

Thermoelectrics (TE) (*cont.*)
 figure of merit, 23
 generators, 24
 PbSe, 24
 PbTe, 24
 polycrystalline SnS-based materials, 53
 pristine and p-type-doped Sns single
  crystals, 54–56
 p-type samples, 31
 p-type semiconductors, 160
 p-type-doped Sns single crystals, 56–59
 Seebeck coefficient, 161
 Seebeck effect, 47
 temperature, 163
 transport properties, 162
 undoped and Hole-Doped polycrystalline
  Sns, 53, 54
 Zintl-Klemm concept, 160
Thermoelectric (TE) materials
 anion-centered tetrahedron, 139
 anisotropy, 19
 crystal structure and phase transition, 138,
  140, 141
 crystallization/melting methods, 11
 electrical conductivity, 17
 electron and phonon transport
  properties, 139
 electron mobility, 16
 extrusion technique, 13
 function of temperature, 18
 generator application, 15
 green energy, 13
 ionic radius and electronegativity, 139
 kinetic coefficients, 20
 lattice thermal conductivity, 100
 longtime thermal stability, 102
 low-temperature materials, 15
 mid-temperature thermoelectric
  materials, 13
 multiphase PbTe, 98
 nanotechnology and low-dimensional
  structures, 14, 15
 novel thermocouple, 20
 phase transition phenomenon, 139
 phase transition, 139
 practical applications, 88
 pressing technique, 12
 resonant levels, 86
 Seebeck coefficient, 16, 17
 SPS technique, 12
 temperature-dependent crystal structure,
  139
 thermal conductivity, 18

thermocouple, 20
thermomechanical stresses, 11
Thomson effect, 308
Tin selenide (SnSe)
 acoustic phonons, 24
 ARPES, 29, 30
 crystal structure, 25, 26
 crystallographic and physical properties, 24
 defect chemistry, 27
 electronic band structure, 28, 29
 electronic transport properties
  (*see* Electronic transport properties)
 extrinsic dopants, 40
 geometric architecting, 24
 lattice thermal conductivity, 40
 low-temperature allotrope, 25
 PbTe compounds, 24
 polycrystalline and single-crystalline, 26, 27
 single-crystalline, 24
 thermal conductivity, 25
 thermal properties (*see* Thermal properties)
 thermoelectric figure of merit, 23
 thermoelectric materials, 23
 thermoelectricity, 42
 ugly-duckling compound, 24
 ZT values, 40, 42
Tin sulfide (SnS)
 BoltzTrap calculation, 50–52
 crystal structure, 48, 50
 crystal structures, 49
 density of pristine, 49
 DFT calculations, 60
 doping efficiency, 60
 effective masses, 51
 electrical transport properties, 56, 58
 electronic band structure, 50–52
 figure of merit, 59, 60
 low thermal conductivity, 52
 polycrystalline, 54
 pristine and p-type-Doped SnS single
  crystals, 54–56
 room temperature concentration, charge
  carriers, 58
 Seebeck coefficient, 47, 48
 single crystals, 55
 SnS-based materials, 48
 square-shaped samples, 59
 synthesis methods, 53
 TE materials, 47
 temperature-dependent electronic
  properties, 57
 thermal conductivity, 47, 58
 thermal properties, 58, 59

undoped and Hole-Doped
polycrystalline, 53
Tin telluride, 73
Transmission electron microscopy (TEM),
148, 186, 208, 260

V
Valence band maxima (VBM), 161, 217
van der Waals (vdW) forces, 8–10
Variable range hopping (VRH), 240
Vickers microhardness tester, 152

W
Wiedemann-Franz law, 6

X
X-ray diffraction (XRD), 107, 141, 260, 261

Z
Zinc-blende phase, 139
1-2-2 Zintl phase
electron-crystal, 159
Mg_3Sb_2 Zintl Compounds, 167, 169, 171
optimizing concentration, charge carriers,
164–166
phonon-glass, 159
solid solution, 166
theoretical investigation, 171, 172
thermoelectric studies, 160
ZT values, 105–133, 278, 291

Printed in the United States
By Bookmasters